Higher-Order Theories of Functionally Graded Circular Cylindrical Beams and Columns

Higher-Order Theories of Functionally Graded Circular Cylindrical Beams and Columns

Xueyang Zhang
Central South University, Changsha, China

Weili Ma
Chang'an University, ShaanXi, China

Zhenliang Hu
University of South China, Hengyang, China

Xianfang Li
Central South University, Changsha, China

ELSEVIER

中南大學出版社
www.csupress.com.cn

Elsevier
Radarweg 29, PO Box 211, 1000 AE Amsterdam, Netherlands
125 London Wall, London EC2Y 5AS, United Kingdom
50 Hampshire Street, 5th Floor, Cambridge, MA 02139, United States

ISBN: 978-0-443-44718-1

For information on all Elsevier publications visit our website at https://www.elsevier.com/books-and-journals

Publisher: Matthew Deans
Acquisitions Editor: Somoshri Banerji
Editorial Project Manager: Glyn Jones
Production Project Manager: Paul Prasad Chandramohan
Cover Designer: Raman Kumar

Typeset by TNQ Tech

Contents

Preface

As one of the most important structural units, cylindrical beams are widely used in lots of engineering fields, and the study of beams with circular cross-sections also has a long history. Generally speaking, the classical beam theories, such as the Euler–Bernoulli beam theory and the Rayleigh beam theory, are the most adopted in the research of cylindrical beams. However, the classical beam theories are not adequate for providing accurate bending, buckling, and vibration results when the radius-to-length ratio of the beam is relatively large. Thus, the transverse shear strain cannot be neglected in order to obtain more accurate solutions.

In this book, a detailed exploration of the mechanical behaviors of functionally graded circular cylindrical beams and columns, including bending, vibration, and stability, is presented, considering factors such as shear deformation, cross-section warping, elastic foundation interaction, nonlocal effects at the nanoscale, etc. With a focus on higher-order shear theory and analytical methods, the book provides derivations and explanations of complex equations of the circular cylindrical structures. To ensure the accuracy of the proposed theory, the calculation results are compared with classical beam theories, existing literature, and finite element simulation results. Several new warp-shape functions are proposed, and the influence of warp-shape functions on the higher-order beam theory is analyzed.

There are seven chapters in this book. Chapter 1 gives an introduction to the classical beam theories and higher-order shear beam theory. Chapters 2–4 present the bending, vibration, and buckling solutions of a functionally graded circular cylindrical beam via higher-order shear beam theory, respectively. Chapter 5 presents the stability solutions while considering the elastic foundations. Chapter 6 deals with mechanical behaviors combined with the nonlocal effects. Chapter 7 gives some advanced studies, such as the combined effects of thermoelasticity, surface elasticity, piezoelectricity, flexoelectricity, and coupling stress.

Owing to the vastness of the literature and the limited abilities of the authors, there may be relevant papers that escaped our search. To these authors, we offer our sincere apologies. Such omissions shall be rectified in a future edition. The authors gratefully acknowledge the support and encouragement of Central South University in carrying out the writing of this book.

Authors

Chapter 1

Introduction

As one of the most common structural elements, long elastic cylinders [1] are frequently encountered in civil, mechanical, aerospace, chemical, and marine engineering, for example, in structures such as steel ropes used in bridges and components of aircraft landing gear. At the micro level, carbon nanotubes and microtubules can be regarded as one dimensional beam like structures with an annular cross section and their mechanical behaviors can be analyzed by using elasticity theories of beams [2,3]. Strictly speaking, for elastic tubes including carbon nanotubes and microtubules, a more adequate description of their mechanical behaviors may be provided by an elastic cylindrical shell model. However, for thick tubes, particularly sufficiently long carbon nanotubes and microtubules, the elastic theory of beams is still appropriate for analyzing their elastic responses under loading [4]. Therefore, the structural analysis of beams with circular cross section (Fig. 1.1) is of great interest not only at the macroscale but also at the micro and nanoscale.

The study of beams with circular cross section has a long history. Generally speaking, the research related to this topic can be classified into two categories. One is based on three dimensional (3D) elasticity theory. In this field, numerous studies have been devoted to the free vibration and wave propagation of cylindrical rods by using 3D elasticity theory [5–11]. Although this approach provides an adequate description of the problem, the 3D analysis contains three spatial coordinates, and the manipulations involved are rather complex. Moreover, for some specific problems, the boundary conditions at both ends cannot be satisfied completely. For example, for axisymmetric vibrations of a free rod of finite length, three of the six boundary conditions are exactly satisfied and the other three are only approximately satisfied [9]. The second approach is to transform a 3D elastic problem into a one-dimensional (1D) elastic problem with the aid of crucial and reasonable hypotheses, which is particularly efficient for a slender beam. Compared to the above 3D analysis, the 1D approach is simpler and more feasible in engineering practice since only one spatial coordinate is involved. So far, many 1D theories for beams [12] have been developed, such as the Euler-Bernoulli beams, Rayleigh beams, Timoshenko beams, shear beams, etc. [13,14].

Higher-Order Theories of Functionally Graded Circular Cylindrical Beams and Columns
https://doi.org/10.1016/B978-0-443-44718-1.00003-7

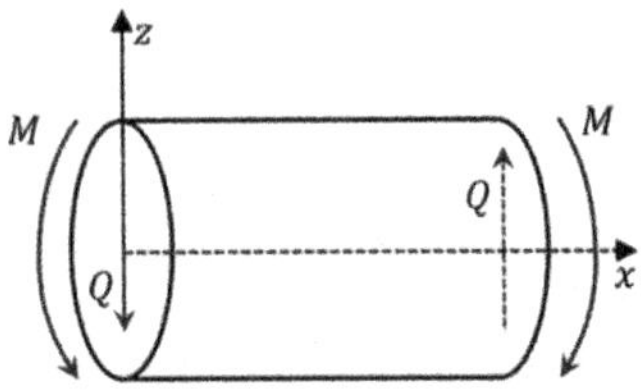

FIGURE 1.1 The schematic of circular cylindrical structures.

1.1 Euler-Bernoulli beam theory

Consider a flexible cylindrical beam with uniform circular cross section of radius R which is deflected under the action of transverse loading. We choose a Cartesian coordinate system such that the x-axis is oriented along the central axis of the beam, and the positive z-axis is directed upward and perpendicular to the x-axis. Meanwhile, polar cylindrical coordinates (x, r, θ) are also used, which are related to (x, y, z) by the following relations

$$y = r\cos\theta,\ z = r\sin\theta. \tag{1.1}$$

The three displacement components in the x, y, and z directions are denoted by u, v, and w, respectively. According to Saint-Venant's principle, it is reasonable to assume that the following stress components vanish: $\sigma_{yy} = \sigma_{zz} = \sigma_{yz} = 0$ [5,15]. In this case, the 3D constitutive equations reduce to the following form:

$$\begin{cases} \sigma_{xx} = E\varepsilon_{xx} \\ \tau_{xy} = G\gamma_{xy} \\ \tau_{xz} = G\gamma_{xz} \end{cases} \tag{1.2}$$

where σ_{jj} and ε_{jj} are the normal stresses and strains, respectively; τ_{jj} and $\gamma_{ij}(i \neq j)$ are the shear stresses and strains, respectively; E and G are the Young's modulus and shear modulus of the beam, respectively. Moreover, the normal and shear strains ε_{jj} and γ_{ij} are respectively expressed by the following relations:

$$\begin{cases} \varepsilon_{jj} = \dfrac{\partial u_j}{\partial x_j} \\ \gamma_{ij} = \dfrac{\partial u_j}{\partial x_i} + \dfrac{\partial u_i}{\partial x_j} \end{cases} \tag{1.3}$$

where $i, j = 1, 2, 3$, correspond to x, y, z, respectively. Using the equilibrium of force, the bending moment M and shear force Q at any position x of the beam can be expressed as

$$M = \int_A \sigma_{xx} z \mathrm{d}A \tag{1.4}$$

$$Q = \int_A \tau_{xz} \mathrm{d}A \tag{1.5}$$

respectively, where A is the area of the cross section.

As is well know, the Euler-Bernoulli theory of beams neglects the effects of rotary inertia and shear deformation of cross sections.

$$\varepsilon_{xx} = z\frac{\partial \psi}{\partial x} = -z\frac{\partial^2 w}{\partial x^2} \tag{1.6}$$

where ψ is the rotation of the cross section at the centroidal axis. Thus, the bending moment M can be obtained as

$$M = \int_A E\varepsilon_{xx} z \mathrm{d}A = -EI\frac{\partial^2 w}{\partial x^2} \tag{1.7}$$

According to the equilibrium equation, we have

$$\frac{\partial Q}{\partial x} = \rho A \frac{\partial^2 w}{\partial t^2} \tag{1.8}$$

$$\frac{\partial M}{\partial x} = Q \tag{1.9}$$

where $A = \pi R^2, I = \pi R^4/4$ are the cross sectional area and its inertia moment of area, respectively. The constitutive equation of the Euler-Bernoulli Beam is

$$EI\frac{\partial^4 w}{\partial x^4} + \rho A \frac{\partial^2 w}{\partial t^2} = q \tag{1.10}$$

Due to this simplification, the Euler-Bernoulli beam theory tends to overestimate both the natural frequencies of free vibration and the critical loads of buckling [16].

1.2 Timoshenko beam theory

The Euler-Bernoulli beam theory was first improved by Rayleigh [1], who considered the contribution of the rotary inertia of cross section, but still ignored shear deformation. Later, by taking the effects of both rotary inertia and shear deformation of cross section (Fig. 1.2) into account, Timoshenko proposed a further improvement of the beam theory [17,18].

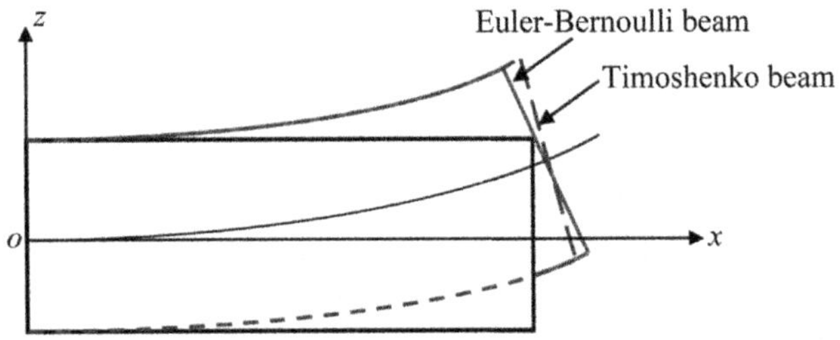

FIGURE 1.2 The schematic of Timoshenko beam.

The normal and shear strains ε_{jj} and γ_{ij} are respectively expressed by the following relations:

$$\varepsilon_{xx} = \frac{\partial u}{\partial x} = z\frac{\partial \psi}{\partial x} \tag{1.11}$$

$$\gamma_{xz} = \frac{\partial w}{\partial x} + \frac{\partial u}{\partial z} = \frac{\partial w}{\partial x} + \psi \tag{1.12}$$

Substituting the above Eqs. (1.11) and (1.12) into Eqs. (1.4) and (1.5), the bending moment M and shear force Q can be written as

$$M = EI\frac{\partial \psi}{\partial x} \tag{1.13}$$

$$Q = \kappa G\left(\frac{\partial w}{\partial x} + \psi\right) \tag{1.14}$$

where κ is the shear correction coefficient. Considering the equilibrium equation of a microelement, we have

$$Q = \frac{\partial M}{\partial x} + \rho I\frac{\partial^2 \psi}{\partial t^2} \tag{1.15}$$

$$\frac{\partial Q}{\partial x} = \rho A\frac{\partial^2 w}{\partial t^2} \tag{1.16}$$

We can also obtain the corresponding coupled governing equations of the Timoshenko beam theory as follows

$$\begin{cases} EI\dfrac{\partial^2 \psi}{\partial x^2} - \kappa AG\left(\dfrac{\partial w}{\partial x} + \psi\right) = \rho I\dfrac{\partial^2 \psi}{\partial t^2} \\ \kappa AG\left(\dfrac{\partial^2 w}{\partial x^2} + \dfrac{\partial \psi}{\partial x}\right) = \rho A\dfrac{\partial^2 w}{\partial t^2} - q \end{cases} \tag{1.17}$$

To date, the Timoshenko beam theory has been widely accepted and used in the analysis of transverse bending and vibration [19–22], as well as in the stability of elastic beams [23,24], particularly for shorter beams. However, in the Timoshenko beam theory, the shear stress on the cross section is assumed uniform, and a shear correction coefficient κ must be introduced to relax the traction-free condition on the surface of the beam. The shear correction coefficient is an inherent part of this theory and cannot be directly determined by the theory itself. There have been numerous discussions about the determination of the shear correction factor for various geometries based on static analysis of elastic beams [25–28].

1.3 Higher-order beam theories

Owing to the limitations of the Euler-Bernoulli and Timoshenko beam theories, it is highly desirable to develop an analytical approach for deriving a governing equation without the need for a shear correction coefficient. It is worth noting that, based on these consideration, more accurate higher-order theories for rectangular beams and even of plates have been formulated [29,30]. These refined beam theories do not contain a shear correction coefficient; moreover, the shear deformation and the rotary inertia as well as the surface boundary conditions are taken into account simultaneously. Nevertheless, for beams with other-shaped cross sections, there is little information on the study of a refined higher-order theory without introduction of the shear correction coefficient, even for a cylindrical beam with circular cross section. It is interesting to point out that for a static analysis, the exact solution to the bending of a cylindrical beam with circular cross section is available only for a cantilevered beam subjected to a tip force without shear correction coefficient [5]. In addition, Wadee et al. [31] employed the energy method and small-strain nonlinear elasticity theory to analyze the buckling of thin-walled elastic tubes with circular cross sections under pure bending. Based on the above considerations, it is necessary to further develop a higher-order beam theory suitable for cylindrical beams with circular cross section, in order to adequately describe their dynamic behaviors.

To analyze the flexural behavior of circular cylindrical beams, it is expedient to introduce the radial displacement and shear strain γ_{xr} over an arbitrary circular cross section, which are defined as:

$$U_r = v\cos\theta + w\sin\theta \tag{1.18}$$

$$\gamma_{xr} = \frac{\partial U_r}{\partial x} + \frac{\partial u}{\partial r} \tag{1.19}$$

respectively. Substituting Eq. (1.18) for U_r into Eq. (1.19) for γ_{xr}, after some manipulations, yields:

$$\gamma_{xr} = \frac{1}{r}\left(z\frac{\partial u}{\partial z} + z\frac{\partial w}{\partial x} + y\frac{\partial u}{\partial y} + y\frac{\partial v}{\partial x}\right) \tag{1.20}$$

When the beam is restricted to transverse motion in the xoz-plane and subjected to loading along the z-direction, the elastic displacement component v can be assumed approximately as $v = v(y, z)$. Hence

$$\gamma_{xr} = \frac{1}{r}\left[z\left(\frac{\partial u}{\partial z} + \frac{\partial w}{\partial x}\right) + y\frac{\partial u}{\partial y}\right] \tag{1.21}$$

Similar to the treatment of a rectangular beam [29], we introduce the rotation of the cross section at the centroidal axis as ψ. In order to guarantee the shear stress to be free at the surface of the cylindrical beam, it is readily shown that the longitudinal displacement u in the x-direction can be expressed as follows:

$$u = z\psi - \frac{z^3 + zy^2}{3R^2}\left(\frac{\partial w}{\partial x} + \psi\right) \tag{1.22}$$

where the deflection w and the rotation angle ψ are functions of the longitudinal coordinate x and time t, and are positive in the z-direction and in a clockwise manner, respectively. Hence from Eqs. (1.21) and (1.22), the shear strain at the cross section takes the form

$$\gamma_{xr} = \frac{z}{rR^2}\left(R^2 - z^2 - y^2\right)\left(\psi + \frac{\partial w}{\partial x}\right) \tag{1.23}$$

From this expression for shear strain, one easily finds

$$\gamma_{xr} = 0, \text{as} \quad y^2 + z^2 = R^2 \tag{1.24}$$

which equivalently means vanishing shear stress $\tau_{xr} = 0$ at the surface because of the constitutive relations. In other words, the shear free surface condition is identically satisfied. Therefore, using the above result, we can analyze the cylindrical beams in which the effect of shear stress is taken into account. Note that the classical Euler-Bernoulli beam theory completely neglects the presence of the shear stress at the cross section. Although the Timoshenko beam theory takes account of the presence of the shear stress at the cross section, the shear stress is assumed uniform and a shear correction factor, which cannot be determined by the theory itself, must be introduced. Here our model is capable of coping with the presence of shear stress at the cross section. Moreover, no additional correction factor is needed.

Furthermore, we made some observations. Although the vertical displacement (deflection) w is a function of the centroidal axis coordinate x, the horizontal displacement u and shear strain γ (or shear stress) are dependent on all three spatial coordinates x, y, and z. This is because the beam has a circular cross section. In the present study, the situation is completely different from that of the Timoshenko and Levinson beams for rectangular beams, where the longitudinal displacement component u and shear stress τ only involve two spatial coordinates x and z, and the remaining spatial coordinate y is suppressed because of the assumption of plane stress.

With these results, the constitutive Eq. (1.2) become

$$\begin{cases} \sigma_{xx} = E\left[z\dfrac{\partial\psi}{\partial x} - \dfrac{z^3+zy^2}{3R^2}\left(\dfrac{\partial^2 w}{\partial x^2}+\dfrac{\partial\psi}{\partial x}\right)\right] \\ \tau_{xy} = -\dfrac{2Gzy}{3R^2}\left(\dfrac{\partial w}{\partial x}+\psi\right) \\ \tau_{xz} = G\left[\dfrac{\partial w}{\partial x}+\psi-\dfrac{3z^2+y^2}{3R^2}\left(\dfrac{\partial w}{\partial x}+\psi\right)\right] \end{cases} \tag{1.25}$$

Substitution of Eq. (1.25) for σ_{xx} and τ_{xz}, respectively, into Eqs. (1.4) and (1.5) leads to

$$\begin{cases} M = \dfrac{1}{9}EI\left(7\dfrac{\partial\psi}{\partial x}-2\dfrac{\partial^2 w}{\partial x^2}\right) \\ Q = \dfrac{2}{3}GA\left(\psi+\dfrac{\partial w}{\partial x}\right) \end{cases} \tag{1.26}$$

In deriving Eq. (1.26), we have used the following integral results

$$\int_A\left(z^2, \frac{z^4+z^2y^2}{R^2}, \frac{z^2}{R^2}\right)\mathrm{d}A = \left(I, \frac{2}{3}I, \frac{1}{4}A\right) \tag{1.27}$$

Since attention is paid on flexural deformation and vibration in the vertical plane, that is, the xoz-plane, then in this case, we neglect the horizontal motion in the xoy-plane. Consequently, the equations of motion for the problem under consideration read

$$\frac{\partial\sigma_{xx}}{\partial x}+\frac{\partial\tau_{xy}}{\partial y}+\frac{\partial\tau_{xz}}{\partial z} = \rho\frac{\partial^2 u}{\partial t^2} \tag{1.28}$$

$$\frac{\partial\tau_{xz}}{\partial x}+\frac{\partial\tau_{yz}}{\partial y}+\frac{\partial\sigma_{zz}}{\partial z} = \rho\frac{\partial^2 w}{\partial t^2} - Z \tag{1.29}$$

where ρ is the mass density of the beam, and Z is the body force per unit volume in the z-direction. In addition, the boundary conditions at the surface of the beam can be stated below:

$$\begin{cases} \sigma_{xy}(x, y_0, z_0) - \dfrac{\mathrm{d}y}{\mathrm{d}z}\sigma_{xz}(x, y_0, z_0) = 0 \\ \sigma_{yz}(x, y_0, z_0) - \dfrac{\mathrm{d}y}{\mathrm{d}z}\sigma_{zz}(x, y_0, z_0) = S(s) \end{cases} \tag{1.30}$$

where y_0 and z_0 satisfy $y_0^2+z_0^2=R^2$, meaning that the point (y_0, z_0) lies at the surface of the beam, S is the resultant of surface traction in the z-direction, and s is a coordinate along the circumference.

After substituting Eq. (1.25) into Eq. (1.28), we multiply both sides of Eq. (1.28) by z and then integrate over the cross sectional area A, resulting in

$$EI\left(\frac{\partial^2\psi}{\partial x^2}-\frac{2}{7}\frac{\partial^3 w}{\partial x^3}\right)-\frac{6GA}{7}\left(\frac{\partial w}{\partial x}+\psi\right)=\rho I\left(\frac{\partial^2\psi}{\partial t^2}-\frac{2}{7}\frac{\partial^3 w}{\partial x\partial t^2}\right) \tag{1.31}$$

Similarly, a straight integration of Eq. (1.29) over the cross sectional area A gives

$$\frac{2}{3}GA\left(\frac{\partial^2 w}{\partial x^2}+\frac{\partial\psi}{\partial x}\right)=\rho A\frac{\partial^2 w}{\partial t^2}-q \tag{1.32}$$

where

$$q=\int_A Z\mathrm{d}A+\oint_{\partial A} S(s)\mathrm{d}s \tag{1.33}$$

q being a distributed force of N/m dimension and being applied at a lateral surface of the beam in the z-direction. We therefore obtain two coupled governing Eqs. (1.31) and (1.32) for dynamic behaviors of cylindrical beams with circular cross section. These two coupled equations are similar to those derived by Levinson [29]. The latter case only applies to rectangular beams. Here we extend the Levinson beam theory to cylindrical beams with circular cross section.

A further simplification of the resulting two coupled equations of motions Eqs. (1.31) and (1.32) can be achieved by introducing a new auxiliary function. Following the technique described in Ref. [32], if we take the deflection w and the rotation ψ in the form

$$w=F-\frac{7EI}{6GA}\frac{\partial^2 F}{\partial x^2}+\frac{7\rho I}{6GA}\frac{\partial^2 F}{\partial t^2} \tag{1.34}$$

$$\psi=\frac{\rho I}{3GA}\frac{\partial^3 F}{\partial x\partial t^2}-\frac{EI}{3GA}\frac{\partial^3 F}{\partial x^3}-\frac{\partial F}{\partial x} \tag{1.35}$$

where F is a new auxiliary function of length dimension, it is readily checked that when substituting Eqs. (1.34) and (1.35) into Eq. (1.31), it is identically fulfilled. On the other hand, when substituting Eqs. (1.34) and (1.35) into Eq. (1.32), it becomes

$$EI\frac{\partial^4 F}{\partial x^4}+\rho A\frac{\partial^2 F}{\partial t^2}-\rho I\left(1+\frac{7E}{6G}\right)\frac{\partial^4 F}{\partial x^2\partial t^2}+\frac{7\rho^2 I}{6G}\frac{\partial^4 F}{\partial t^4}=q \tag{1.36}$$

Once the solution to the governing equation, F is obtained through appropriate boundary conditions, all the physical quantities of interest including the deflection w, rotation ψ, bending moment M, and shear force Q are obtainable. For instance, w and ψ are given by Eqs. (1.34) and (1.35), respectively. Remembering Eq. (1.26), we can give the bending moment M and shear force Q in terms of F as follows

$$\begin{cases} M = -EI\dfrac{\partial^2 F}{\partial x^2} \\ Q = -EI\dfrac{\partial^3 F}{\partial x^3} + \rho I\dfrac{\partial^3 F}{\partial x \partial t^2} \end{cases} \tag{1.37}$$

For the Timoshenko beams, by using a similar procedure, if taking

$$w = F - \frac{EI}{\kappa GA}\frac{\partial^2 F}{\partial x^2} + \frac{\rho I}{\kappa GA}\frac{\partial^2 F}{\partial t^2} \tag{1.38}$$

$$\psi = -\frac{\partial F}{\partial x} \tag{1.39}$$

We immediately find that the coupled governing Eq. (1.17) can be converted to a single governing equation as follows:

$$EI\frac{\partial^4 F}{\partial x^4} + \rho A\frac{\partial^2 F}{\partial t^2} - \rho I\left(1 + \frac{E}{\kappa G}\right)\frac{\partial^4 F}{\partial x^2 \partial t^2} + \frac{\rho^2 I}{\kappa G}\frac{\partial^4 F}{\partial t^4} = q \tag{1.40}$$

In this case, M and Q are still given by Eq. (1.37).

Comparing Eq. (1.36) with Eq. (1.40), we can find that two equations are identical if $\kappa = 6/7$ [17]. However, it is emphasized that the expressions for ψ in terms of F are not consistent since $\psi = \psi_T + Q/3GA$, where ψ_T stands for the rotation given by Eq. (1.39) according to the Timoshenko beam theory. Due to this reason, there is a discrepancy between two solutions derived from the present theory and the Timoshenko beam theory with $\kappa = 6/7$ for some specific beams, as will be viewed in the following.

1.4 Background of functionally graded (FG) circular cylindrical structures

Functionally graded materials (FGMs) have attracted considerable scientific and technological interest in recent years due to their outstanding performance in the high temperature environment [33]. Typically, FGMs are fabricated from ceramic and metal; they show advantages in withstanding thermal shocks and maintaining structural integrity, since the ceramic component provides higher temperature resistance, while the metallic component prevents the structure from cracking. Considerable research has been done for the manufacture of FGMs. The material properties of FGMs can be designed by changing the volume fraction of two or more components, and the material properties may vary along arbitrary direction like plates or beams thickness, radial direction of cylinders, etc.

Numerous theoretical have been conducted on FGM beams or rods. For example, Sankar presented an elasticity solution for a thickness-wise functionally graded beam subjected to transverse loads [34]. Aydogdu investigated the free vibration of simply supported thickness-wise FGM beam [35]. Li formulated a unified approach to analyze the static and dynamic behavior of thickness-wise functionally graded beams using the Euler-Bernoulli and Timoshenko theories of beams [32]. For axially graded beams, some approximate methods and exact analytic approaches have been developed to deal with free vibration and buckling problems of Euler-Bernoulli and Timoshenko beams [36–38]. For a functionally graded beam with bi-directional gradient such as the thickness- and length-directions, some works have been conducted. Simsek [39] dealt with free and forced vibration of Timoshenko beams with various ends. Nguyen et al. analyzed the vibration of a bi-dimensional functionally graded Timoshenko beam excited by a moving load [40]. Furthermore, Ghayesh [41] gave an analysis of nonlinear vibration analysis of axially graded shear-deformable tapered beams. Sun and Li [42] analyzed free vibration of axially loaded functionally graded Timoshenko beams with nonuniform cross sections. Tang and Ding [43] studied nonlinear vibration of a bi-directional functionally graded beam subjected to hygro-thermal loads. Using isogeometric analysis, Nguyen et al. provided an efficient computational approach for analyzing functionally graded plate structures [44–46].

A typical FGM is a particulate composite with a prescribed distribution of volume fractions of constituent phases. For example, several numerical examples are given to show the effects of the gradient and the warping shapes on the flexure, buckling loads, and natural frequencies of radially graded tubes. For radially graded tubes, we suppose the material properties to change in the following power-law form

$$Y(r) = Y_i + (Y_o - Y_i)\left(\frac{r - R_i}{R_o - R_i}\right)^N \tag{1.41}$$

whereY_i and Y_o are the corresponding material properties on the inner and the outer surfaces, respectively, and N is the power-law gradient index describing the volume fraction change of both constituents. When $N = 1$, the material properties exhibit a linear variation on the radial coordinate. For other values of N, Fig. 1.3 shows the variation of material properties for $R_i/R_o = 0.5$. In the present study, a radially graded tube with a length to radius ratio $L/R_o = 10$, and a Poisson's ratio $\nu = 0.3$ is used, unless otherwise stated, similarly, aluminum and zirconia are selected as the materials for the inner and outer surfaces, respectively, the material properties of which are [47]

$$\begin{cases} \text{Al}: E_i = 70\ \text{GPa}, G_i = 27\ \text{GPa}, \rho_i = 2702\ \text{kg/m}^3 \\ \text{ZrO}_2: E_o = 200\ \text{GPa}, G_o = 77\ \text{GPa}, \rho_o = 5700\ \text{kg/m}^3 \end{cases} \tag{1.42}$$

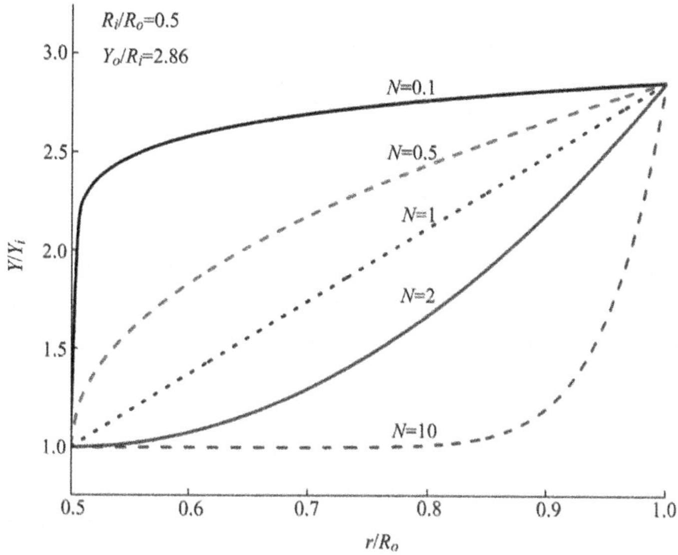

FIGURE 1.3 Variation of material properties with $Y_o/Y_i = 2.86$ and $R_i/R_o = 0.5$ when taking different gradient indexes N.

1.5 Scope of the book

The book provides a detailed exploration of the mechanical behaviors of functionally graded circular cylindrical beams and columns based on a higher order shear deformation beam theory. A brief description of the book's layout is given to provide a brief overview of what will be discussed.

Chapter 2 presents a new approach for analyzing transverse flexure of circular cylindrical beams or tubes with arbitrary radial nonhomogeneity. A third-order shear deformation beam theory for radially graded beams or tubes are proposed. Coupled governing equations for the deflection and rotation about the neutral axis of cross section are derived from equilibrium equations and then converted to a single fourth-order partial differential governing equation.

Chapter 3 investigates the transverse vibration of functionally graded circular cylindrical beams and tubes with radial nonhomogeneity, as well as the tubes subjected to axial tensile and compressive forces. Using the shear stress free condition at the surface of the cylinder, coupled governing equations for deflection and rotation angle are derived. The frequency equations for free flexural vibration of radially graded cylinder and hollow cylinders with various boundary conditions are obtained.

Chapter 4 analyzes the stability of circular cylindrical columns/beams composed of functionally gradient materials, as well as tubes subjected to axial tensile and compressive forces, where shear deformation is taken into account. In addition, a transition parameter to describe the direction of axial load from

Engesser's hypothesis to Haringx's one is introduced, then a unified method is presented to determine the critical load.

Chapter 5 presents and discusses several recently suggested foundation models (Winkle model, Pasternak model, Vlasov model, Kerr model and several viscoelastic models). The structural stability of clamped-free beam-columns resting on elastic foundations under a nonconservative force is investigated. The higher-order beam theory is developed by considering the warping of the cross section, rotary inertia, and axial force. As an example, the characteristic equation of dynamic stability of rectangular/circular cantilevers on a Winkler-Pasternak foundation is obtained for a generalized follower force. Critical loads for divergence instability and flutter instability are evaluated.

Chapter 6 studies the behaviors of flexural waves traveling in carbon nanotubes (CNTs) and the transverse free vibration of functionally graded (FG) nanotubes with radial inhomogeneity in a free space and embedded in an elastic matrix. An exact higher-order model for analyzing dynamic behavior of nonlocal elastic beams with circular cross section is proposed. The frequency equations of free vibration of FG nanotubes are obtained for various ends. The influences of the nonlocal, geometric, and material parameters on the frequencies are discussed.

Chapter 7 explores some advanced topics (thermoelastic coupling effects, surface effects, the coupled stress effects as well as the piezoelectric and flexoelectric effects) in the context of the classical and higher-order beam theories. The governing equations of the beam structures combined with these physical effects are derived briefly.

References

[1] Weaver JW, Timoshenko SP, Young DH. Vibration problems in engineering. John Wiley and Sons; 1991.

[2] Kis A, Kasas S, Babić B, et al. Nanomechanics of microtubules. Phys Rev Lett 2002;9 (24):248101.

[3] Wong EW, Sheehan PE, Lieber CM. Nanobeam mechanics: Elasticity, strength, and toughness of nanorods and nanotubes. Science 1997;277(5334):1971–5.

[4] Ru CQ. Elastic models for carbon nanotubes[M]//Encyclopedia of nanoscience and nanotechnology2. American Scientific Publishers; 2004. p. 731–44.

[5] Love AEH. A treatise on the mathematical theory of elasticity. Courier Corporation; 1944.

[6] Sokolnikoff IS. Mathematical theory of elasticity. New York: Robert E. Krieger Publishing Company; 1983.

[7] Cheung YK, Wu CI. Free vibrations of thick, layered cylinders having finite length with various boundary conditions. J Sound Vib 1972;24(2):189–200.

[8] Hutchinson JR. Transverse vibrations of beams, exact versus approximate solutions. J ApplMech 1981;48(4):923–8.

[9] Hutchinson JR. Axisymmetric vibrations of a free finite-length rod. J AcoustSoc Am 1972;51(1B):233–40.

[10] Leissa AW, So J. Accurate vibration frequencies of circular cylinders from three-dimensional analysis. J AcoustSoc Am 1995;98(4):2136–41.
[11] Leissa AW, So J. Comparisons of vibration frequencies for rods and beams from one-dimensional and three-dimensional analyses. J AcoustSoc Am 1995;98(4):2122–35.
[12] Karnovsky IA, Lebed O. Formulas for structural dynamics: Tables, graphs and solutions. New York: McGraw-Hill Professional; 2000.
[13] HanS M, Benaroya H, Wei T. Dynamics of transversely vibrating beams using four engineering theories. J Sound Vib 1999;225(5):935–88.
[14] Li XF, Yu ZW, Zhang H. Free vibration of shear beams with finite rotational inertia. J Constr Steel Res 2011;67(10):1677–83.
[15] Asaro R, Lubarda V. Mechanics of solids and materials. Cambridge University Press; 2006.
[16] Huang Y, Wu JX, Li XF, et al. Higher-order theory for bending and vibration of beams with circular cross section. J Eng Math 2013;80(1):91–104.
[17] Timoshenko SP. LXVI. On the correction for shear of the differential equation for transverse vibrations of prismatic bars41. London: Edinburgh Dublin Philosophical Magazine and Journal of Science; 1921. p. 744–6.
[18] Timoshenko SP. X. On the transverse vibrations of bars of uniform cross-section, 43. London: Edinburgh Dublin Philosophical Magazine and Journal of Science; 1922. p. 125–31.
[19] Esmailzadeh E, Ghorashi M. Vibration analysis of a Timoshenko beam subjected to a travelling mass. J Sound Vib 1997;199(4):615–28.
[20] Krawczuk M, Palacz M, Ostachowicz W. The dynamic analysis of a cracked Timoshenko beam by the spectral element method. J Sound Vib 2003;264(5):1139–53.
[21] Salarieh H, Ghorashi M. Free vibration of Timoshenko beam with finite mass rigid tip load and flexural–torsional coupling. Int J MechSci 2006;48(7):763–79.
[22] Li XF, Wang BL. Vibrational modes of Timoshenko beams at small scales. Appl Phys Lett 2009;94(10).
[23] Blaauwendraad J. Shear in structural stability: On the Engesser–Haringxdiscord. J ApplMech 2010;77(3):031005.
[24] Le Grognec P, Nguyen QH, Hjiaj M. Exact buckling solution for two-layer Timoshenko beams with interlayer slip. Int J Solid Struct 2012;49(1):143–50.
[25] Cowper GR. The shear coefficient in Timoshenko's beam theory. J ApplMech 1966;33 (2):335–40.
[26] Hutchinson JR. Shear coefficients for Timoshenko beam theory. J ApplMech 2001;68 (1):87–92.
[27] Kaneko T. On Timoshenko's correction for shear in vibrating beams. J Phys Appl Phys 1975;8(16):1927.
[28] Yu W, Hodges DH. Generalized Timoshenko theory of the variational asymptotic beam sectional analysis. J Am Helicopter Soc 2005;50(1):46–55.
[29] Levinson M. A new rectangular beam theory. J Sound Vib 1981;74(1):81–7.
[30] Reddy JN. A simple higher-order theory for laminated composite plates. J ApplMech 1984;51(4):745–52.
[31] Wadee MK, Wadee MA, Bassom AP, et al. Longitudinally inhomogeneous deformation patterns in isotropic tubes under pure bending. Proc R Soc A 2006;462(2067):817–38.
[32] Li XF. A unified approach for analyzing static and dynamic behaviors of functionally graded Timoshenko and Euler–Bernoulli beams. J Sound Vib 2008;318(4–5):1210–29.
[33] Suresh S. Graded materials for resistance to contact deformation and damage. Science 2001;292(5526):2447–51.

[34] Sankar BV. An elasticity solution for functionally graded beams. Compos SciTechnol 2001;61(5):689–96.
[35] Aydogdu M, Taskin V. Free vibration analysis of functionally graded beams with simply supported edges. Mater Des 2007;28(5):1651–6.
[36] Huang Y, Li XF. A new approach for free vibration of axially functionally graded beams with non-uniform cross-section. J Sound Vib 2010;329(11):2291–303.
[37] Li XF, Kang YA, Wu JX. Exact frequency equations of free vibration of exponentially functionally graded beams. ApplAcoust 2013;74(3):413–20.
[38] Tang AY, Wu JX, Li XF, et al. Exact frequency equations of free vibration of exponentially non-uniform functionally graded Timoshenko beams. Int J MechSci 2014;89:1–11.
[39] Simsek M. Bi-directional functionally graded materials (BDFGMs) for free and forced vibration of Timoshenko beams with various boundary conditions. Compos Struct 2015;133:968–78.
[40] Nguyen DK, Nguyen QH, Tran TT, et al. Vibration of bi-dimensional functionally graded Timoshenko beams excited by a moving load. ActaMech 2017;228(1):141–55.
[41] Ghayesh MH. Nonlinear vibration analysis of axially functionally graded shear-deformable tapered beams. Appl Math Model 2018;59:583–96.
[42] Sun DL, Li XF. Initial value method for free vibration of axially loaded functionally graded Timoshenko beams with nonuniform cross section. Mech Base Des Struct Mach 2019;47 (1):102–20.
[43] Tang Y, Ding Q. Nonlinear vibration analysis of a bi-directional functionally graded beam under hygro-thermal loads. Compos Struct 2019;225:111076.
[44] Nguyen HX, Nguyen TN, Abdel-Wahab M, et al. A refined quasi-3D isogeometric analysis for functionally graded microplates based on the modified couple stress theory. Comput Methods ApplMechEng 2017;313:904–40.
[45] Nguyen NT, Hui D, Lee J, et al. An efficient computational approach for size-dependent analysis of functionally graded nanoplates. Comput Methods ApplMechEng 2015;297:191–218.
[46] Nguyen TN, Ngo TD, Nguyen-Xuan H. A novel three-variable shear deformation plate formulation: theory and Isogeometricimplementation. Comput Methods ApplMechEng 2017;326:376–401.
[47] Vel SS, Batra RC. Three-dimensional exact solution for the vibration of functionally graded rectangular plates. J Sound Vib 2004;272(3–5):703–30.

Chapter 2

Bending of higher-order functionally graded circular cylindrical beams

2.1 Introduction

Cylinders with an annular cross-section are widely used in many engineering fields from the macro scale to the micro scale, such as steel pipes, carbon nanotubes, microtubules, etc. As one of the most important structure units, cylindrical beams have gained increasing numbers of applications in many areas such as the mechanical, civil, aerospace, and marine engineering sectors in recent years. In these applications, the tubes or pipes in service are loaded by complex mechanical loading which directly affects the structural integrity, reliability, and stability of structures. In contrast to the Euler–Bernoulli and Timoshenko theories of beams, a third-order shear deformation beam theory for radially graded beams or tubes is proposed accounting for warping, shear deformation, and rotational moment of inertia of the cross-section. Coupled governing equations for deflection and rotation about the neutral axis of the cross-section are derived from equilibrium equations and are then converted into a single fourth-order partial differential governing equation. The deflection and stress distribution for various ends are derived explicitly.

2.2 Assumption and governing equations

2.2.1 Cylinder cylindrical beams

Consider a functionally graded cylindrical beam-column with uniform circular cross-section of radius R which is deflected under the action of transverse loading or bending moment [1]. In this work, the material properties of the circular cylindrical beam are assumed to vary arbitrarily along the radial direction. For such a structure, we choose a Cartesian coordinate system such that the x-axis is oriented in the neutral axis of the cylindrical beam, the positive z-axis is directed upward and perpendicular to the x-axis, and u, v, and w are the corresponding elastic displacement components in the x, y, and z

Higher-Order Theories of Functionally Graded Circular Cylindrical Beams and Columns
https://doi.org/10.1016/B978-0-443-44718-1.00007-4

directions, respectively. Meanwhile, we also adopt polar cylindrical coordinates (x, r, θ) and the corresponding displacement components (u, u_r, u_θ), which satisfy the following relations (Fig. 2.1):

$$\begin{cases} y = r\cos\theta, z = r\sin\theta \\ u_r = v\cos\theta + w\sin\theta \end{cases} \tag{2.1}$$

When the beam is restricted to transverse motion in the xoz-plane, the elastic displacement component v in the y-direction can be assumed approximately independent of x, and the deflection w in the z-direction is dependent on x, $w = w(x, t)$. To satisfy the shear-stress-free condition at the surface of the cylindrical beam, the longitudinal displacement u in the x-direction can be expressed as follows:

$$u = z\psi - \frac{z^3 + zy^2}{3R^2}\left(\frac{\partial w}{\partial x} + \psi\right) \tag{2.2}$$

where ψ is the rotation of the cross-section at the centroidal axis. The deflection w and the rotation ψ are functions of the longitudinal coordinate x and time t and are positive in the z-direction and in a clockwise manner, respectively. As a result, using the kinematic relations the normal strain ε_{xx} and shear strains $\gamma_{xy}, \gamma_{xz}, \gamma_{xr}$ can be expressed as follows:

$$\varepsilon_{xx} = \frac{\partial u}{\partial x} = z\frac{\partial \psi}{\partial x} - \frac{z^3 + zy^2}{3R^2}\left(\frac{\partial^2 w}{\partial x^2} + \frac{\partial \psi}{\partial x}\right) \tag{2.3}$$

$$\gamma_{xy} = \frac{\partial u}{\partial y} + \frac{\partial v}{\partial x} = -\frac{2zy}{3R^2}\left(\frac{\partial w}{\partial x} + \psi\right) \tag{2.4}$$

$$\gamma_{xz} = \frac{\partial u}{\partial z} + \frac{\partial w}{\partial x} = \frac{\partial w}{\partial x} + \psi - \frac{3z^2 + y^2}{3R^2}\left(\frac{\partial w}{\partial x} + \psi\right) \tag{2.5}$$

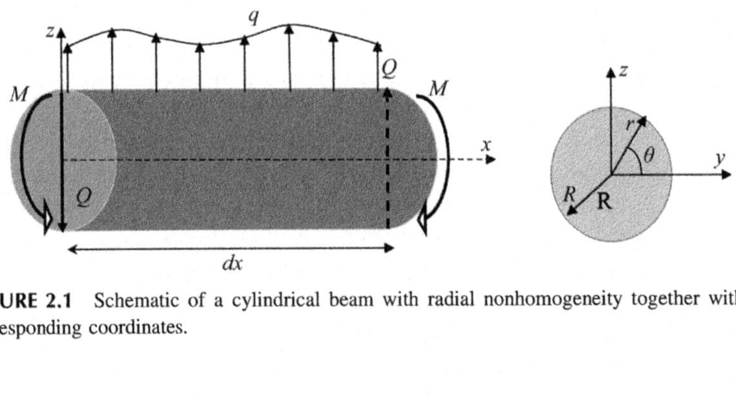

FIGURE 2.1 Schematic of a cylindrical beam with radial nonhomogeneity together with the corresponding coordinates.

$$\gamma_{xr} = \frac{\partial u_r}{\partial x} + \frac{\partial u}{\partial r} = \frac{z}{rR^2}\left(R^2 - z^2 - y^2\right)\left(\psi + \frac{\partial w}{\partial x}\right) \tag{2.6}$$

According to Saint-Venant's principle, the following stress components are assumed to vanish in a beam: $\sigma_{yy} = \sigma_{zz} = \tau_{yz} = 0$ [2]. In this case, the 3D constitutive equations reduce to the following form:

$$\sigma_{xx} = E(r)\varepsilon_{xx} = E(r)\left[z\frac{\partial \psi}{\partial x} - \frac{z^3 + zy^2}{3R^2}\left(\frac{\partial^2 w}{\partial x^2} + \frac{\partial \psi}{\partial x}\right)\right] \tag{2.7}$$

$$\tau_{xy} = G(r)\gamma_{xy} = -\frac{2G(r)zy}{3R^2}\left(\frac{\partial w}{\partial x} + \psi\right) \tag{2.8}$$

$$\tau_{xz} = G(r)\gamma_{xz} = G(r)\left[\frac{\partial w}{\partial x} + \psi - \frac{3z^2 + y^2}{3R^2}\left(\frac{\partial w}{\partial x} + \psi\right)\right] \tag{2.9}$$

where σ_{xx}, τ_{xy}, and τ_{xz} are the normal stress and the shear stresses, respectively; and *E*(*r*) and *G*(*r*) are the Young's modulus and shear modulus of the cylindrical beam dependent on *r*, respectively. In addition, from Eq. (2.6), we can see that the shear-stress-free surface condition is completely satisfied owing to $\tau_{xr} = G(r)\gamma_{xr}$. Therefore, the above relations allow us to analyze flexural behavior as well as the transverse vibration of cylindrical beams where the shear stress at the cross-section is taken into account.

From the equilibrium of internal forces, the bending moment *M* and shear force *Q* at any position *x* of the beam are

$$\begin{cases} M = \int_A \sigma_{xx} z \mathrm{d}A \\ Q = \int_A \tau_{xz} \mathrm{d}A \end{cases} \tag{2.10}$$

respectively, where *A* stands for the area of the cross-section. Note that in the Timoshenko beam theory, the shear stress is assumed to be uniform and a shear correction factor must be introduced, which cannot be directly determined by the model itself. Here the uniform stress assumption is abandoned. Instead, the presence of shear stress is acknowledged, and it is required to satisfy the shear-stress-free surface condition. Moreover, it is unnecessary to introduce an additional shear correction factor in our model. Substitution of Eqs. (2.7) and (2.9) for σ_{xx} and τ_{xz}, respectively, into the above integrals results in

$$M = \widetilde{E}_2\frac{\partial \psi}{\partial x} - E_4\frac{\partial^2 w}{\partial x^2} \tag{2.11}$$

$$Q = \widetilde{G}_0\left(\psi + \frac{\partial w}{\partial x}\right) \tag{2.12}$$

where

$$\widetilde{E}_2 = E_2 - E_4, \widetilde{G}_0 = G_0 - G_2$$
$$E_2 = \pi \int_0^R E(r) r^3 \mathrm{d}r, E_4 = \frac{\pi}{3R^2} \int_0^R E(r) r^5 \mathrm{d}r \tag{2.13}$$
$$G_0 = 2\pi \int_0^R G(r) r \mathrm{d}r, G_2 = \frac{4\pi}{3R^2} \int_0^R G(r) r^3 \mathrm{d}r$$

Because the cylindrical beam is deflected in the z-direction subjected to transverse loading or bending moment, we then neglect the motion along the y-direction. Thus, with reference to flexural deformation in the xoz-plane, only two equations of motion are involved:

$$\frac{\partial \sigma_{xx}}{\partial x} + \frac{\partial \tau_{xy}}{\partial y} + \frac{\partial \tau_{xz}}{\partial z} = 0 \tag{2.14}$$

$$\frac{\partial \tau_{xz}}{\partial x} + \frac{\partial \tau_{yz}}{\partial y} + \frac{\partial \sigma_{zz}}{\partial z} = 0 \tag{2.15}$$

where $\rho(r)$ is the mass density of the beam depending on r.

After substituting Eqs. (2.7)–(2.9) into Eq. (2.14), both sides of Eq. (2.14) are multiplied by z and then integrated over the cross-sectional area A, resulting in

$$\widetilde{E}_2 \frac{\partial^2 \psi}{\partial x^2} - E_4 \frac{\partial^3 w}{\partial x^3} - \widetilde{G}_0 \left(\frac{\partial w}{\partial x} + \psi \right) = 0 \tag{2.16}$$

Similarly, direct integration of Eq. (2.15) over the cross-sectional area A gives

$$\widetilde{G}_0 \left(\frac{\partial^2 w}{\partial x^2} + \frac{\partial \psi}{\partial x} \right) = -q \tag{2.17}$$

where q is the distributed force with units of N/m, applied along the centroidal axis of the beam in the z-direction. We therefore obtain two coupled governing equations [Eqs. (2.16) and (2.17)] for the bending of functionally graded circular cylindrical beams.

A further simplification of the resulting two coupled equations [Eqs. (2.16) and (2.17)] can be achieved by introducing a new auxiliary function. This is accomplished by expressing the deflection w and the rotation ψ in the form

$$w = F - \frac{\widetilde{E}_2}{\widetilde{G}_0} \frac{\partial^2 F}{\partial x^2} \tag{2.18}$$

$$\psi = -\frac{E_4}{\widetilde{G}_0} \frac{\partial^3 F}{\partial x^3} - \frac{\partial F}{\partial x} \tag{2.19}$$

where F is a new auxiliary function of length dimension, it has been readily confirmed that when substituting Eqs. (2.18) and (2.19) into Eq. (2.16), it is completely fulfilled. On the other hand, when substituting Eqs. (2.18) and (2.19) into Eq. (2.17), we get

$$E_2 \frac{\partial^4 F}{\partial x^4} = q \tag{2.20}$$

It should be noted that a similar governing equation has been obtained for a rectangular functionally graded Timoshenko beam in Ref. [3]. As pointed out above, the introduction of a shear correction factor is inevitable in the Timoshenko beam, but now, no additional factor is needed for a functionally graded circular cylindrical beam. For a circular cylindrical beam, the above governing equation has not been derived, even for a homogeneous case.

Once the solution to the governing equation, F, is obtained through appropriate boundary conditions, all the physical quantities of interest including the deflection w, rotation ψ, bending moment M, and shear force Q can be obtained. For instance, w and ψ are given by Eqs. (2.18) and (2.19), respectively. Referring to Eqs. (2.11) and (2.12), the bending moment M and shear force Q can be expressed in terms of F as follows

$$\begin{cases} M = -E_2 \dfrac{\partial^2 F}{\partial x^2} \\ Q = -E_2 \dfrac{\partial^3 F}{\partial x^3} \end{cases} \tag{2.21}$$

Consider a special case of homogeneous cylindrical beams with constant material properties: $E(r) = E$, $G(r) = G$, and $\rho(r) = \rho$. We have

$$\begin{cases} E_2 = EI, \widetilde{E}_2 = \dfrac{7}{9} EI, \widetilde{G}_0 = \dfrac{2}{3} GA \\ \rho_2 = \rho I, \widetilde{\rho}_2 = \dfrac{7}{9} \rho I, \rho_0 = \rho A \end{cases} \tag{2.22}$$

where $I = \pi R^4/4$ is the inertia moment of the cross-sectional area, and the above-derived governing equation then reduces to

$$EI \frac{\partial^4 F}{\partial x^4} = q \tag{2.23}$$

2.2.2 Cylindrical tube beams

The bending of a straight cylindrical tube with radially varying material properties over an annular cross-section has been considered [4], and the geometry of a radially graded hollow cylinder is shown in Fig. 2.2. We denote its inner and outer radii as R_i and R_o, respectively, and the Young's modulus $E(r)$ and shear modulus $G(r)$ of the tube are dependent on r, r being the radius,

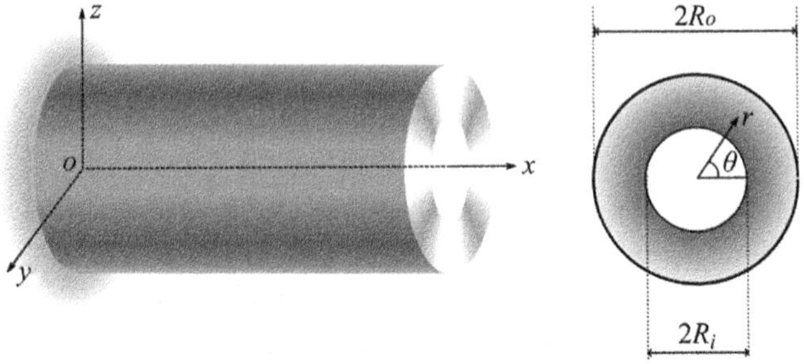

FIGURE 2.2 Schematic of a circular cylindrical beam with radial gradient together with the corresponding coordinates.

i.e., $R_i \le r \le R_o$. In the following, we choose a Cartesian coordinate system and a cylindrical polar coordinate system (Fig. 2.2), where the x-axis coincides with the centroidal axis of the tube, y, z are related to r, θ by

$$\begin{cases} y = r\cos\theta \\ z = r\sin\theta \end{cases} \tag{2.24}$$

Consider a straight hollow tube subjected to a transverse force or distributed loading in the xoz-plane and a bending moment perpendicular to the xoz-plane, the tube is deflected in the xoz-plane. We denote three displacement components by (u, u_r, u_θ) according to the coordinates (x, r, θ), and (u, v, w) according to the coordinates (x, y, z), respectively. It is clear that they are related by the following relation:

$$u_r = v\cos\theta + w\sin\theta \tag{2.25}$$

In polar coordinates, based on Eqs. (2.24) and (2.25), the relationship between the shear strain and the displacement vector can be expressed as

$$\gamma_{xr} = \frac{\partial u_r}{\partial x} + \frac{\partial u}{\partial r} = \frac{1}{r}\left(y\frac{\partial v}{\partial x} + z\frac{\partial w}{\partial x} + y\frac{\partial u}{\partial y} + z\frac{\partial u}{\partial z}\right) \tag{2.26}$$

Consider a simple case, that is, the displacement component v in the y-direction is independent of x. The shear strain can be rewritten as

$$\gamma_{xr} = \frac{1}{r}\left[z\left(\frac{\partial w}{\partial x} + \frac{\partial u}{\partial z}\right) + y\frac{\partial u}{\partial y}\right] \tag{2.27}$$

If taking the shear strain γ_{xr} of the form

$$y\frac{\partial u}{\partial y} + z\left(\frac{\partial w}{\partial x} + \frac{\partial u}{\partial z}\right) = z\frac{\left(y^2 + z^2 - R_i^2\right)\left(y^2 + z^2 - R_o^2\right)}{y^2 + z^2}\zeta(x) \tag{2.28}$$

where $\zeta(x)$ is an unknown function, it is evident that the shear-stress-free boundary conditions

$$\begin{cases} \tau_{xr}(x,r) = 0 \\ r = R_i, R_o \end{cases} \tag{2.29}$$

are automatically satisfied at the inner and outer surface of the tube owing to the shear Hooke's law

$$\tau_{xr}(x,r) = G(x)\gamma_{xr}(x,r) \tag{2.30}$$

To obtain the expression for $\zeta(x)$, u is separated into two parts

$$u(x,y,z) = u_1(x,z) + \left(\frac{1}{3}y^2 - \frac{R_i^2 R_o^2}{y^2 + z^2}\right) z\zeta(x) \tag{2.31}$$

Substituting Eq. (2.31) into Eq. (2.28), we find

$$\frac{\partial u_1}{\partial z} = -\frac{\partial w}{\partial x} + \left[z^2 - \left(R_i^2 + R_o^2\right)\right]\zeta(x) \tag{2.32}$$

Integrating both sides of Eq. (2.32) with respect to z yields

$$u_1 = \frac{1}{3}z^3\zeta(x) - z\left[\left(R_i^2 + R_o^2\right)\zeta(x) + \frac{\partial w}{\partial x}\right] \tag{2.33}$$

Then, we introduce a new function ψ representing the rotation angle of u_1, ψ can be expressed as

$$\psi(x) = -\left(R_i^2 + R_o^2\right)\zeta(x) - \frac{\partial w}{\partial x} \tag{2.34}$$

After combining Eq. (2.34) and Eq. (2.31), $\zeta(x)$ can be eliminated immediately, then we get an expression for the displacement component u below

$$u = z\psi + z\frac{3R_i^2R_o^2 - \left(y^2 + z^2\right)^2}{3\left(R_i^2 + R_o^2\right)\left(y^2 + z^2\right)}\left(\frac{\partial w}{\partial x} + \psi\right) \tag{2.35}$$

With the above expression for u, the normal strain ε_{xx} and shear strains γ_{xy}, γ_{xz}, and γ_{xr} are immediately obtained as follows

$$\varepsilon_{xx} = z\frac{\partial \psi}{\partial x} + z\frac{3R_i^2R_o^2 - \left(y^2 + z^2\right)^2}{3\left(R_i^2 + R_o^2\right)\left(y^2 + z^2\right)}\left(\frac{\partial^2 w}{\partial x^2} + \frac{\partial \psi}{\partial x}\right) \tag{2.36}$$

$$\gamma_{xy} = -2yz\frac{3R_i^2R_o^2 + \left(y^2 + z^2\right)^2}{3\left(R_i^2 + R_o^2\right)\left(y^2 + z^2\right)^2}\left(\frac{\partial w}{\partial x} + \psi\right) \tag{2.37}$$

$$\gamma_{xz} = \left[1 + \frac{3R_i^2R_o^2\left(y^2 - z^2\right) - \left(y^2 + 3z^2\right)\left(y^2 + z^2\right)^2}{3\left(R_i^2 + R_o^2\right)\left(y^2 + z^2\right)^2}\right]\left(\frac{\partial w}{\partial x} + \psi\right) \tag{2.38}$$

$$\gamma_{xr} = \frac{z}{r}\left[1 - \frac{R_i^2 R_o^2 + \left(y^2+z^2\right)^2}{\left(R_i^2+R_o^2\right)\left(y^2+z^2\right)}\right]\left(\frac{\partial w}{\partial x}+\psi\right) \tag{2.39}$$

Based on the assumption of beams, the stress components σ_{yy}, σ_{zz}, and τ_{yz} are considered to vanish in the whole beam [2]. In addition, the shear stresses vanish on the inner and outer surfaces of a tube. This is clearly seen from the following relationships:

$$\sigma_{xx} = E(r)\varepsilon_{xx} = E(r)\left[z\frac{\partial \psi}{\partial x} + z\frac{3R_i^2R_o^2 - \left(y^2+z^2\right)^2}{3\left(R_i^2+R_o^2\right)\left(y^2+z^2\right)}\left(\frac{\partial^2 w}{\partial x^2}+\frac{\partial \psi}{\partial x}\right)\right] \tag{2.40}$$

$$\tau_{xy} = G(r)\gamma_{xy} = -2G(r)yz\frac{3R_i^2R_o^2 + \left(y^2+z^2\right)^2}{3\left(R_i^2+R_o^2\right)\left(y^2+z^2\right)^2}\left(\frac{\partial w}{\partial x}+\psi\right) \tag{2.41}$$

$$\tau_{xz} = G(r)\gamma_{xz} = G(r)\left[1 + \frac{3R_i^2R_o^2\left(y^2-z^2\right) - \left(y^2+3z^2\right)\left(y^2+z^2\right)^2}{3\left(R_i^2+R_o^2\right)\left(y^2+z^2\right)^2}\right]\left(\frac{\partial w}{\partial x}+\psi\right) \tag{2.42}$$

$$\tau_{xr} = G(r)\frac{z}{r}\left[1 - \frac{R_i^2R_o^2 + \left(y^2+z^2\right)^2}{\left(R_i^2+R_o^2\right)\left(y^2+z^2\right)}\right]\left(\frac{\partial w}{\partial x}+\psi\right) \tag{2.43}$$

The above relations allow us to analyze the flexural behavior as well as the flexural vibration of straight radially graded cylindrical tubes, where the shear stresses vanish on the inner and outer surfaces of the tube. According to the Timoshenko beam theory, the shear stress on the cross-section is assumed to be uniform, and a shear correction coefficient must be introduced to relax the traction-free condition on the surface of the beam [5]. The shear correction coefficient is an inherent part of this theory and cannot be directly determined by the theory itself. Here the uniform stress assumption is abandoned. The shear stress is assumed to satisfy the shear-stress-free surface condition, making the shear correction coefficient unnecessary. Due to the equilibrium of internal force, the bending moment M and shear force Q at any position x are

$$M = \int_A \sigma_{xx} z \mathrm{d}A \tag{2.44}$$

$$Q = \int_A \tau_{xz} \, \mathrm{d}A \tag{2.45}$$

respectively, where A is the cross-sectional area. Substituting Eqs. (2.40) and (2.42) into the above integrals, one gets

$$M = \widetilde{E}_2 \frac{\partial \psi}{\partial x} - E_4 \frac{\partial^2 w}{\partial x^2} \tag{2.46}$$

$$Q = \widetilde{G}_0 \left(\psi + \frac{\partial w}{\partial x} \right) \tag{2.47}$$

where

$$\begin{cases} \widetilde{E}_2 = E_2 - E_4 \\ \widetilde{G}_0 = G_0 - G_2 \end{cases} \tag{2.48}$$

$$\begin{cases} E_2 = \pi \int_{R_i}^{R_o} E(r) r^3 \mathrm{d}r \\ E_4 = \dfrac{\pi}{3\left(R_i^2 + R_o^2\right)} \int_{R_i}^{R_o} E(r)\left(r^4 - 3R_i^2 R_o^2\right) r \mathrm{d}r \end{cases} \tag{2.49}$$

$$\begin{cases} G_0 = 2\pi \int_{R_i}^{R_o} G(r) r \mathrm{d}r \\ G_2 = \dfrac{4\pi}{3\left(R_i^2 + R_o^2\right)} \int_{R_i}^{R_o} G(r) r^3 \mathrm{d}r \end{cases} \tag{2.50}$$

Since our attention is focused on a hollow tube subjected to a transverse force or distributed loading in the *xoz*-plane and a bending moment perpendicular to the *xoz*-plane, the bending of the tube takes place in the *xoz*-plane. Thus, the motion in the *y*-direction is neglected, so the remaining two equations of motion read

$$\frac{\partial \sigma_{xx}}{\partial x} + \frac{\partial \tau_{xy}}{\partial y} + \frac{\partial \tau_{xz}}{\partial z} = 0 \tag{2.51}$$

$$\frac{\partial \tau_{xz}}{\partial x} + \frac{\partial \tau_{yz}}{\partial y} + \frac{\partial \sigma_{zz}}{\partial z} = 0 \tag{2.52}$$

Now we substitute Eqs. (2.40)–(2.42) into Eq. (2.51), multiply both sides by z and then integrate over the cross-section. After some manipulations we obtain

$$\widetilde{E}_2 \frac{\partial^2 \psi}{\partial x^2} - E_4 \frac{\partial^3 w}{\partial x^3} - \widetilde{G}_0 \left(\frac{\partial w}{\partial x} + \psi \right) = 0 \tag{2.53}$$

Then we perform integration on both sides of Eq. (2.52) over the cross-section, obtaining

$$\widetilde{G}_0 \left(\frac{\partial^2 w}{\partial x^2} + \frac{\partial \psi}{\partial x} \right) = -q \tag{2.54}$$

where q is a distributed lateral loading along the centroidal axis in the *xoz*-plane. Thus, one obtains two coupled equations [Eqs. (2.53) and (2.54)] governing the dynamic behaviors of radially graded cylindrical tubes.

A further simplification can be achieved by introducing a new auxiliary function of F with length dimension, which satisfies

$$E_2 \frac{\partial^4 F}{\partial x^4} = q \tag{2.55}$$

Choosing the deflection w and the rotation ψ in the form

$$w = F - \frac{\widetilde{E}_2}{\widetilde{G}_0} \frac{\partial^2 F}{\partial x^2} \tag{2.56}$$

$$\psi = -\frac{E_4}{\widetilde{G}_0} \frac{\partial^3 F}{\partial x^3} - \frac{\partial F}{\partial x} \tag{2.57}$$

We can see that after substituting Eqs. (2.56) and (2.57) into Eqs. (2.53) and (2.54), they are completely fulfilled. Therefore, we only need to determine the auxiliary function F through appropriate boundary conditions. For convenience, we also provide bending moment M and shear force Q in terms of F. This is done by substituting Eqs. (2.56) and (2.57) into Eqs. (2.46) and (2.47). The bending moment M and shear force Q can be expressed as

$$M = -E_2 \frac{\partial^2 F}{\partial x^2} \tag{2.58}$$

$$Q = -E_2 \frac{\partial^3 F}{\partial x^3} \tag{2.59}$$

The remaining task is to solve the governing equation [Eq. (2.55)] for the unknown F under suitable boundary conditions, which can be provided by the deflection w, rotation ψ, bending moment M, and shear force Q.

According to the governing equation for static bending of circular cylindrical beams or tubes with radial nonhomogeneity, the governing equation reduces to the following expression:

$$E_2 \frac{d^4 F}{dx^4} = q(x) \tag{2.60}$$

2.3 Analytical solution of several boundary conditions

A general solution to Eqs. (2.23) and (2.60) is

$$F(x) = \frac{1}{6E_2} \int_0^x (x-s)^3 q(s) ds + A_1 x^3 + A_2 x^2 + A_3 x + A_4 \tag{2.61}$$

where $A_i (i = 1, \ldots, 4)$ are unknown constants to be obtained through appropriate boundary conditions. In the following, we consider two typical cases: cantilever beams and simply supported beams. For other cases, the exact solution is also easily obtained and so is not included here.

2.3.1 Circular cylindrical beams

2.3.1.1 Cantilever circular cylindrical beam subjected to uniform distributed loading

Let us first consider a cantilever beam with circular cross-section subjected to uniform distributed force q. Hence, the boundary conditions can be written as

$$\begin{cases} w(0) = 0 \\ \psi(0) = 0 \\ M(L) = 0 \\ Q(L) = 0 \end{cases} \tag{2.62}$$

for a cantilever with the free end at $x = L$ and the clamped end at $x = 0$. Based on the above boundary conditions, the coefficients A_j in Eq. (2.61) can be easily obtained. After calculation, the final expression for $F(x)$ is

$$F(x) = \frac{q}{24E_2}\left(x^4 - 4Lx^3 + 6L^2x^2 + \frac{24E_4L}{\widetilde{G}_0}x + \frac{12\widetilde{E}_2L^2}{\widetilde{G}_0}\right) \tag{2.63}$$

With $F(x)$ at hand, we can obtain all physical quantities of interest as follows:

$$\begin{cases} u = -\dfrac{q}{6E_2}\left[\left(x^3 - 3Lx^2 + 3L^2x\right)z + \dfrac{6E_4xz}{\widetilde{G}_0} - \dfrac{2E_2(x-L)\left(z^3 + zy^2\right)}{\widetilde{G}_0R^2}\right] \\ w = \dfrac{q}{24E_2}\left[\left(x^4 - 4Lx^3 + 6L^2x^2\right) + \dfrac{12}{\widetilde{G}_0}\left(2E_2L - \widetilde{E}_2x\right)x\right] \\ \psi = -\dfrac{q}{6E_2}\left[\left(x^3 - 3Lx^2 + 3L^2x\right) + \dfrac{6E_4x}{\widetilde{G}_0}\right] \end{cases} \tag{2.64}$$

and

$$\begin{cases} \sigma_{xx} = -\dfrac{qE(r)}{E_2}\left[\dfrac{1}{2}\left(x^2 - 2Lx + L^2\right)z + \dfrac{E_4z}{\widetilde{G}_0} - \dfrac{E_2\left(z^3 + zy^2\right)}{3\widetilde{G}_0R^2}\right] \\ \tau_{xy} = -\dfrac{2qG(r)}{3\widetilde{G}_0R^2}(x-L)zy \\ \tau_{xz} = \dfrac{qG(r)}{3\widetilde{G}_0R^2}(x-L)\left(3R^2 - 3z^2 - y^2\right) \end{cases} \tag{2.65}$$

From the above, it is observed that the contribution of flexural behavior is composed of two parts, one arising from the bending moment and the other from shear force. Letting $\widetilde{G}_0$ tend to infinity, equivalently to the fact that the beam has a sufficient high reduced shear rigidity, the above results reduce to those for Euler–Bernoulli functionally graded beams of circular cross-section.

2.3.1.2 Simply supported circular cylindrical beam subjected to a concentrated force

Next, we consider a simply supported circular cylindrical beam subjected to a concentrated force ($P_y = 0,\ P_z = P$) at the mid-point of the beam. From symmetry, one can write the following conditions:

$$\begin{cases} w(0) = 0 \\ M(0) = 0 \\ Q\left(\dfrac{L}{2}\right) = \dfrac{P}{2} \\ \psi\left(\dfrac{L}{2}\right) = 0 \end{cases} \tag{2.66}$$

Based on these conditions, we obtain $F(x)$ as follows:

$$F(x) = -\frac{P}{12E_2}x^3 + \frac{P}{2E_2}\left(\frac{L^2}{8} + \frac{E_4}{\widetilde{G}_0}\right)x, 0 \le x \le \frac{L}{2} \tag{2.67}$$

With this $F(x)$, we can obtain all physical quantities of interest. For brevity, we only give those for $x \in [0, L/2]$ below. Those for $x \in [L/2, L]$ can be obtained directly by symmetry

$$\begin{cases} u = \dfrac{P}{4E_2}\left(x^2 - \dfrac{L^2}{4}\right)z - \dfrac{P}{6\widetilde{G}_0}\dfrac{z^3 + zy^2}{R^2} \\ w = -\dfrac{P}{12E_2}x^3 + \dfrac{P}{2E_2}\left(\dfrac{L^2}{8} + \dfrac{E_2}{\widetilde{G}_0}\right)x \\ \psi = \dfrac{P}{4E_2}\left(x^2 - \dfrac{L^2}{4}\right) \end{cases} \tag{2.68}$$

and

$$\begin{cases} \sigma_{xx} = \dfrac{PE(r)}{2E_2}xz \\ \tau_{xy} = -\dfrac{PG(r)}{3\widetilde{G}_0R^2}zy \\ \tau_{xz} = \dfrac{PG(r)}{6\widetilde{G}_0R^2}\left(3R^2 - 3z^2 - y^2\right) \end{cases} \tag{2.69}$$

2.3.2 Cylindrical tube beams

2.3.2.1 Cantilever tube beam subjected to uniform distributed loading

Here we consider a cantilever tube beam subjected to uniform distributed loading q. In this case, the boundary conditions for a cantilever tube beam with the free end at $x = L$ and the clamped end at $x = 0$ are stated as

$$\begin{cases} w(0) = 0 \\ \psi(0) = 0 \\ M(L) = 0 \\ Q(L) = 0 \end{cases} \tag{2.70}$$

where L is the length of the tube beam. Using the above boundary conditions, it is a simple matter to determine A_j in Eq. (2.61). Omitting the details, one gets the final expression for $F(x)$ as follows:

$$F(x) = \frac{q}{24E_2}\left(x^4 - 4Lx^3 + 6L^2x^2 + \frac{24E_4L}{\widetilde{G}_0}x + \frac{12\widetilde{E}_2L^2}{\widetilde{G}_0}\right) \tag{2.71}$$

With $F(x)$ at hand, we can obtain all physical quantities of interest as follows:

$$\begin{cases} u = -\dfrac{qz}{6E_2}\left[x^3 - 3Lx^2 + 3L^2x + \dfrac{6E_4x}{\widetilde{G}_0} - \dfrac{6R_i^2R_o^2 - 2\left(y^2+z^2\right)^2}{\left(R_i^2+R_o^2\right)\left(y^2+z^2\right)}\dfrac{E_2(L-x)}{\widetilde{G}_0}\right] \\ w = \dfrac{q}{24E_2}\left[x^4 - 4Lx^3 + 6L^2x^2 + \dfrac{12}{\widetilde{G}_0}\left(2E_2L - \widetilde{E}_2x\right)x\right] \\ \psi = -\dfrac{q}{6E_2}\left(x^3 - 3Lx^2 + 3L^2x + \dfrac{6E_4x}{\widetilde{G}_0}\right) \end{cases} \tag{2.72}$$

and

$$\begin{cases} \sigma_{xx} = -\dfrac{qzE(r)}{2E_2}\left[x^2 - 2Lx + L^2 + \dfrac{2E_4}{\widetilde{G}_0} + \dfrac{3R_i^2R_o^2 - \left(y^2+z^2\right)^2}{3\left(R_i^2+R_o^2\right)\left(y^2+z^2\right)}\dfrac{2E_2}{\widetilde{G}_0}\right] \\ \tau_{xy} = \dfrac{2qG(r)yz}{\widetilde{G}_0}\dfrac{3R_i^2R_o^2 + \left(y^2+z^2\right)^2}{3\left(R_i^2+R_o^2\right)\left(y^2+z^2\right)^2}(x-L) \\ \tau_{xz} = -\dfrac{qG(r)}{\widetilde{G}_0}\left[1 + \dfrac{3R_i^2R_o^2\left(y^2-z^2\right) - \left(y^2+3z^2\right)\left(y^2+z^2\right)^2}{3\left(R_i^2+R_o^2\right)\left(y^2+z^2\right)^2}\right](x-L) \\ \tau_{xr} = -\dfrac{qG(r)z}{\widetilde{G}_0r}\left[1 - \dfrac{R_i^2R_o^2 + \left(y^2+z^2\right)^2}{\left(R_i^2+R_o^2\right)\left(y^2+z^2\right)}\right](x-L) \end{cases} \tag{2.73}$$

From the above, it can be observed that the contribution of flexural behavior is composed of two parts, one arising from the bending moment and the other from shear force. Letting $\widetilde{G}_0$ tend to infinity, equivalently to the fact that the beam has sufficiently high reduced shear rigidity, this means shear locking. Thus, the above results reduce to those for Euler–Bernoulli radially graded beams of annular cross-section, in which there is uniform vanishing shear stress, as expected. On the other hand, if setting $R_i = 0$, a hollow tube reduces to a solid cylinder. In this case, the above stresses reduce to those for a solid cylinder subjected to uniform transverse loading [1].

2.3.2.2 *Simply supported tube beam subjected to a concentrated force*

Next, we consider a simply supported tube beam subjected to a concentrated force ($P_y = 0$, $P_z = P$) at the mid-span of the beam. Due to symmetry, one can write the following conditions:

$$\begin{cases} w(0) = 0 \\ M(0) = 0 \\ Q\left(\dfrac{L}{2}\right) = \dfrac{P}{2} \\ \psi\left(\dfrac{L}{2}\right) = 0 \end{cases} \tag{2.74}$$

Based on these conditions, we can obtain $F(x)$ as follows:

$$F(x) = -\frac{P}{12E_2}x^3 + \frac{P}{2E_2}\left(\frac{L^2}{8} + \frac{E_4}{\widetilde{G}_0}\right)x, 0 \le x \le \frac{L}{2} \tag{2.75}$$

With this $F(x)$, we can obtain all physical quantities of interest. To save space, we only provide expressions for $x \in [0, L/2]$ below. Those for $x \in [L/2, L]$ can be obtained directly by symmetry.

$$u = z\psi + z\frac{3R_i^2R_o^2 - \left(y^2 + z^2\right)^2}{3\left(R_i^2 + R_o^2\right)\left(y^2 + z^2\right)}\left(\partial w \partial x + \psi\right) \tag{2.76}$$

$$\begin{cases} u = \dfrac{Pz}{4E_2}\left(x^2 - \dfrac{L^2}{4}\right) + \dfrac{Pz}{6\widetilde{G}_0}\dfrac{3R_i^2R_o^2 - \left(y^2 + z^2\right)^2}{\left(R_i^2 + R_o^2\right)\left(y^2 + z^2\right)} \\ w = -\dfrac{P}{12E_2}x^3 + \dfrac{P}{2E_2}\left(\dfrac{L^2}{8} + \dfrac{E_2}{\widetilde{G}_0}\right)x \\ \psi = \dfrac{P}{4E_2}\left(x^2 - \dfrac{L^2}{4}\right) \end{cases} \tag{2.77}$$

and

$$\begin{cases} \sigma_{xx} = \dfrac{xzPE(r)}{2E_2} \\ \tau_{xy} = -\dfrac{yzPG(r)}{\widetilde{G}_0} \dfrac{3R_i^2R_o^2 + (y^2+z^2)^2}{3(R_i^2+R_o^2)(y^2+z^2)^2} \\ \tau_{xz} = \dfrac{PG(r)}{2\widetilde{G}_0}\left[1 + \dfrac{3R_i^2R_o^2(y^2-z^2) - (y^2+3z^2)(y^2+z^2)^2}{3(R_i^2+R_o^2)(y^2+z^2)^2}\right] \\ \tau_{xr} = \dfrac{zPG(r)}{2\widetilde{G}_0 r}\left[1 - \dfrac{R_i^2R_o^2 + (y^2+z^2)^2}{(R_i^2+R_o^2)(y^2+z^2)}\right] \end{cases} \tag{2.78}$$

2.4 Bending behaviors analysis

2.4.1 Cylinder cylindrical beams

Numerical computations are carried out for a circular cylindrical beam with radial nonhomogeneity. To show the effect of nonhomogeneity on deflections, stress distribution, and natural frequencies, we here adopt material properties in the following power-law form

$$P(r) = P_c + (P_s - P_c)\left(\frac{r}{R}\right)^N \tag{2.79}$$

where P_c and P_s are the corresponding material properties of the center and the outer surface, respectively, and N is the power-law gradient parameter describing the volume fraction change of both constituents. In the following calculations, a cylindrical beam with a length to radius ratio of $L/R = 10$ is used, and two materials, aluminum and zirconia, are selected, the material properties of which are [6]:

$$\begin{cases} \text{Al}: E_c = 70\ \text{GPa}, G_c = 27\ \text{GPa}, \rho_c = 2702\ \text{kg/m}^3 \\ \text{ZrO}_2: E_s = 200\ \text{GPa}, G_s = 77\ \text{GPa}, \rho_s = 5700\ \text{kg/m}^3 \end{cases} \tag{2.80}$$

Moreover, it is assumed that it is rich in aluminum at the center and rich in zirconia at the surface.

As a representative example, consider a cantilever functionally graded beam with circular cross-section where uniform distributed force q is applied at the longitudinal axis. The variation of dimensionless deflection $w(x)E_c/q$ with dimensionless coordinate x/L is plotted in Fig. 2.3 for different N values.

It can be observed that the deflection increases with the gradient index N rising, implying that different volume fractions can produce different deflections, although the chosen materials remain unchanged. Furthermore, the distribution of axial stress $\sigma_{xx}(0, 0, z)$ at the clamped end versus

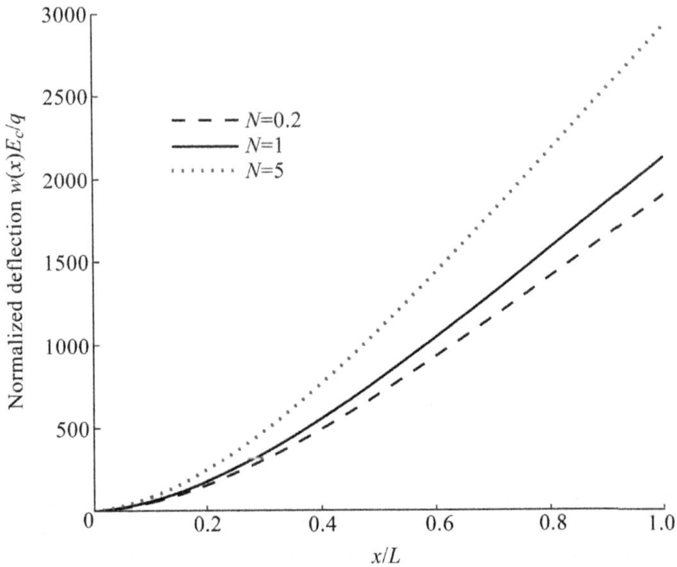

FIGURE 2.3 Normalized deflection $w(x)E_c/q$ versus dimensionless coordinate x/L for different gradient indexes of a cantilever functionally graded circular cylindrical beam subjected to uniformly distributed loading.

z/R is displayed in Fig. 2.4. From Fig. 2.4, σ_{xx} can be seen to vary nearly linearly along the diameter direction from the bottom to the top for $N = 0.2$. Moreover, the stress σ_{xx} in magnitude reaches a maximum at $z = \pm R$, which coincides with the classical results. However, the maximum σ_{xx} in magnitude at $z = \pm R$ is increased when N becomes larger. Fig. 2.5 shows a variation of the normalized maximum stress $\sigma_{xx}(x, 0, R)A/qL$ versus x/L. On the other hand, we also plot the distribution of the shear stress σ at the clamped end versus z/R in Fig. 2.6. Unlike the distribution of σ_{xx}, the shear stress σ_{xz} is greatly influenced by the gradient index, which mainly occurs in the region near the center. To be precise, the effects of gradient index N on the distribution of σ_{xz} are slight for the larger radius ($0.7 \leq |z/R|$). For $|z/R| < 0.7$, the shear stress is sensitive to the gradient index. In particular, the shear stress remains almost unchanged and approximately takes a constant in $|z/R| < 0.7$ for $N = 2.5$. For other values of N, the shear stress takes either a local minimum or a local maximum. This feature is also provided for a functionally graded beam with rectangular cross-section in Ref. [7].

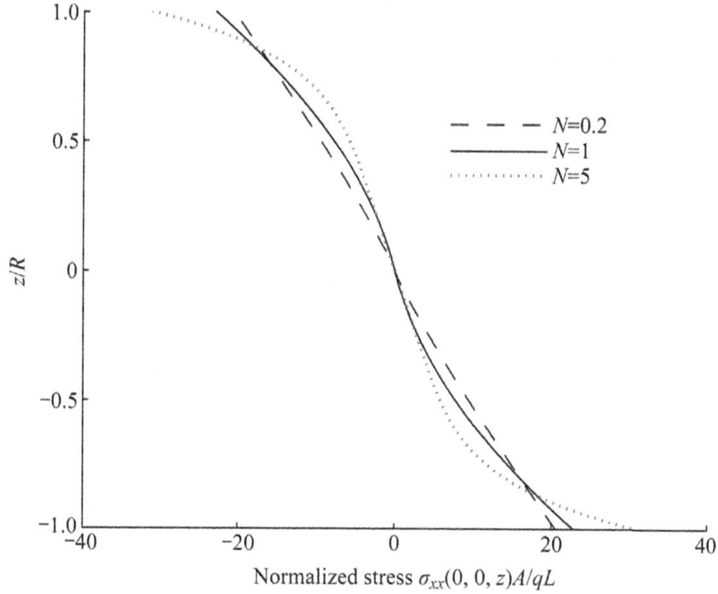

FIGURE 2.4 The distribution of the normal stress $\sigma_{xx}(0, 0, z)\ A/qL$ with dimensionless coordinate z/R at the clamped end of a cantilever functionally graded circular cylindrical beam subjected to uniformly distributed loading.

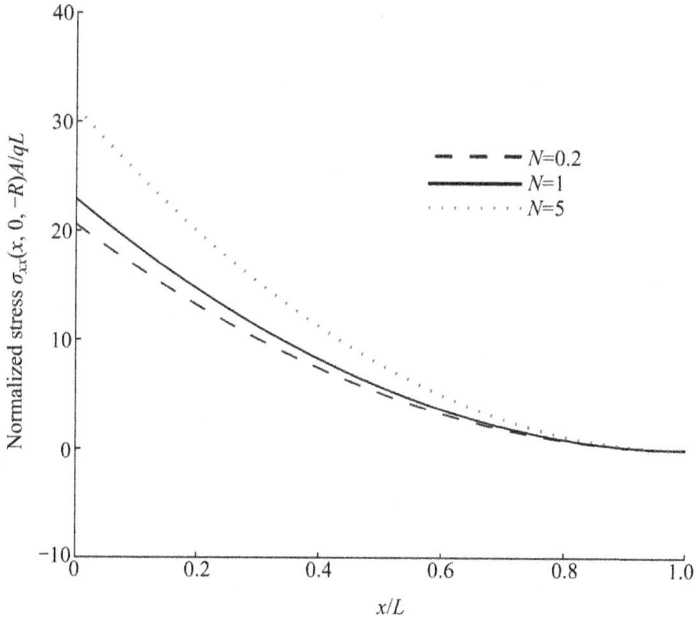

FIGURE 2.5 The distribution of the normal stress $\sigma_{xx}(x,0,-R)A/qL$ with dimensionless coordinate x/L for a cantilever functionally graded circular cylindrical beam subjected to uniformly distributed loading.

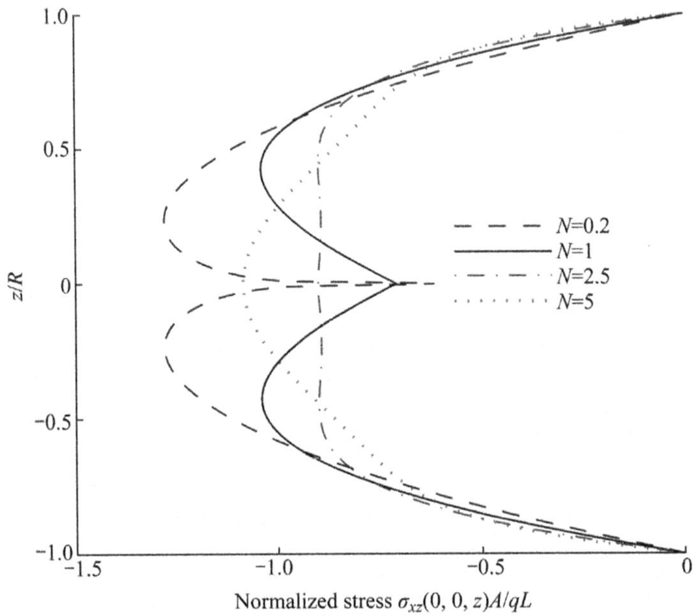

FIGURE 2.6 The distribution of the normal stress $\sigma_{xz}(0, 0, z)\ A/qL$ with dimensionless coordinate z/R at the clamped end of a cantilever functionally graded circular cylindrical beam subjected to uniformly distributed loading.

2.4.2 Cylindrical tube beams

To study the effects of the gradient parameter on flexure, we take a cantilever tube beam subjected to uniform distributed loading as an example. In the following calculation, the material properties of the tube are assumed to obey the power law

$$Y(r) = Y_i + (Y_o - Y_i)\left(\frac{r - R_i}{R_o - R_i}\right)^N \tag{2.81}$$

where Y_i and Y_o are the corresponding material properties on the inner and outer surfaces, respectively, and N is the power-law gradient index describing the volume fraction change of both constituents involved. When $N = 1$, the material properties exhibit a linear variation on the radial coordinate. For other values of N, Fig. 2.7 shows the variation of material properties for $R_i/R_o = 0.5$. In the present study, a radially graded tube with a length to radius ratio of $L/R_o = 10$, $\nu = 0.3$ is used, unless otherwise stated, and similarly, aluminum and zirconia are selected as the inner and outer surface materials, respectively, and the material properties in Eq. (2.80) are chosen [6].

As a representative example, consider a radially graded cantilever beam with annular cross-section subjected to uniform distributed load q. The variation of dimensionless deflection $w(x)E_iI/qL^4$ with the dimensionless axial coordinate x/L is plotted in Fig. 2.8 for different values of N, where I denotes

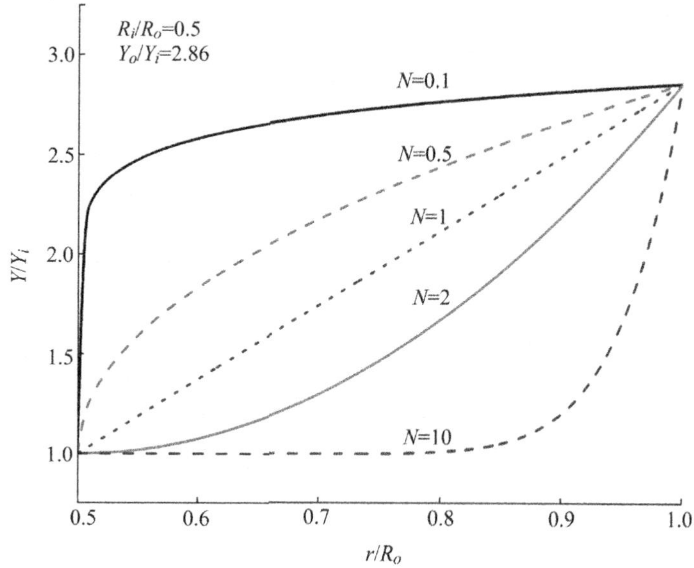

FIGURE 2.7 Variation of material properties with $Y_o/Y_i = 2.86$ and $R_i/R_o = 0.5$ when taking different gradient indexes, N.

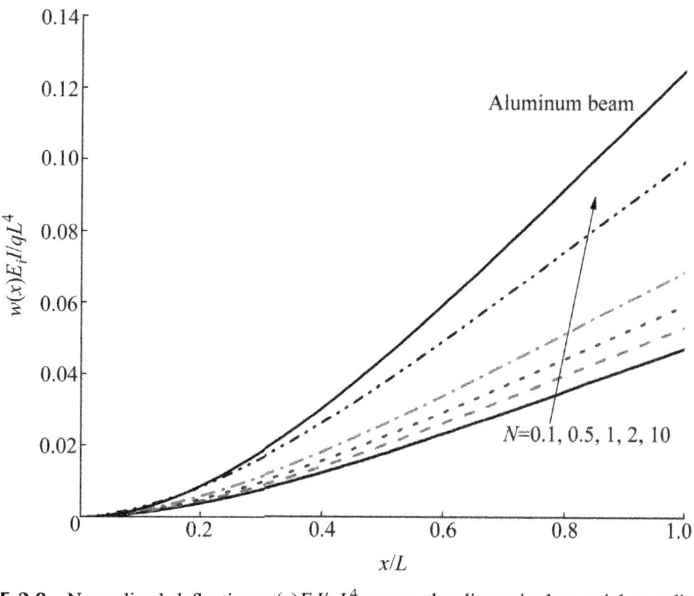

FIGURE 2.8 Normalized deflection $w(x)E_iI/qL^4$ versus the dimensionless axial coordinate x/L for different gradient indexes of a cantilever functionally graded tube beam subjected to uniformly distributed loading with $L/R_o = 10$.

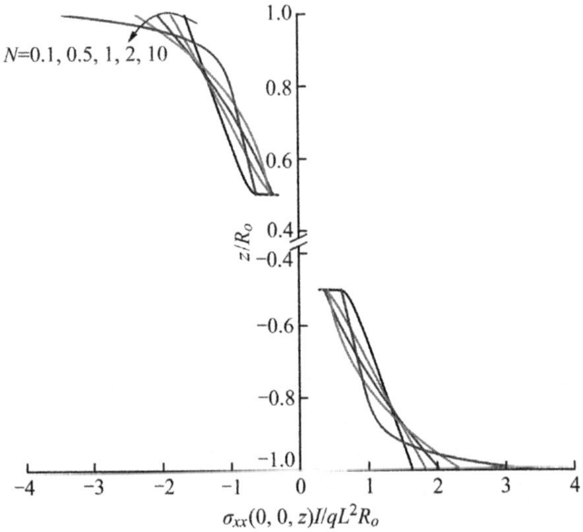

FIGURE 2.9 Effect of the gradient index on the distribution of the normal stress $\sigma_{xx}(0, 0, z)I/qL^2R_o$ at the clamped end of a radially graded cantilever tube beam subjected to uniformly distributed loading with $L/R_o = 10$.

the rotational moment of inertia of the cross-section, i.e., $I = \pi\left(R_o^4 - R_i^4\right)/4$. It can be observed that the deflection increases as N increases, implying that different volume fractions lead to different deflections, although the chosen materials remain unchanged. For comparison, we also display the variation of the deflection for a homogeneous cantilever tube beam made of aluminum. Clearly, due to the contribution of zirconia with a higher Young's modulus in a functionally graded tube, the radially graded tube beam seems to become stiffer, and the deflection is lower. When taking $N = 10$ or more, the deflection gradually gets closer to that for a homogeneous cantilever aluminum tube beam. It is easily understood that as N increases, the volume fraction of aluminum increases at the inner surface of a radially graded tube also, and the properties of aluminum become dominant, as shown in Fig. 2.8.

Furthermore, the distribution of the axial stress $\sigma_{xx}(0,0,z)$ at the clamped end versus z/R_o is displayed in Fig. 2.9. The stress σ_{xx} in magnitude reaches a maximum at $z = \pm R_o$, which coincides with the classical results. However, the maximum σ_{xx} in magnitude at $z = \pm R_o$ increases when N becomes larger. The top and bottom positions correspond to the maximum compressive and tensile stress, respectively.

For shear stress τ_{xz}, we plot the distribution profile of $\tau_{xz}L/q$ over the cross-section of the clamped end in Figs. 2.10–2.12. Clearly, the shear stress τ_{xz} arrives at its maximum at the midplane, in agreement with the classical result. The location of the maximum shear stress in the midplane depends on the gradient index. For example, when the gradient index N increases, the location of the maximum shear stress shifts from the inner surface to the outer surface,

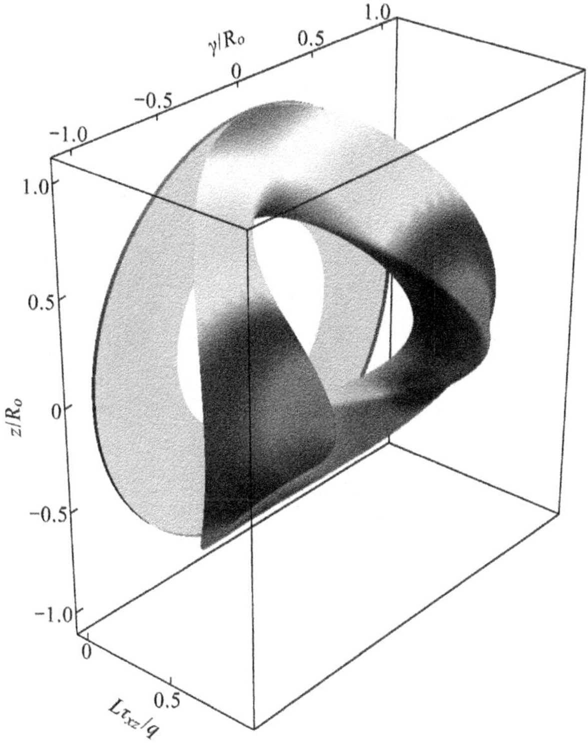

FIGURE 2.10 The distribution of the shear stress $\tau_{xz}(0, y, z)L/q$ over the clamped end of a radially graded cantilever tube beam subjected to uniformly distributed loading ($N = 0.1$).

as shown in Figs. 2.10–2.12. For comparison, for a homogeneous cantilever tube beam made of aluminum, the distribution of the shear stress is presented in Fig. 2.13. The gradient index has a remarkable effect on the stress distribution. Finally, we point out that the shear stress τ_{xz} does not vanish on the inner and outer surfaces except for the top and bottom positions. Contrary to that, the shear stress τ_{xr} always remains at zero on the inner and outer surfaces, as can be seen from Eqs. (2.73) and (2.78).

To verify the effectiveness of the present approach, the finite element method (FEM) was used to examine the accuracy of the numerical results. The commercial software MSC Patran/Nastran was chosen for the FEM analysis, and a 3D Hex8 element with eight nodes was applied. Table 2.1 shows the numerical results of a dimensionless deflection $w(x)E_iI/qL^4$ based on the FEM and the present approach. When adopting the FEM, three different meshes, with the numbers of elements being 9000, 16,000, and 64,800, respectively, were taken for the same model, and they are denoted as FEM-1, FEM-2, and FEM-3 in Table 2.1, respectively. As a representative example, the meshing model of FEM-2 is shown

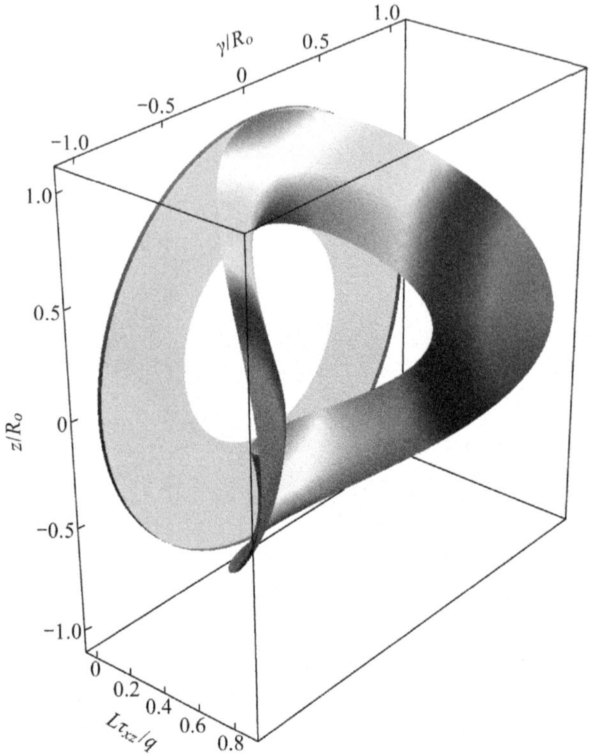

FIGURE 2.11 The distribution of the shear stress $\tau_{xz}(0, y, z)L/q$ over the clamped end of a radially graded cantilever tube beam subjected to uniformly distributed loading ($N = 1$).

in Fig. 2.14. To save space, those of FEM-1 and FEM-3 are not displayed here. Additionally, to model the gradient variation, the Al/ZrO_2 tube was partitioned into k equal parts along the radial direction, with k being a positive integer (Fig. 2.15). In each part, the material is assumed to be homogeneous and have the same material properties, and the total properties are piecewise constants, where the elastic displacement and stresses are required to be continuous across the interface of two adjacent parts.

From the relative errors listed in Table 2.1, one can see that as the number of elements grows, the FEM results converge well. All the relative errors between the latter two FEM results and present ones are less than 3%. This confirms the accuracy of the present results and their agreement with the finite element results. Therefore, the subsequent FEM model in this study uses the same element size as in FEM-2. It is interesting to point out that based on FEM-2, the maximum dimensionless lateral displacement vE_iI/qL^4 is found to be 0.000264, which is far less than the maximum of dimensionless deflection $w(x)E_iI/qL^4$ at 0.06034. Thus, comparing with the flexural

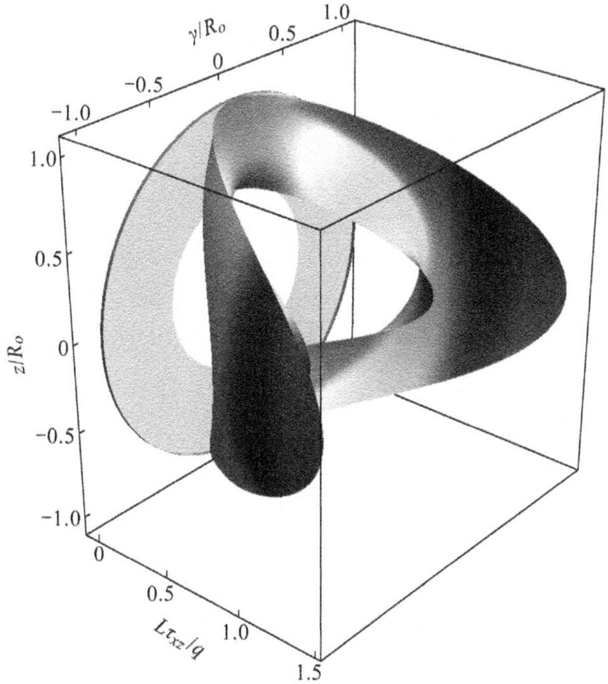

FIGURE 2.12 The distribution of the shear stress $\tau_{xz}(0,\ y,\ z)L/q$ over the clamped end of a radially graded cantilever tube beam subjected to uniformly distributed loading ($N = 10$).

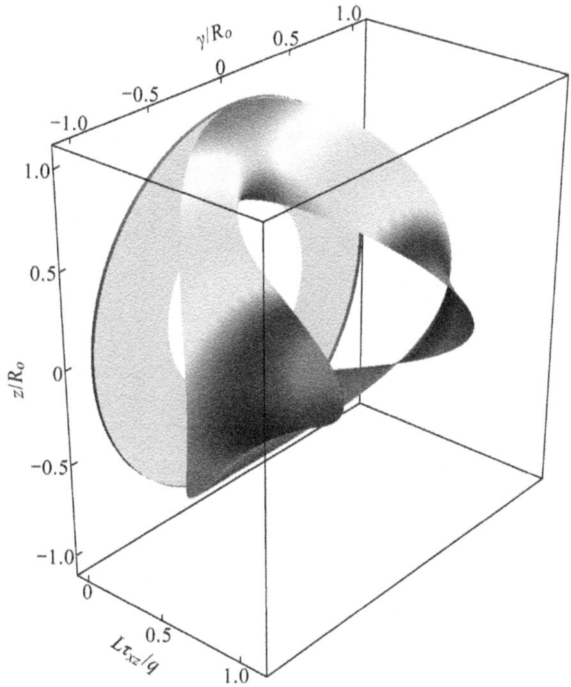

FIGURE 2.13 The distribution of the shear stress $\tau_{xz}(0,\ y,\ z)L/q$ over the clamped end of a radially graded cantilever tube beam subjected to uniformly distributed loading: homogeneous aluminum tube beam.

TABLE 2.1 Dimensionless deflection $w(x)E_iI/qL^4$ for a functionally graded cantilever tube beam with $L/R_o = 10$ subjected to uniformly distributed loading.

x/L	0.1	0.2	0.3	0.4	0.5	0.6	0.7	0.8	0.9	1.0
Present	0.001629	0.005062	0.009857	0.015690	0.022280	0.028640	0.036740	0.044260	0.051810	0.059310
FEM-1	0.001398	0.004852	0.009705	0.015660	0.022430	0.029670	0.03721	0.045000	0.052800	0.060590
Errors-1	14.18%	4.15%	1.54%	0.19%	0.67%	3.60%	1.28%	1.67%	1.91%	2.16%
FEM-2	0.001589	0.004978	0.009730	0.015690	0.022350	0.029470	0.037210	0.044570	0.052550	0.060340
Errors-2	2.46%	1.66%	1.29%	0.00%	0.31%	2.90%	1.28%	0.70%	1.43%	1.74%
FEM-3	0.001594	0.004987	0.009755	0.015610	0.022250	0.029420	0.036960	0.044450	0.05204	0.059840
Errors-3	2.15%	1.48%	1.03%	0.51%	0.13%	2.72%	0.60%	0.43%	0.44%	0.89%

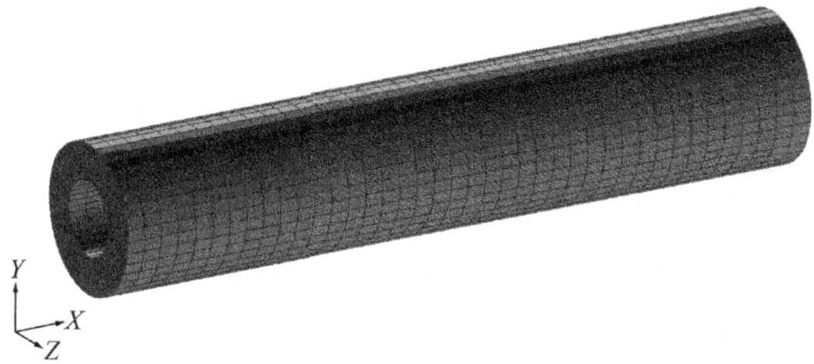

FIGURE 2.14 The meshing model used in the FEM.

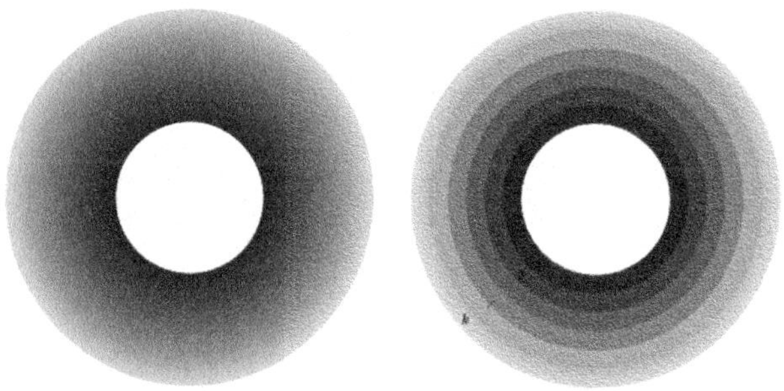

FIGURE 2.15 The continuously varying (left) and piecewise constant (right) material properties along the radial direction.

deflection w, one can completely ignore the displacement component v. This also indicates that the assumption of the lateral displacement v being independent of the variable x holds true.

References

[1] Huang Y, Li XF. Bending and vibration of circular cylindrical beams with arbitrary radial nonhomogeneity. Int J Mech Sci 2010;52(4):595–601.

[2] Love AEH. A treatise on the mathematical theory of elasticity. Courier Corporation; 1944.

[3] Li XF. A unified approach for analyzing static and dynamic behaviors of functionally graded Timoshenko and Euler–Bernoulli beams. J Sound Vib 2008;318(4–5):1210–29.

[4] Ma WL, Jiang ZC, Lee KY, et al. A refined beam theory for bending and vibration of functionally graded tube-beams. Compos Struct 2020;236:111878.

[5] Elishakoff I, Kaplunov J, Nolde E. Celebrating the centenary of Timoshenko's study of effects of shear deformation and rotary inertia. Appl Mech Rev 2015;67(6):060802.

[6] Vel SS, Batra RC. Three-dimensional exact solution for the vibration of functionally graded rectangular plates. J Sound Vib 2004;272(3–5):703–30.

[7] Apetre NA, Sankar BV, Ambur DR. Analytical modeling of sandwich beams with functionally graded core. J Sandw Struct Mater 2008;10(1):53–74.

Chapter 3

Vibration of higher-order Functionally Graded circular cylindrical beams

3.1 Introduction

Solid and hollow cylinders with circular cross-sections are one of the most frequently encountered engineering structures. In nature, the trunk of a tree may be understood as representative of a circular cylindrical beam with radial nonhomogeneity. In addition, carbon nanoropes and carbon nanotubes may be taken as solid and hollow circular cylindrical beams, respectively, when using a continuum mechanics approach. For such circular cylindrical structures, there has been a very long history of research. In contrast to the classical Euler–Bernoulli and Timoshenko theories of beams, a refined beam theory or third-order shear deformation beam theory for radially graded tubes is proposed, where warping, shear deformation, and rotational moment of inertia of the cross-section are all considered. The shear correction coefficient is not needed. Coupled governing equations for deflection and rotation about the neutral axis of the cross-section are derived from equilibrium equations, and then converted into a single fourth-order partial differential governing equation. The frequency equations for free flexural vibration of radially graded cylindrical beams and hollow cylinders with clamped–clamped, pinned–pinned, and clamped–free ends are obtained and the natural frequencies calculated for different power-law gradients and various length-to-thickness ratios. The effects of radial gradient on the stress distribution and natural frequencies are analyzed in detail.

3.2 Assumption and governing equations

3.2.1 Circular cylindrical beams and tubes

In the preceding chapter, we investigated the bending behavior of functionally graded cylindrical beams and tube beams with uniform circular cross-section subjected to a transverse load or bending moment, see Sections 2.2.1 and 2.2.2,

Higher-Order Theories of Functionally Graded Circular Cylindrical Beams and Columns
https://doi.org/10.1016/B978-0-443-44718-1.00006-2

respectively. Here, the vibration behavior of the cylinder (tube) beam is analyzed. Similarly, the cylindrical beam is deflected in the z-direction not in the y-direction and subjected to a transverse load or bending moment, we then neglect the motion along the y-direction and derive two governing equations in the xoz-plane:

$$\frac{\partial \sigma_{xx}}{\partial x}+\frac{\partial \tau_{xy}}{\partial y}+\frac{\partial \tau_{xz}}{\partial z}=\rho(r)\frac{\partial^2 u}{\partial t^2} \tag{3.1}$$

$$\frac{\partial \tau_{xz}}{\partial x}+\frac{\partial \tau_{yz}}{\partial y}+\frac{\partial \sigma_{zz}}{\partial z}=\rho(r)\frac{\partial^2 w}{\partial t^2} \tag{3.2}$$

where $\rho\,(r\,)$ is the mass density of the beam depending on r.

After substituting Eqs. (2.7–2.9) into Eq. (3.1), we multiply both sides of Eq. (3.1) by z and then integrate over the cross-sectional area A, resulting in

$$\widetilde{E}_2\frac{\partial^2 \psi}{\partial x^2}-E_4\frac{\partial^3 w}{\partial x^3}-\widetilde{G}_0\left(\frac{\partial w}{\partial x}+\psi\right)=\widetilde{\rho}_2\frac{\partial^2 \psi}{\partial t^2}-\rho_4\frac{\partial^3 w}{\partial x \partial t^2} \tag{3.3}$$

where

$$\begin{aligned}\widetilde{\rho}_2&=\rho_2-\rho_4\\ \rho_2&=\pi\int_0^R \rho(r)r^3\mathrm{d}r\\ \rho_4&=\frac{\pi}{3R^2}\int_0^R \rho(r)r^5\mathrm{d}r\end{aligned} \tag{3.4}$$

for a cylinder beam,

$$\begin{cases}\rho_2=\pi\displaystyle\int_{R_i}^{R_o} \rho(r)r^3\mathrm{d}r\\ \rho_4=\dfrac{\pi}{3\left(R_i^2+R_o^2\right)}\displaystyle\int_{R_i}^{R_o} \rho(r)\left(r^4-3R_i^2R_o^2\right)r\mathrm{d}r\end{cases} \tag{3.5}$$

for a cylindrical tube beam.

Similarly, a straight integration of Eq. (3.2) over the cross-sectional area A gives

$$\widetilde{G}_0\left(\frac{\partial^2 w}{\partial x^2}+\frac{\partial \psi}{\partial x}\right)=\rho_0\frac{\partial^2 w}{\partial t^2}-q, \tag{3.6}$$

where

$$\rho_0=2\pi\int_0^R \rho(r)r\mathrm{d}r \tag{3.7}$$

for a cylindrical beam,

$$\rho_0 = 2\pi \int_{R_i}^{R_o} \rho(r) r dr \tag{3.8}$$

for a cylindrical tube beam. Here, q is the distributed force of N/m dimension and applied at a principal axis of the beam in the z-direction. We therefore obtain two coupled governing Eqs. (3.3) and (3.6) for the bending of functionally graded circular cylindrical beams.

A further simplification of the resulting two coupled Eqs. (3.3) and (3.6) can be achieved by introducing a new auxiliary function. This can be done by expressing the deflection w and the rotation ψ in the form

$$w = F - \frac{\widetilde{E}_2}{\widetilde{G}_0}\frac{\partial^2 F}{\partial x^2} + \frac{\widetilde{\rho}_2}{\widetilde{G}_0}\frac{\partial^2 F}{\partial t^2} \tag{3.9}$$

$$\psi = \frac{\rho_4}{\widetilde{G}_0}\frac{\partial^3 F}{\partial x \partial t^2} - \frac{E_4}{\widetilde{G}_0}\frac{\partial^3 F}{\partial x^3} - \frac{\partial F}{\partial x} \tag{3.10}$$

where F is a new auxiliary function of length dimension, which is readily checked when substituting Eqs. (3.9) and (3.10) into Eq. (3.10) that it is identically fulfilled. For convenience, we also give bending moment M and shear force Q in terms of F. This is done by substituting Eqs. (3.61) and (3.62) into Eqs. (3.48) and (3.49). The bending moment M and shear force Q can be expressed as

$$M = -E_2\frac{\partial^2 F}{\partial x^2} + \frac{E_2\rho_4 - E_4\rho_2}{\widetilde{G}_0}\frac{\partial^4 F}{\partial x^2 \partial t^2} \tag{3.11}$$

$$Q = \rho_2\frac{\partial^3 F}{\partial x \partial t^2} - E_2\frac{\partial^3 F}{\partial x^3} \tag{3.12}$$

The remaining task is to solve the governing Eq. (3.67) for the unknown F under suitable boundary conditions, which can be provided by the deflection w, rotation ψ, bending moment M, and shear force Q. On the other hand, when substituting Eqs. (3.9) and (3.10) into Eq. (3.6), we get

$$E_2\frac{\partial^4 F}{\partial x^4} + \rho_0\frac{\partial^2 F}{\partial t^2} - \left(\rho_2 + \frac{\rho_0\widetilde{E}_2}{\widetilde{G}_0}\right)\frac{\partial^4 F}{\partial x^2 \partial t^2} + \frac{\widetilde{\rho}_2\rho_0}{\widetilde{G}_0}\frac{\partial^4 F}{\partial t^4} = q \tag{3.13}$$

It should be noted that a similar governing equation has been obtained for a rectangular functionally graded Timoshenko beam [1]. As pointed out above, the introduction of a shear correction factor is inevitable in the Timoshenko beam, but now, no additional factor is needed for a functionally graded circular cylindrical beam.

Once the solution to the governing equation, F, is obtained through proper boundary conditions, all the physical quantities of interest including the deflection w, rotation ψ, bending moment M, and shear force Q are obtainable. For instance, w and ψ are given by Eqs. (3.9) and (3.10), respectively. Remembering Eqs. (3.11) and (3.12), we can give the bending moment M and shear force Q in terms of F as follows:

$$\begin{cases} M = -E_2 \dfrac{\partial^2 F}{\partial x^2} \\ Q = -E_2 \dfrac{\partial^3 F}{\partial x^3} + \rho_2 \dfrac{\partial^3 F}{\partial x \partial t^2} \end{cases} \tag{3.14}$$

Consider a special case of homogeneous cylindrical beams with constant material properties: $E(r) = E, G(r) = G, \rho(r) = \rho$. In this case, we have

$$\begin{cases} E_2 = EI \\ \widetilde{E}_2 = \dfrac{7}{9} EI \\ \widetilde{G}_0 = \dfrac{2}{3} GA \\ \rho_2 = \rho I \\ \widetilde{\rho}_2 = \dfrac{7}{9} \rho I \\ \rho_0 = \rho A \end{cases} \tag{3.15}$$

where $I = \pi R^4/4$ is the inertia moment of the cross-sectional area, and the above-derived governing equation then reduces to

$$EI \frac{\partial^4 F}{\partial x^4} + \rho A \frac{\partial^2 F}{\partial t^2} - \rho I \left(1 + \frac{7E}{6G}\right) \frac{\partial^4 F}{\partial x^2 \partial t^2} + \frac{7\rho^2 I}{6G} \frac{\partial^4 F}{\partial t^4} = q \tag{3.16}$$

3.2.2 Cylindrical tube beams with axial force

In order to establish an exact higher-order shear beam theory for radially graded hollow cylinders subjected to transverse and axial loads, we consider the free flexural vibration of a functionally graded cylindrical tube with annular cross-section under tensile or compressive axial force, as shown in Fig. 3.1. The hollow cylinder's material properties are designed to vary along the radial direction, i.e., they are dependent on the radial coordinate. For the convenience of later analysis, two coordinate systems will be used. One is a Cartesian coordinate system and the other is the polar cylindrical coordinates. The x-axis of the Cartesian coordinate system (x, y, z) is oriented along the centroidal axis of the undeformed tube, and the origin O is placed at the center of the cross-section at a reference position of the tube. We also introduce the polar cylindrical coordinate system (x, r, θ) with the same origin and the same

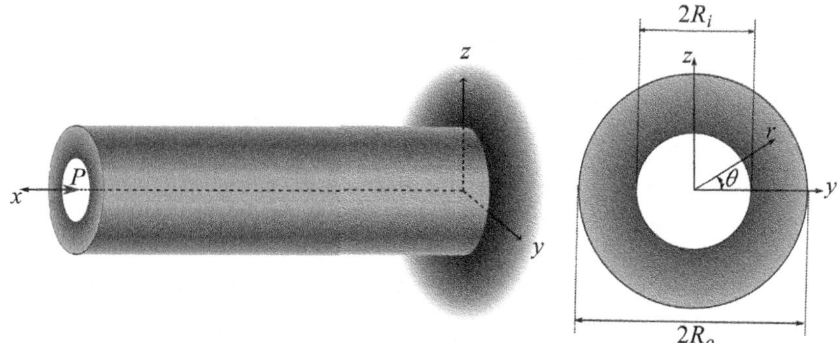

FIGURE 3.1 Schematic of an axially compressed cylindrical beam with annular cross-section and radial nonhomogeneity together with the corresponding coordinates.

x-axis. It is clear that between both coordinate systems, the following relationships hold:

$$\begin{cases} y = r\cos\theta \\ z = r\sin\theta \\ r = \sqrt{y^2 + z^2} \end{cases} \tag{3.17}$$

In the following, (u, v, w) are denoted as displacement components in (x, y, z) axes, respectively, when Cartesian coordinates are used, and (u, u_r, u_θ) are defined as displacement components along (x, r, θ) directions according to the polar cylindrical coordinates, respectively. Similar to the analyses of rectangular beams [2–4] and circular beams [5,6], the displacement component $u(x, y, z, t)$ along the x-axis is assumed to take the form

$$u(x, y, z, t) = -z\frac{\partial w}{\partial x} + f(y, z)\left(\frac{\partial w}{\partial x} + \psi\right) \tag{3.18}$$

where the function $f(y, z)$ represents the cross-sectional warping shape; $w(x, t)$ stands for the deflection of transverse bending of the midplane; $\psi(x, t)$ denotes the cross-sectional rotation at the midplane (here the positive direction of ψ is opposite to $\partial w / \partial x$). According to the usual treatment of beams [7], we neglect the displacement component v along the y-direction and obtain the shear strain

$$\gamma_{xr} = \sin\theta\frac{\partial w}{\partial x} + \frac{\partial u}{\partial r} \tag{3.19}$$

from which we find

$$\begin{cases} \gamma_{xr} = \dfrac{1}{r}\left(z\dfrac{\partial f}{\partial z} + y\dfrac{\partial f}{\partial y}\right)\left(\dfrac{\partial w}{\partial x} + \psi\right) \\ \sigma_{xr} = G(r)\gamma_{xr} \end{cases} \tag{3.20}$$

where $G(r)$ denotes the radius-dependent shear modulus of the functionally graded tube. Note that in Eq. (3.18), if allowing the annular cross-section after deformation to remain plane and perpendicular to the neutral surface, i.e., $f(y,z) = 0$, the classical Euler–Bernoulli beam theory is reduced. By taking $f(y,z) = z$, the Timoshenko beam theory can be retrieved in this case, the annular cross-section after deformation is no longer perpendicular to the midplane, but still remains plane. For the latter case, a shear correction factor is required, which cannot be derived from the theory itself [8,9]. Some research has been carried out into prismatic beams with a circular cross-section. Huang and Li [5,6] proposed a warping shape function of the circular cross-section. Recently, Ma et al. [10,11] extended the above results to radially graded tubes using the warping shape of the rational function

$$f(y,z) = z\left[1 + \frac{3R_i^2R_o^2 - r^4}{3(R_i^2 + R_o^2)r^2}\right] \tag{3.21}$$

As is well known, the warping shape may be represented by various functions. A variety of warping shape functions for rectangular functionally graded beams can be found in the literature, some of which are collected in Refs. [3,4]. For circular cylindrical beam-columns, three different warping shape functions were presented recently in Refs. [12,13]. However, for bending of a hollow cylinder or cylindrical tubes, in addition to the above-mentioned warping shape of rational function, other alternative warping shapes may be given. For example, one may assume the following inverse trigonometric functions as cross-sectional warping shape function:

$$f(y,z) = z\left[1 + \frac{s_1(r) + s_2(r)}{2r\sqrt{R_i^2 + R_o^2}}\right] \tag{3.22}$$

where

$$\begin{cases} s_1(r) = (R_i^2 + R_o^2)\sin^{-1}\sqrt{\dfrac{r^2 - R_i^2}{R_o^2 - R_i^2}} - 2R_iR_o\tan^{-1}\left(\dfrac{R_o}{R_i}\sqrt{\dfrac{r^2 - R_i^2}{R_o^2 - r^2}}\right) \\ s_2(r) = \sqrt{(R_o^2 - r^2)(r^2 - R_i^2)} \end{cases} \tag{3.23}$$

Then, by substituting Eq. (3.22) into Eq. (3.20), one can infer that $\gamma_{xr} = 0$ on the cylindrical tube's inner and outer surfaces automatically. This implies that $\sigma_{xr} = 0$ on the surfaces: $r = R_i$ and $r = R_o$ in view of Hooke's law, $\sigma_{xr} = G(r)\gamma_{xr}$.

With the above axial displacement u in Eq. (3.18) together with kinematic equations, it is easy to obtain the expression of the axial strain ε_{xx} and the shear strain γ_{xz} below

$$\begin{cases} \varepsilon_{xx} = -z\dfrac{\partial^2 w}{\partial x^2} + f(y,z)\left(\dfrac{\partial^2 w}{\partial x^2} + \dfrac{\partial \psi}{\partial x}\right) \\ \gamma_{xz} = \dfrac{\partial f}{\partial z}\left(\dfrac{\partial w}{\partial x} + \psi\right) \end{cases} \tag{3.24}$$

Furthermore, the associated stress components of interest can be derived as follows:

$$\begin{cases} \sigma_{xx} = E(r)\varepsilon_{xx} \\ \sigma_{xz} = G(r)\dfrac{\partial f}{\partial z}\left(\dfrac{\partial w}{\partial x} + \psi\right) \end{cases} \tag{3.25}$$

In the following section, we focus on the dynamic behavior of a radially graded tube under an axial load and derive a governing differential equation. As stated in the foregoing section, the shear stress component is automatically relaxed on the cylindrical tubes' surfaces for the warping shape functions, i.e., $\sigma_{xr} = 0$. Since the tube's bending vibration is restricted in the xoz-plane, the equation of motion in the y-direction does not need to be considered, one has

$$\frac{\partial V}{\partial x} = m_0\frac{\partial^2 w}{\partial t^2} \tag{3.26}$$

$$\frac{\partial M}{\partial x} - V - P\frac{\partial w}{\partial x} = -m_2\frac{\partial^3 w}{\partial t^2 \partial x} + m_{2f}\left(\frac{\partial^3 w}{\partial t^2 \partial x} + \frac{\partial^2 \psi}{\partial t^2}\right) \tag{3.27}$$

with

$$\begin{cases} m_0 = \displaystyle\int_A \rho(r)\mathrm{d}A \\ m_2 = \displaystyle\int_A z^2\rho(r)\mathrm{d}A \\ m_{2f} = \displaystyle\int_A zf(y,z)\rho(r)\mathrm{d}A \end{cases} \tag{3.28}$$

where A is for the cross-sectional area, P is the axial force (here it is particularly noted that $P > 0$ means compressive, $P < 0$ means tensile), M is the bending moment, and V is the transverse shear force perpendicular to the undeformed column. In deriving Eq. (3.27), we have used the expression (3.18) for u. In the above, one can derive the expression of the bending moment and shear force as follows

$$\begin{cases} M = \int_A z\sigma_{xx}\mathrm{d}A \\ V = \int_A \sigma_{xz}\mathrm{d}A - P\dfrac{\partial w}{\partial x} \end{cases} \tag{3.29}$$

Next, substituting Eqs. (3.25) and (3.24) into Eq. (3.29) leads to

$$\begin{cases} M = (EI)^*\left(\dfrac{\partial^2 w}{\partial x^2} + \dfrac{\partial \psi}{\partial x}\right) - (EI)_0\dfrac{\partial^2 w}{\partial x^2} \\ V = (GA)^*\left(\dfrac{\partial w}{\partial x} + \psi\right) - P\dfrac{\partial w}{\partial x} \end{cases} \tag{3.30}$$

where

$$\begin{cases} (EI)^* = \int_A zf(y,z)E(r)\mathrm{d}A \\ (EI)_0 = \int_A z^2E(r)\mathrm{d}A \\ (GA)^* = \int_A G(r)\dfrac{\partial f}{\partial z}\mathrm{d}A \end{cases} \tag{3.31}$$

Substituting Eq. (3.30) into Eqs. (3.26) and (3.27), one immediately obtains a system of differential equations

$$\begin{aligned} &(EI)^*\left(\frac{\partial^3 w}{\partial x^3} + \frac{\partial^2 \psi}{\partial x^2}\right) - (EI)_0\frac{\partial^3 w}{\partial x^3} - (GA)^*\left(\frac{\partial w}{\partial x} + \psi\right) \\ &= -m_2\frac{\partial^3 w}{\partial t^2\partial x} + m_{2f}\left(\frac{\partial^3 w}{\partial t^2\partial x} + \frac{\partial^2 \psi}{\partial t^2}\right) \end{aligned} \tag{3.32}$$

$$(GA)^*\left(\frac{\partial^2 w}{\partial x^2} + \frac{\partial \psi}{\partial x}\right) - P\frac{\partial^2 w}{\partial x^2} = m_0\frac{\partial^2 w}{\partial t^2} \tag{3.33}$$

Since the above system includes two unknown functions (deflection and rotation of the cross-section), it is inconvenient to analyze the dynamic behavior of a functionally graded beam-column. For further simplification, we introduce an unknown auxiliary function F with length dimension to convert the above system into a single equation for the introduced new function. To this end, we express w and ψ in terms of F as follows:

$$w = F - \frac{(EI)^*}{(GA)^*}\frac{\partial^2 F}{\partial x^2} + \frac{m_{2f}}{(GA)^*}\frac{\partial^2 F}{\partial t^2} \tag{3.34}$$

$$\psi = -\frac{\partial F}{\partial x} - \frac{(EI)_0 - (EI)^*}{(GA)^*}\frac{\partial^3 F}{\partial x^3} + \frac{m_2 - m_{2f}}{(GA)^*}\frac{\partial^3 F}{\partial x \partial t^2} \tag{3.35}$$

With Eqs. (3.34) and (3.35) at hand, we find that Eqs. (3.32) and (3.33) are automatically fulfilled only if the unknown function F satisfies the following single governing differential equation:

$$\left[(EI)_0 - \frac{P(EI)^*}{(GA)^*}\right]\frac{\partial^4 F}{\partial x^4} + P\frac{\partial^2 F}{\partial x^2} + m_0\frac{\partial^2 F}{\partial t^2} - \left[m_2 + \frac{m_0(EI)^* - Pm_{2f}}{(GA)^*}\right]\frac{\partial^4 F}{\partial x^2 \partial t^2} + \frac{m_0 m_{2f}}{(GA)^*}\frac{\partial^4 F}{\partial t^4} = 0 \tag{3.36}$$

Consequently, the solution procedure can be greatly simplified and the remaining task is to obtain an appropriate solution F by solving the final governing differential [Eq. (3.36)] under proper boundary conditions. To facilitate our analysis, it is suitable to give the expression of bending moment M as well as shear force V in terms of F, one gets:

$$\begin{aligned} M &= -(EI)_0\frac{\partial^2 F}{\partial x^2} + \frac{m_0(EI)^* - m_{2f}(EI)_0}{(GA)^*}\frac{\partial^4 F}{\partial x^2 \partial t^2} \\ V &= -\left[(EI)_0 - \frac{P(EI)^*}{(GA)^*}\right]\frac{\partial^3 F}{\partial x^3} + \left[m_2 - \frac{m_{2f}P}{(GA)^*}\right]\frac{\partial^3 F}{\partial x \partial t^2} - P\frac{\partial F}{\partial x} \end{aligned} \tag{3.37}$$

In closing this section, we consider a special case of neglecting shear deformation, i.e., $f = 0$ and sufficiently large shear modulus G. In this case, we find $m_{2f} = 0$, $(GA)^* \rightarrow \infty$, and Eqs. (3.32) and (3.33) can be converted to

$$(EI)_0\frac{\partial^4 w}{\partial x^4} + P\frac{\partial^2 w}{\partial x^2} + m_0\frac{\partial^2 w}{\partial t^2} - m_2\frac{\partial^4 w}{\partial t^2 \partial x^2} = 0 \tag{3.38}$$

This is the governing equation for dynamic behavior of functionally graded Rayleigh beams, in which rotational inertia is considered, but the influence of shear deformation is excluded. If further omitting the rotational inertia of the cross-section, the well-known governing equation for dynamic behavior of axially loaded radially graded Euler–Bernoulli beams is recovered.

3.3 Foundation of free vibration

3.3.1 Cylindrical beams and tubes

In this section, the free vibration of radially graded cylindrical beams with an annular cross-section is analyzed. For this purpose, the introduced auxiliary function F may take the form

$$F(x,t)=f(x)\mathrm{e}^{\mathrm{i}\omega t} \tag{3.39}$$

where ω is the circular frequency of free vibration. Substituting Eq. (3.39) into Eq. (3.31), in the absence of applied loading, i.e., $q=0$, one can get

$$E_2\frac{\mathrm{d}^4 f}{\mathrm{d}x^4}+\left(\rho_2+\frac{\rho_0\widetilde{E}_2}{\widetilde{G}_0}\right)\omega^2\frac{\mathrm{d}^2 f}{\mathrm{d}x^2}-\rho_0\omega^2\left(1-\frac{\widetilde{\rho}_2}{\widetilde{G}_0}\omega^2\right)f=0 \tag{3.40}$$

This is an ordinary differential equation with constant coefficients. By using a standard approach, we can solve the above equation and give a general solution to Eq. (3.40) as

$$f(x)=C_1\cos(\lambda_1 x)+C_2\sin(\lambda_1 x)+C_3\cosh(\lambda_2 x)+C_4\sinh(\lambda_2 x) \tag{3.41}$$

for frequencies ω less than the cut-off frequency ω_c, and

$$f(x)=C_1\cos(\lambda_3 x)+C_2\sin(\lambda_3 x)+C_3\cos(\lambda_4 x)+C_4\sin(\lambda_4 x) \tag{3.42}$$

for frequencies ω larger than the cut-off frequency ω_c, where the cut-off frequency $\omega_c=\sqrt{\widetilde{G}_0/\widetilde{\rho}_2}$, $C_j (j=1, \ldots, 4)$ are unknown constants, and

$$\lambda_{1,2}=\omega\sqrt{\frac{1}{2}\left[\sqrt{\left(\frac{\rho_2}{E_2}+\frac{\rho_0\widetilde{E}_2}{\widetilde{G}_0 E_2}\right)^2+\frac{4\rho_0}{E_2}\left(\frac{1}{\omega^2}-\frac{1}{\omega_c^2}\right)}\pm\left(\frac{\rho_2}{E_2}+\frac{\rho_0\widetilde{E}_2}{\widetilde{G}_0 E_2}\right)\right]} \tag{3.43}$$

and

$$\lambda_{3,4}=\omega\sqrt{\frac{1}{2}\left[\pm\sqrt{\left(\frac{\rho_2}{E_2}+\frac{\rho_0\widetilde{E}_2}{\widetilde{G}_0 E_2}\right)^2+\frac{4\rho_0}{E_2}\left(\frac{1}{\omega^2}-\frac{1}{\omega_c^2}\right)}+\left(\frac{\rho_2}{E_2}+\frac{\rho_0\widetilde{E}_2}{\widetilde{G}_0 E_2}\right)\right]} \tag{3.44}$$

An expression for $f(x)$ for $\omega=\omega_c$ is omitted here. In the above expressions for $f(x)$, the sine and cosine terms in Eqs. (3.41) and (3.42) correspond to flexural waves propagating in the hollow tube, while the other terms represent nonpropagating fields or evanescent components.

Combining the above expressions with the relevant end conditions, the natural frequencies of free vibration can be determined. In what follows we discuss three frequently encountered typical cases: simply supported beams, doubly clamped beams, and cantilever beams. Moreover, for convenience, we only focus on free vibration of lower frequencies. Or rather, we only study the case of Eq. (3.41). For $f(x)$ given by Eq. (3.42), an analogous treatment can be done. To this end, it is expedient to give expressions for $w, \psi, M,$ and Q in terms of the $\lambda_{1,2}$. This is done by substituting Eq. (3.41) into Eqs. (3.61)–(3.64). After some algebra, one finally obtains

$$w = (1+\Pi_1)[C_1 \cos(\lambda_1 x) + C_2 \sin(\lambda_1 x)] + (1-\Pi_2)[C_3 \cosh(\lambda_2 x) + C_4 \sinh(\lambda_2 x)] \tag{3.45}$$

$$\begin{aligned}\psi = &\lambda_1(1-\Lambda_1)[C_1 \sin(\lambda_1 x) - C_2 \cos(\lambda_1 x)] \\ &- \lambda_2(1+\Lambda_2)[C_3 \sinh(\lambda_2 x) + C_4 \cosh(\lambda_2 x)]\end{aligned} \tag{3.46}$$

$$\begin{aligned}M = &\left(E_2 + \frac{E_2\rho_4 - E_4\rho_2}{\widetilde{G}_0}\right)\{\lambda_1^2[C_1 \cos(\lambda_1 x) + C_2 \sin(\lambda_1 x)] \\ &- \lambda_2^2[C_3 \cosh(\lambda_2 x) + C_4 \sinh(\lambda_2 x)]\}\end{aligned} \tag{3.47}$$

$$\begin{aligned}Q = &-\lambda_1\left(E_2\lambda_1^2 - \rho_2\omega^2\right) \\ &\left\{C_1 \sin(\lambda_1 x) - C_2 \cos(\lambda_1 x) + \frac{\lambda_2(\Pi_2+\Lambda_2)}{\lambda_1(\Pi_1+\Lambda_1)}[C_3 \cosh(\lambda_2 x) + C_4 \sinh(\lambda_2 x)]\right\}\end{aligned} \tag{3.48}$$

where

$$\begin{cases} \Pi_1 = \dfrac{\widetilde{E}_2\lambda_1^2 - \widetilde{\rho}_2\omega^2}{\widetilde{G}_0} \\ \Pi_2 = \dfrac{\widetilde{E}_2\lambda_2^2 + \widetilde{\rho}_2\omega^2}{\widetilde{G}_0} \end{cases} \tag{3.49}$$

$$\begin{cases} \Lambda_1 = \dfrac{E_4\lambda_1^2 - \rho_4\omega^2}{\widetilde{G}_0} \\ \Lambda_2 = \dfrac{E_4\lambda_2^2 + \rho_4\omega^2}{\widetilde{G}_0} \end{cases} \tag{3.50}$$

Using the above results for the deflection, rotation, bending moment, and shear force, Eqs. (3.45–3.48), under appropriate boundary conditions at both ends of a tube beam, we can easily determine the natural frequencies of free vibration of a tube beam.

3.3.2 Cylindrical tube beams with axial force

This section is concerned with the dynamic behavior of axially loaded functionally graded material (FGM) cylindrical tubes. For this purpose, the expression for the auxiliary function F for harmonic vibration takes the following form

$$F(x,t) = f(x)e^{i\omega t} \tag{3.51}$$

where $f(x)$ is an unknown function to be determined and ω is the circular frequency. Substituting Eq. (3.51) into Eq. (3.36), by suppressing the time factor, one obtains an ordinary differential equation for $f(x)$ as follows:

$$b_0 \frac{d^4 f}{dx^4} + 2b_1 \frac{d^2 f}{dx^2} - b_2 f = 0 \tag{3.52}$$

in which

$$b_0 = (EI)_0 - \frac{P(EI)^*}{(GA)^*} \tag{3.53}$$

$$b_1 = \frac{1}{2}\left\{\left[m_2 + \frac{m_0(EI)^* - Pm_{2f}}{(GA)^*}\right]\omega^2 + P\right\} \tag{3.54}$$

$$b_2 = m_0\omega^2\left[1 - \frac{m_{2f}\omega^2}{(GA)^*}\right] \tag{3.55}$$

For simplicity, in the following, only lower frequencies of vibration are analyzed. It suffices to solve Eq. (3.52) by applying a standard approach. After some algebra, one easily obtains the general solution as

$$f(x) = C_1 \cos(\lambda_1 x) + C_2 \sin(\lambda_1 x) + C_3 \cosh(\lambda_2 x) + C_4 \sinh(\lambda_2 x) \tag{3.56}$$

with

$$\lambda_{1,2} = \sqrt{\frac{\sqrt{b_1^2 + b_0 b_2} \pm b_1}{b_0}} \tag{3.57}$$

In Eq. (3.56), the hyperbolic sine and hyperbolic cosine functions indicate nonpropagating fields or evanescent components, and the other two related to trigonometric functions represent bending waves' propagation. It is worth pointing out that for free vibration at higher frequencies, where each term in Eq. (3.56) becomes trigonometric and the hyperbolic functions vanish, an analogous analysis can be done but is omitted here for brevity. Based on the given boundary conditions, circular frequencies can be readily determined from the above expression together with the relevant end conditions.

Next, four frequently encountered typical cases are considered, similar to the previous section. For this purpose, one should provide $w, \psi, M,$ and Q first. This is done by substituting Eq. (3.56) into the governing equations. One then has the following expressions

$$\begin{aligned} w = {} & (1 + \Pi_1)[C_1 \cos(\lambda_1 x) + C_2 \sin(\lambda_1 x)] + (1 - \Pi_2) \\ & [C_3 \cosh(\lambda_2 x) + C_4 \sinh(\lambda_2 x)] \end{aligned} \tag{3.58}$$

$$\begin{aligned} \psi = {} & \lambda_1(1 - \Lambda_1)[C_1 \sin(\lambda_1 x) - C_2 \cos(\lambda_1 x)] - \lambda_2(1 + \Lambda_2) \\ & [C_3 \sinh(\lambda_2 x) + C_4 \cosh(\lambda_2 x)] \end{aligned} \tag{3.59}$$

$$M = M_0\left\{\lambda_1^2[C_1\cos(\lambda_1 x) + C_2\sin(\lambda_1 x)] - \lambda_2^2[C_3\cosh(\lambda_2 x) + C_4\sinh(\lambda_2 x)]\right\} \tag{3.60}$$

$$V = -\lambda_1\lambda_2^*[C_1\sin(\lambda_1 x) - C_2\cos(\lambda_1 x)] - \lambda_2\lambda_1^*[C_3\sinh(\lambda_2 x) + C_4\cosh(\lambda_2 x)] \tag{3.61}$$

where

$$\begin{cases} \Pi_1 = \dfrac{(EI)^*\lambda_1^2 - m_{2f}\omega^2}{(GA)^*} \\ \Pi_2 = \dfrac{(EI)^*\lambda_2^2 + m_{2f}\omega^2}{(GA)^*} \end{cases} \tag{3.62}$$

$$\begin{cases} \Lambda_1 = \dfrac{(EI)_0\lambda_1^2 - m_2\omega^2}{(GA)^*} - \Pi_1 \\ \Lambda_2 = \dfrac{(EI)_0\lambda_2^2 + m_2\omega^2}{(GA)^*} - \Pi_2 \end{cases} \tag{3.63}$$

$$\begin{cases} \lambda_1^* = b_0\lambda_1^2 - \dfrac{m_0(EI)^*}{(GA)^*}\omega^2 \\ \lambda_2^* = b_0\lambda_2^2 + \dfrac{m_0(EI)^*}{(GA)^*}\omega^2 \end{cases} \tag{3.64}$$

$$M_0 = (EI)_0 + \frac{m_0(EI)^* - m_{2f}(EI)_0}{(GA)^*}\omega^2 \tag{3.65}$$

Using the above results, we apply the relationships in Eqs. (3.58)–(3.61) to obtain the frequency equations.

3.4 Vibration analysis of circular cylindrical beams

3.4.1 Simply supported tube beams

Consider a simply supported tube beam of length L with the boundary conditions:

$$\begin{cases} w(0) = 0 \\ M(0) = 0 \\ w(L) = 0 \\ M(L) = 0 \end{cases} \tag{3.66}$$

By applying the boundary conditions in Eqs. (3.66)–(3.45) and Eq. (3.47), respectively, one easily gets $C_1 = C_3 = 0$, and C_2 and C_4 satisfy the following algebraic equations

$$(1+\Pi_1)C_2\sin(\lambda_1 L)+(1-\Pi_2)C_4\sinh(\lambda_2 L)=0 \tag{3.67}$$

$$\lambda_1^2 C_2\sin(\lambda_1 L)-\lambda_2^2 C_4\sinh(\lambda_2 L)=0 \tag{3.68}$$

Due to the existence of a nontrivial solution, the determinant of the coefficient matrix has to vanish, which results in

$$\sin(\lambda_1 L)\sinh(\lambda_2 L)=0 \tag{3.69}$$

Solving the above equation, recalling Eq. (3.43), we derive the natural frequencies, which result in the following equation

$$\omega L\sqrt{\frac{1}{2}\left[\sqrt{\left(\frac{\rho_2}{E_2}+\frac{\rho_0\widetilde{E}_2}{\widetilde{G}_0E_2}\right)^2+\frac{4\rho_0}{E_2}\left(\frac{1}{\omega^2}-\frac{1}{\omega_c^2}\right)}+\left(\frac{\rho_2}{E_2}+\frac{\rho_0\widetilde{E}_2}{E_2\widetilde{G}_0}\right)\right]} \tag{3.70}$$

$$=n\pi, n=1,2,3,\ldots$$

The natural frequencies can be given explicitly by finding the roots of Eq. (3.70)

$$\omega_n=\frac{\omega_c}{\sqrt{2}}\sqrt{1+\eta\left(\frac{n\pi}{L}\right)^2-\sqrt{\left[1+\eta\left(\frac{n\pi}{L}\right)^2\right]^2-\frac{4E_2\widetilde{\rho}_2}{\widetilde{G}_0\rho_0}\left(\frac{n\pi}{L}\right)^4}} \tag{3.71}$$

where

$$\eta=\frac{\rho_2}{\rho_0}+\frac{\widetilde{E}_2}{\widetilde{G}_0} \tag{3.72}$$

The natural frequencies can be exactly determined by solving Eq. (3.71). Note that the lower positive roots are selected. With the natural frequencies, the mode shape of a simply supported tube beam corresponding to the n-th natural frequency reads

$$w=\sin\left(\frac{n\pi x}{L}\right) \tag{3.73}$$

This result coincides with the classical one, and it turns out that the mode shapes of a simply supported radially graded tube beam are independent of the gradient index. Nonetheless, according to Eq. (3.71), the natural frequencies are clearly dependent on the gradient index and material properties.

To verify the accuracy of the presented theory, the first three natural frequencies of a simply supported radially graded cylindrical tube made of Al/ZrO_2 are calculated based on the presented theory. The material properties obey power law (Eq. 2.81). Numerical results of dimensionless natural frequencies $\Omega_n = L^2\sqrt{\rho_o A\omega_n^2/E_o I}$ based on the present approach and finite element method (FEM) are listed in Table 3.1 for the simplest linear gradient, $N=1$. Here we discuss the free vibration of a simply supported radially

TABLE 3.1 Dimensionless frequencies Ω_n of a simply supported radially graded tube beam with $R_i/R_o = 0.5$, $N = 1$, $v = 0.3$.

	$L/R_o = 20$		$L/R_o = 10$		$L/R_o = 5$	
n	Present	FEM	Present	FEM	Present	FEM
1	9.7057	9.7001	9.1405	9.1793	7.6461	7.7656
2	36.5620	36.7170	30.5845	31.0625	21.0083	21.5269
3	75.6586	76.4260	56.6664	57.9680	34.7390	35.2628

TABLE 3.2 The number of elements and DOFs of a simply supported radially graded tube beam with $R_i = 1$, $R_i/R_o = 0.5$.

	$L/R_o = 20$	$L/R_o = 10$	$L/R_o = 5$
Elements	32,000	16,000	8000
DOFs	105,138	52,338	25,938

graded tube beam and compare our numerical results with those based on the FEM. As can be seen in Table 3.1, our numerical results agree well with those obtained from the FEM, in which the numbers of the elements and degrees of freedom (DOFs) used in the FEM model are listed in Table 3.2. The comparison reveals satisfactory agreement. A slight difference is possible due to the relatively rough partition. If more layers are partitioned, it is anticipated that more accurate results of the FEM can be achieved. Figs. 3.2–3.4 show the finite element results of the first three mode shapes of a simply supported tube. In addition, from Eq. (3.73) we clearly conclude that the mode shapes of simply supported radially graded tube beams are independent of the gradient index, which is essentially different from the natural frequencies.

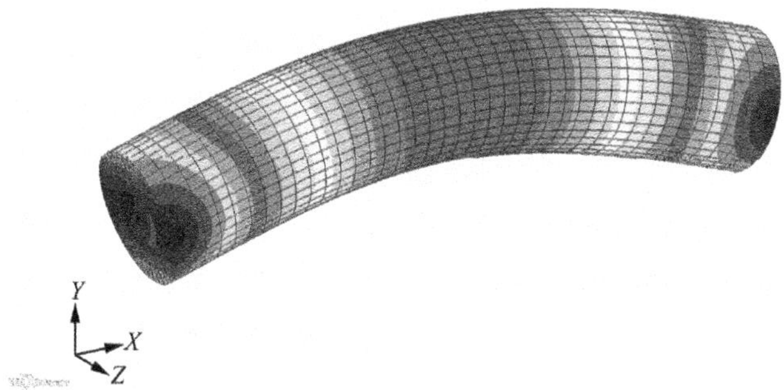

FIGURE 3.2 The first three mode shapes for a simply supported tube beam with $L/R_o = 10$ in FEM results, first mode shape.

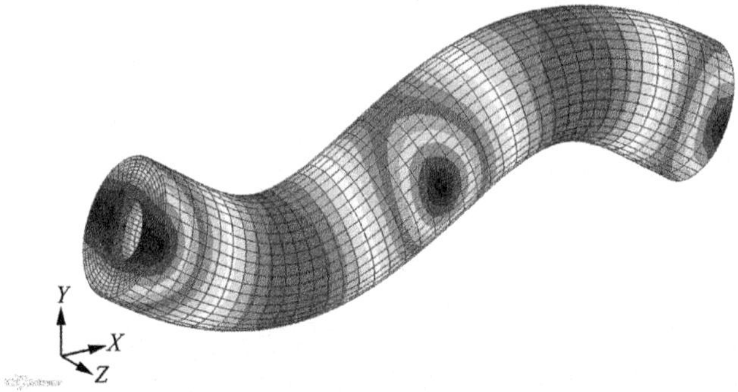

FIGURE 3.3 The first three mode shapes for a simply supported tube beam with $L/R_o = 10$ in FEM results, second mode shape.

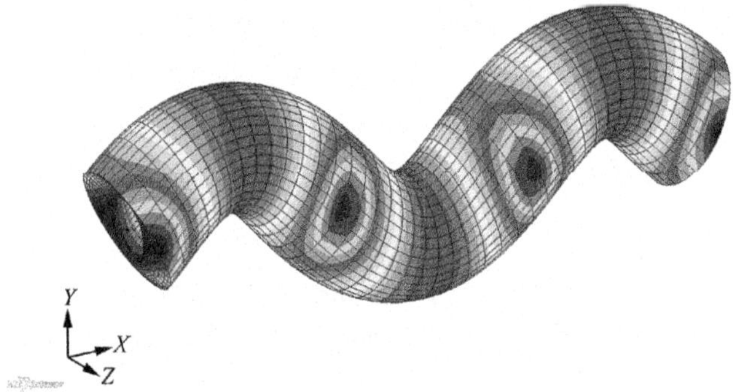

FIGURE 3.4 The first three mode shapes for a simply supported tube beam with $L/R_o = 10$ in FEM results, third mode shape.

3.4.2 Clamped tube beams

For a doubly clamped tube beam of length L, the boundary conditions can be written as

$$\begin{cases} w(0) = 0 \\ \psi(0) = 0 \\ w(L) = 0 \\ \psi(L) = 0 \end{cases} \tag{3.74}$$

Substituting the first two boundary conditions in Eq. (3.74) into Eqs. (3.61) and (3.62) yields

$$\begin{cases} C_3 = -\dfrac{1+\Pi_1}{1-\Pi_2}C_1 \\ C_4 = -\dfrac{\lambda_1(1-\Lambda_1)}{\lambda_2(1+\Lambda_2)}C_2 \end{cases} \tag{3.75}$$

In addition, using the third boundary condition in Eq. (3.74) along with Eq. (3.75), we have

$$C_2 = -\frac{\cos(\lambda_1 L) - \cosh(\lambda_2 L)}{\sin(\lambda_1 L) - \Psi \sinh(\lambda_2 L)}C_1 \tag{3.76}$$

with

$$\Psi = \frac{\lambda_1(1-\Lambda_1)(1-\Pi_2)}{\lambda_2(1+\Lambda_2)(1+\Pi_1)} \tag{3.77}$$

Finally, applying the remaining boundary condition in Eq. (3.74), one obtains the frequency equation

$$2 - 2\cos(\lambda_1 L)\cosh(\lambda_2 L) - \left(\Psi - \frac{1}{\Psi}\right)\sin(\lambda_1 L)\sinh(\lambda_2 L) = 0 \tag{3.78}$$

Once the natural frequencies are determined by solving the above frequency, Eq. (3.78), the corresponding mode shape can be given by

$$w = \cos(\lambda_1 x) - \cosh(\lambda_2 x) - \frac{\cos(\lambda_1 L) - \cosh(\lambda_2 L)}{\sin(\lambda_1 L) - \Psi \sinh(\lambda_2 L)}[\sin(\lambda_1 x) - \Psi \sinh(\lambda_2 x)] \tag{3.79}$$

Table 3.3 also shows a comparison of the first three dimensionless frequencies Ω_n between our results with those obtained from the FEM for a doubly clamped tube beam. In Figs. 3.5–3.7, the finite element results of the first three modes shapes for a clamped tube beam are displayed.

TABLE 3.3 Dimensionless frequencies Ω_n of a clamped tube beam with $R_i/R_o = 0.5$, $N = 1$, $v = 0.3$.

	$L/R_o = 20$		$L/R_o = 10$		$L/R_o = 5$	
n	Present	FEM	Present	FEM	Present	FEM
1	20.6660	20.7593	17.0964	17.2690	11.4105	11.6524
2	51.9404	52.3566	37.9044	38.6744	22.3890	23.5231
3	92.2070	93.3804	62.0131	63.8464	35.3381	37.2219

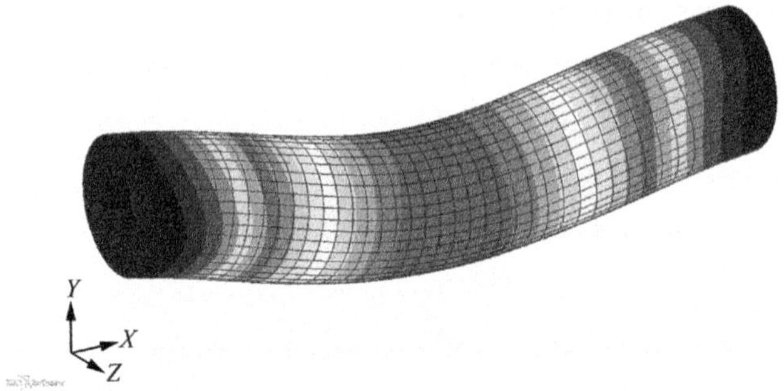

FIGURE 3.5 The first three mode shapes for a clamped tube beam with $L/R_o = 10$ in FEM results, first mode shape.

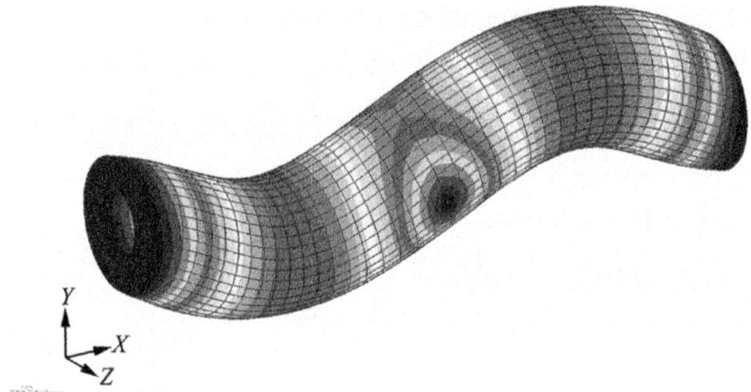

FIGURE 3.6 The first three mode shapes for a clamped tube beam with $L/R_o = 10$ in FEM results, second mode shape.

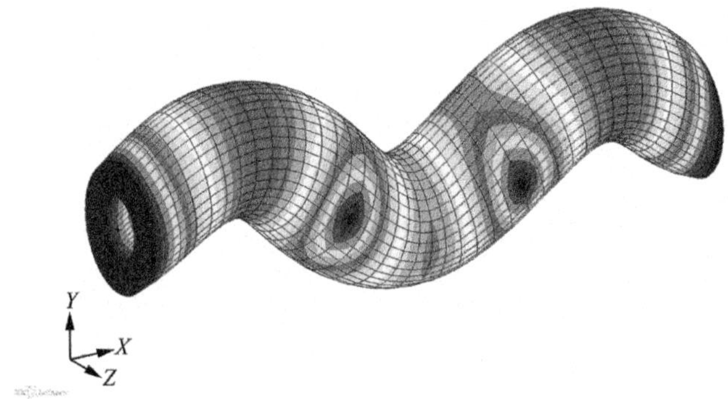

FIGURE 3.7 The first three mode shapes for a clamped tube beam with $L/R_o = 10$ in FEM results, third mode shape.

3.4.3 Cantilever tube beams

In this case, the boundary conditions of a cantilever tube beam read

$$\begin{cases} w(0) = 0 \\ \psi(0) = 0 \\ M(L) = 0 \\ Q(L) = 0 \end{cases} \tag{3.80}$$

According to the first two boundary conditions in Eq. (3.80), one gets the same C_3 and C_4 as in Eq. (3.75). Furthermore, application of the third boundary condition gives

$$C_2 = -\frac{\lambda_1(1+\Lambda_2)\Psi\cos(\lambda_1 L)+\lambda_2(1-\Lambda_1)\cosh(\lambda_2 L)}{\Psi[\lambda_1(1+\Lambda_2)\sin(\lambda_1 L)+\lambda_2(1-\Lambda_1)\sinh(\lambda_2 L)]}C_1 \tag{3.81}$$

With Eq. (3.75), Eq. (3.81), and the last boundary condition in Eq. (3.80) at hand, we can derive the frequency equation of a radially graded cantilever tube beam as follows:

$$\frac{1-\Lambda_1}{1+\Lambda_2}\left[\left(\frac{\Pi_2+\Lambda_2}{\Pi_1+\Lambda_1}+\frac{\lambda_2}{\lambda_1\Psi}\right)\cos(\lambda_1 L)\cosh(\lambda_2 L)+\right.$$
$$\left.\left(\frac{\lambda_2}{\lambda_1}-\frac{\Pi_2+\Lambda_2}{(\Pi_1+\Lambda_1)\Psi}\right)\sin(\lambda_1 L)\sinh(\lambda_2 L)\right]+1+\frac{\lambda_2(\Pi_2+\Lambda_2)(1-\Lambda_1)^2}{\Psi\lambda_1(\Pi_1+\Lambda_1)(1+\Lambda_2)^2}=0 \tag{3.82}$$

The corresponding mode shape can be given by

$$w=\cos(\lambda_1 x)-\cosh(\lambda_2 x)$$
$$-\frac{\lambda_1(1+\Lambda_2)\Psi\cos(\lambda_1 L)+\lambda_2(1-\Lambda_1)\cosh(\lambda_2 L)}{\Psi[\lambda_1(1+\Lambda_2)\sin(\lambda_1 L)+\lambda_2(1-\Lambda_1)\sinh(\lambda_2 L)]}[\sin(\lambda_1 x)-\Psi\sinh(\lambda_2 x)] \tag{3.83}$$

Similarly, in Table 3.4 we give a comparison of the dimensionless frequencies between our numerical results and those obtained from the FEM for a radially graded cantilever tube with $R_{in}/R_{out}=0.5$, $N=1$, and $\nu=0.3$. Figs. 3.8–3.10 show the finite element results of the first three mode shapes of a radially graded cantilever tube.

Based on the characteristic Eqs. (3.71), (3.78), and (3.82), dimensionless natural frequencies of mode Ω_n are calculated for classical end conditions, and the results are provided in Tables 3.5–3.7. As can be seen in Tables 3.5–3.7, regardless of the end conditions, the first two dimensionless natural frequencies increase with increasing N, arriving at its maximum at a certain N value, and decrease as N continues to rise. Nonetheless, from the third-order dimensionless frequencies, they always drop as the power-law index N increases. Therefore, material properties play a dominant role in determining natural frequencies and their trends. In other words, natural frequencies are sensitive to the volume fraction of constituents of each phase.

TABLE 3.4 Dimensionless frequencies Ω of a radially graded cantilever tube beam with $R_i/R_o = 0.5$, $N = 1$, $v = 0.3$.

	$L/R_o = 20$		$L/R_o = 10$		$L/R_o = 5$	
n	Present	FEM	Present	FEM	Present	FEM
1	3.5007	3.5034	3.4047	3.4133	3.0945	3.3713
2	20.7216	20.7654	17.6878	17.8208	12.2091	12.3668
3	53.6195	53.9399	40.9382	41.5966	25.5707	26.1074

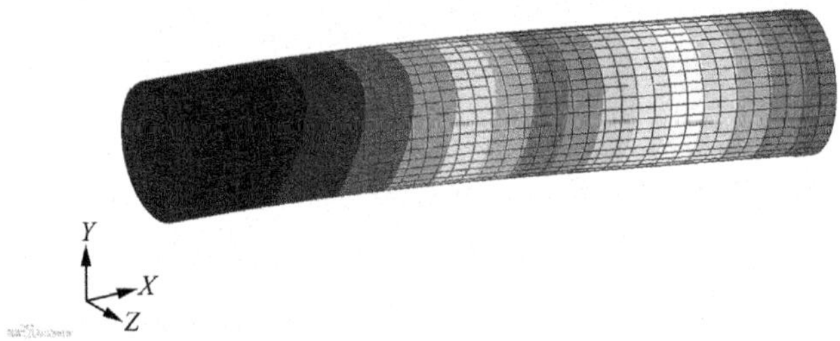

FIGURE 3.8 The first three mode shapes for a cantilever tube beam with $L/R_o = 10$ in FEM results, first mode shape.

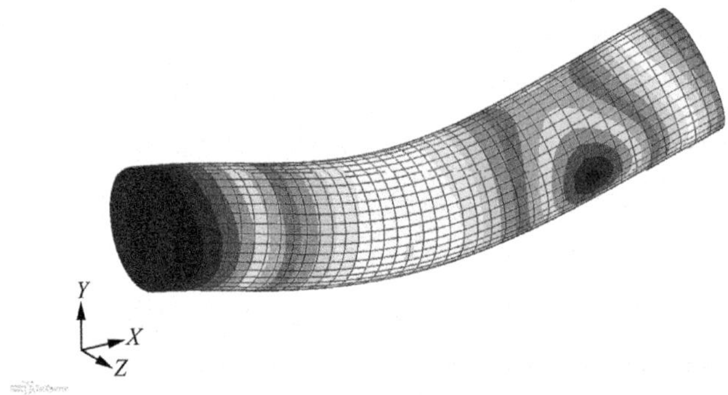

FIGURE 3.9 The first three mode shapes for a cantilever tube beam with $L/R_o = 10$ in FEM results, second mode shape.

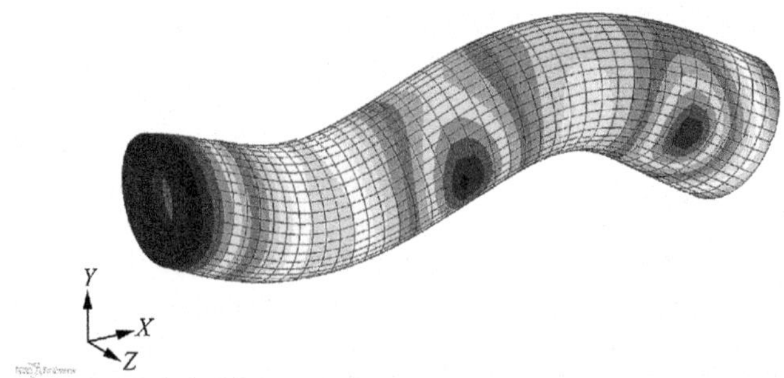

FIGURE 3.10 The first three mode shapes for a cantilever tube beam with $L/R_o = 10$ in FEM results, third mode shape.

TABLE 3.5 Dimensionless natural frequencies Ω_n of a simply supported functionally graded tube beam with $L/R_o = 20$, $\nu = 0.3$.

N	1	2	3	4	5	6
0.1	9.6951	36.6625	76.2160	123.7781	176.0439	230.9249
0.5	9.7303	36.7113	76.1077	123.2787	174.9362	229.0379
1	9.7057	36.5620	75.6586	122.3378	173.3405	226.6658
5	9.2679	34.8820	72.1064	116.4790	164.8996	215.4774
10	8.9657	33.7924	69.9714	113.2095	160.4895	209.9520

TABLE 3.6 Dimensionless natural frequencies Ω_n of a clamped functionally graded tube beam with $L/R_o = 20$, $\nu = 0.3$.

N	1	2	3	4	5	6
0.1	20.7626	52.4913	93.6300	140.4777	190.8816	260.4101
0.5	20.7675	52.3197	93.0550	139.2963	188.9609	240.7467
1	20.6660	51.9404	92.2070	137.8225	186.7600	237.7533
5	19.7020	49.4448	87.6771	130.9443	177.3402	225.6760
10	19.1005	48.0339	85.3216	127.5958	172.9746	220.2788

TABLE 3.7 Dimensionless natural frequencies Ω_n of a cantilever functionally graded tube beam with $L/R_o = 20$, $\nu = 0.3$.

N	1	2	3	4	5	6
0.1	3.4959	20.7747	54.0182	96.7249	145.4473	197.6921
0.5	3.5080	20.8061	53.9408	96.3141	144.4905	196.0227
1	3.5007	20.7216	53.6195	95.5628	143.1424	193.9546
5	3.3437	19.7673	51.0924	90.9604	136.1338	184.3368
10	3.2325	19.1488	49.5801	88.4164	132.5099	179.6305

3.5 Vibration analysis of a cylinder beam with axial force

Next, we consider cylindrical tubes with four typical boundary conditions: hinged–hinged (HH), clamped–free (CF), clamped–clamped (CC), and clamped–hinged (CH) ends, respectively. For the above four different ends, we have

$$\begin{cases} \mathrm{HH}: w(0)=0, M(0)=0, w(L)=0, M(L)=0 \\ \mathrm{CF}: w(0)=0, \psi(0)=0, M(L)=0, V(L)=0 \\ \mathrm{CC}: w(0)=0, \psi(0)=0, w(L)=0, \psi(L)=0 \\ \mathrm{CH}: w(0)=0, \psi(0)=0, w(L)=0, M(L)=0 \end{cases} \tag{3.84}$$

3.5.1 Hinged–hinged (HH) ends

First, for HH tubes, applying Eqs. (3.58) and (3.60) to the boundary conditions results in

$$(1+\Pi_1)C_1+(1-\Pi_2)C_3=0 \tag{3.85}$$

$$\begin{aligned}&(1+\Pi_1)[C_1\cos(\lambda_1 L)+C_2\sin(\lambda_1 L)]+(1-\Pi_2)\\ &[C_3\cosh(\lambda_2 L)+C_4\sinh(\lambda_2 L)]=0\end{aligned} \tag{3.86}$$

$$\lambda_1^2 C_1-\lambda_2^2 C_3=0 \tag{3.87}$$

$$\lambda_1^2[C_1\cos(\lambda_1 L)+C_2\sin(\lambda_1 L)]-\lambda_2^2[C_3\cosh(\lambda_2 L)+C_4\sinh(\lambda_2 L)]=0 \tag{3.88}$$

Solving the above equations, one easily finds the frequency equation as

$$\sin(\lambda_1 L)=0 \tag{3.89}$$

from which, one can find an explicit expression for the natural frequencies below

$$\omega_n=\frac{\omega_c}{\sqrt{2}}\sqrt{1+\eta\left(\frac{n\pi}{L}\right)^2-\sqrt{\left[1+\eta\left(\frac{n\pi}{L}\right)^2\right]^2-\frac{4}{m_0\omega_c^2}\left(\frac{n\pi}{L}\right)^2\left[b_0\left(\frac{n\pi}{L}\right)^2-P\right]}} \tag{3.90}$$

where

$$\begin{cases} \omega_c=\sqrt{\dfrac{(GA)^*}{m_{2f}}} \\ \eta=\dfrac{m_2}{m_0}+\dfrac{m_0(EI)^*-Pm_{2f}}{m_0(GA)^*} \end{cases} \tag{3.91}$$

With Eq. (3.90) at hand, one easily gets the natural frequencies, and the n-th order mode shape can be obtained subsequently, which reads

$$w = \sin\left(\frac{n\pi x}{L}\right) \tag{3.92}$$

3.5.2 Clamped–free (CF) ends

For CF tubes, the application of Eqs. (3.58)–(3.61) to the boundary conditions results in

$$(1+\Pi_1)C_1 + (1-\Pi_2)C_3 = 0 \tag{3.93}$$

$$\lambda_1(1-\Lambda_1)C_2 + \lambda_2(1+\Lambda_2)C_4 = 0 \tag{3.94}$$

$$\lambda_1^2[C_1\cos(\lambda_1 L) + C_2\sin(\lambda_1 L)] - \lambda_2^2[C_3\cosh(\lambda_2 L) + C_4\sinh(\lambda_2 L)] = 0 \tag{3.95}$$

$$\lambda_1\lambda_2^*[C_1\sin(\lambda_1 L) - C_2\cos(\lambda_1 L)] + \lambda_2\lambda_1^*[C_3\sinh(\lambda_2 L) + C_4\cosh(\lambda_2 L)] = 0 \tag{3.96}$$

By applying a similar approach and omitting the details, one finally gets the frequency equation

$$\begin{aligned} 1 &+ \frac{\lambda_2\lambda_1^*(1-\Lambda_1)^2}{\Psi\lambda_1\lambda_2^*(1+\Lambda_2)^2} + \frac{\lambda_2}{\lambda_1}\left[\frac{1-\Lambda_1}{1+\Lambda_2} - \frac{\lambda_1^*(1+\Pi_1)}{\lambda_2^*(1-\Pi_2)}\right]\sin(\lambda_1 L)\sinh(\lambda_2 L) \\ &+ \frac{1-\Lambda_1}{1+\Lambda_2}\left(\frac{\lambda_2}{\Psi\lambda_1} + \frac{\lambda_1^*}{\lambda_2^*}\right)\cos(\lambda_1 L)\cosh(\lambda_2 L) = 0 \end{aligned} \tag{3.97}$$

where

$$\Psi = \frac{\lambda_1(1-\Lambda_1)(1-\Pi_2)}{\lambda_2(1+\Lambda_2)(1+\Pi_1)} \tag{3.98}$$

In this case, one gets the mode shape as

$$\begin{aligned} w = {} & \cos(\lambda_1 x) - \cosh(\lambda_2 x) - \\ & \frac{\lambda_1(1+\Lambda_2)\Psi\cos(\lambda_1 L) + \lambda_2(1-\Lambda_1)\cosh(\lambda_2 L)}{\Psi[\lambda_1(1+\Lambda_2)\sin(\lambda_1 L) + \lambda_2(1-\Lambda_1)\sinh(\lambda_2 L)]}[\sin(\lambda_1 x) - \Psi\sinh(\lambda_2 x)] \end{aligned} \tag{3.99}$$

3.5.3 Clamped–clamped (CC) ends

For CC tubes, the application of Eqs. (3.58) and (3.59) to the boundary conditions results in

$$(1+\Pi_1)C_1 + (1-\Pi_2)C_3 = 0 \tag{3.100}$$

$$\lambda_1(1-\Lambda_1)C_2+\lambda_2(1+\Lambda_2)C_4=0 \tag{3.101}$$

$$\begin{aligned}&(1+\Pi_1)[C_1\cos(\lambda_1 L)+C_2\sin(\lambda_1 L)]\\&\quad+(1-\Pi_2)[C_3\cosh(\lambda_2 L)+C_4\sinh(\lambda_2 L)]=0\end{aligned} \tag{3.102}$$

$$\begin{aligned}&\lambda_1(1-\Lambda_1)[C_1\sin(\lambda_1 L)-C_2\cos(\lambda_1 L)]\\&\quad-\lambda_2(1+\Lambda_2)[C_3\sinh(\lambda_2 L)+C_4\cosh(\lambda_2 L)]=0\end{aligned} \tag{3.103}$$

By omitting the details, one finally gets the frequency equation as

$$1-\cos(\lambda_1 L)\cosh(\lambda_2 L)-\frac{1}{2}\left(\Psi-\frac{1}{\Psi}\right)\sin(\lambda_1 L)\sinh(\lambda_2 L)=0 \tag{3.104}$$

In this case, the corresponding mode shape can be given by

$$w=\cos(\lambda_1 x)-\cosh(\lambda_2 x)-\frac{\cos(\lambda_1 L)-\cosh(\lambda_2 L)}{\sin(\lambda_1 L)-\Psi\sinh(\lambda_2 L)}[\sin(\lambda_1 x)-\Psi\sinh(\lambda_2 x)] \tag{3.105}$$

3.5.4 Clamped–hinged (CH) ends

For CH tubes, the application of Eqs. (3.58)–(3.60) to the boundary conditions results in

$$(1+\Pi_1)C_1+(1-\Pi_2)C_3=0 \tag{3.106}$$

$$\lambda_1(1-\Lambda_1)C_2+\lambda_2(1+\Lambda_2)C_4=0 \tag{3.107}$$

$$\begin{aligned}&(1+\Pi_1)[C_1\cos(\lambda_1 L)+C_2\sin(\lambda_1 L)]\\&\quad+(1-\Pi_2)[C_3\cosh(\lambda_2 L)+C_4\sinh(\lambda_2 L)]=0\end{aligned} \tag{3.108}$$

$$\lambda_1^2[C_1\cos(\lambda_1 L)+C_2\sin(\lambda_1 L)]-\lambda_2^2[C_3\cosh(\lambda_2 L)+C_4\sinh(\lambda_2 L)]=0 \tag{3.109}$$

By omitting the details, one finally gets the frequency equation as

$$\sin(\lambda_1 L)\cosh(\lambda_2 L)-\Psi\cos(\lambda_1 L)\sinh(\lambda_2 L)=0 \tag{3.110}$$

Once the natural frequencies are determined by solving the above frequency, Eq. (3.110), the corresponding mode shape can be given by

$$w=\cos(\lambda_1 x)-\cosh(\lambda_2 x)-\frac{\cos(\lambda_1 L)-\cosh(\lambda_2 L)}{\sin(\lambda_1 L)-\Psi\sinh(\lambda_2 L)}[\sin(\lambda_1 x)-\Psi\sinh(\lambda_2 x)] \tag{3.111}$$

Note that the mode shape of the CH tubes in form is identical to that of the CC tubes, but with different natural frequencies, and their frequency equations are not the same, as illustrated by Eqs. (3.104) and (3.110), respectively. This conclusion is consistent with the classic model [14]. For the above-obtained frequency equations, if the gradient and axial load are neglected, these frequency equations reduce to those for transverse free vibration of homogeneous tubes based on the third-order shear deformation beam theory [10].

Finally, we consider two limiting cases. The first is to take $f(z) = 0$ and the other is to take $f(z) = z$. The former corresponds to the Euler–Bernoulli beam theory and the latter to the Timoshenko beam theory with introduction of the shear correction factor. For the former case, it can be done by setting $(GA)^* \to \infty$, $m_2 = 0$, giving

$$\tilde{\lambda}_{1,2} = \sqrt{\sqrt{\frac{P^2}{4(EI)_0^2} + \frac{m_0\omega^2}{(EI)_0}} \pm \frac{P}{2(EI)_0}} \tag{3.112}$$

Under such circumstances, the frequency equations for the above-mentioned four boundary conditions, which are listed in Table 3.8, can be reduced. In particular, when considering the case of homogeneous beams, these frequency equations are identical to those derived in Refs. [15,16].

For the latter case, when the gradient is absent, we have

$$\lambda_{1,2} = \sqrt{\frac{\sqrt{\left[\rho I\left(1 + \frac{EA - P}{\kappa GA}\right)\omega^2 + P\right]^2 + 4\rho A\omega^2 EI\left(1 - \frac{P}{\kappa GA}\right)\left(1 - \frac{\rho I\omega^2}{\kappa GA}\right)}}{2EI\left(1 - \frac{P}{\kappa GA}\right)} \pm \frac{\rho I\omega^2\left(1 + \frac{EA - P}{\kappa GA}\right) + P}{2EI\left(1 - \frac{P}{\kappa GA}\right)}} \tag{3.113}$$

TABLE 3.8 Frequency equations for radially graded Euler–Bernoulli tubes.

Boundary conditions	Frequency equations
HH	$\sin(\tilde{\lambda}_1 L) = 0$
CF	$2 + \tilde{\lambda}_2^2 - \tilde{\lambda}\frac{1}{\lambda}{}^{2}{}_1\tilde{\lambda}_2 \sin(\tilde{\lambda}_1 L)\sinh(\tilde{\lambda}_2 L) + \tilde{\lambda}_2^4 + \tilde{\lambda}\frac{1}{\lambda}{}^{4}{}_1^2\tilde{\lambda}_2^2 \cos(\tilde{\lambda}_1 L)\cosh(\tilde{\lambda}_2 L) = 0$
CC	$1 - \cos(\tilde{\lambda}_1 L)\cosh(\tilde{\lambda}_2 L) - \tilde{\lambda}_1^2 - \tilde{\lambda}\frac{1}{2\lambda}{}^{2}{}_1\tilde{\lambda}_2 \sin(\tilde{\lambda}_1 L)\sinh(\tilde{\lambda}_2 L) = 0$
CH	$\sin(\tilde{\lambda}_1 L)\cosh(\tilde{\lambda}_2 L) - \tilde{\lambda}\frac{1}{\lambda}{}_2 \cos(\tilde{\lambda}_1)\sinh(\tilde{\lambda}_2 L) = 0$

TABLE 3.9 Frequency equations for radially graded Timoshenko tubes.

Boundary conditions	Frequency equations
HH	$\sin(\lambda_1 L) = 0$
CF	$1 + (1 + \Pi_1)\lambda_2^2\lambda_1 \dot{-} (1 - \Pi_2)\lambda_1^2\lambda_2 * + \frac{\lambda_2}{\lambda_1}\left[1 - \lambda_1 * (1 + \Pi_1 \tfrac{)}{\lambda_2} *\right.$ $\left.(1 - \Pi_2)\right]\sin(\lambda_1 L)\sinh(\lambda_2 L) + (\frac{(1 + \Pi_1)\lambda_2^2}{(1 - \Pi_2)\lambda_1^2} + \frac{\lambda_1^*}{\lambda_2^*})\cos(\lambda_1 L)\cosh(\lambda_2 L) = 0$
CC	$1 - \cos(\lambda_1 L)\cosh(\lambda_2 L) - \frac{1}{2}(\lambda_1 (1 - \Pi_2 \tfrac{)}{\lambda_2}(1 + \Pi_1) - \lambda_2 (1 + \Pi_1 \tfrac{)}{\lambda_1}$ $(1 - \Pi_2))\sin(\lambda_1 L)\sinh(\lambda_2 L) = 0$
CH	$\sin(\lambda_1 L)\cosh(\lambda_2 L) - \lambda_1 (1 - \Pi_2 \tfrac{)}{\lambda_2}(1 + \Pi_1)\cos(\lambda_1 L)\sinh(\lambda_2 L) = 0$

where κ is the shear correction factor. Thus, we can give the frequency equations for radially graded Timoshenko tubes in Table 3.9. Furthermore, by setting the axial load to zero, $P = 0$, one finds that these frequency equations reduce to those for uniform circular cylindrical beams within the framework of the Timoshenko beam theory [17].

Numerical examples are given to show the effects of the gradient and the warping shape on the buckling loads and natural frequencies of radially graded tubes. For radially graded tubes, we assume that the material properties vary according to the following power-law form

$$Y(r) = Y_i + (Y_o - Y_i)\frac{r^N - R_i^N}{R_o^N - R_i^N} \tag{3.114}$$

where Y_i and Y_o represent the material properties, such as Young's modulus E, shear modulus G, and mass density ρ. Here the subscript i indicates the inner surface, while o denotes the outer surface. We denote the gradient index as N to present the volume fraction of two kinds of constituents. Two different materials, aluminum and zirconia, were chosen as the inner and outer components of the gradient, respectively, and their material properties are as follows:

$$\begin{cases} \text{Al}: E_i = 70\ \text{GPa}, G_i = 27\ \text{GPa}, \rho_i = 2702\ \text{kg/m}^3 \\ \text{ZrO}_2: E_o = 200\ \text{GPa}, G_o = 77\ \text{GPa}, \rho_o = 5700\ \text{kg/m}^3 \end{cases} \tag{3.115}$$

In the following, we examine the influence of shear deformation along with warping shape on the natural frequencies of the radially graded tube beams. For this, the numerical results of dimensionless natural frequency $\Omega = L^2\sqrt{\rho_o A\omega^2/E_o I}$ for four typical boundary conditions are listed in Tables 3.10–3.13. Clearly, the influence of the gradient index on the natural frequencies is pronounced. Here we can see that as the gradient index N increases, the dimensionless natural frequencies decrease, regardless of the end conditions. Moreover, the dimensionless natural frequency Ω becomes larger for longer radially graded tubes. It should be remembered that the fact that Ω becomes larger does not imply an increase in the frequency since the normalized factor contains the factor of square of the tube length, similar to the buckling loads described above. For brevity, we do not display the practical frequencies and only show the influence of the gradient index on Ω.

Note that the above numerical results are evaluated from a radially graded tube beam without axial load. Figs. 3.12–3.15 show the influence of axial load $P = PL^2/(\mu\pi)^2 E_o I$ on the nondimensional fundamental frequency Ω_1 for a graded tube beam with $R_i/R_o = 0.5$, $L/R_o = 10$. For simplicity, we only present the dependence of the nondimensional fundamental frequency Ω_1 on the axial load P. It can be clearly seen that a tensile force increases the vibration frequencies, whereas a compressive force decreases them. This conclusion is in agreement with that for classical Euler–Bernoulli beams [15,16]. This is easily understood because the beam-column appears to become stiffer when it is subjected to an axial tensile force, whereas it appears to be softer under axial compression. Therefore, the natural frequencies are viewed as increasing or decreasing in the case of the invariant of beam-column's mass. In particular, as compared with a homogeneous beam-column made of zirconia, the Ω_1–P interaction for $N = -8$ is very close to that of the homogeneous zirconia beam-column. This is because, when taking $N = -8$, the material properties are close to those of the outer surface material, zirconia, as shown in Fig. 3.11.

On the contrary, if choosing the gradient index $N = 10$, implying that the volume fraction of zirconia is lower than that of aluminum, then aluminum is rich in the radially graded tube, so the Ω_1–P interaction curve has a large deviation from the curve of the homogeneous zirconia beam-column. Furthermore, it is anticipated that the Ω_1–P interaction curve is closer to that of the Ω_1–P interaction curve of a homogeneous aluminum due to the same reason. In addition, it can be seen from Figs. 3.12–3.15 that the natural frequencies are more sensitive to a greater gradient index. A greater gradient index N causes a larger Ω_1 value for tensile force, and a lower Ω_1 value under compression. It is interesting to point out that there exists a state at a certain small tensile force, which corresponds to the same intersection position of all frequency–load curves, in which the dimensionless frequency is independent of the gradient index N, as shown in Figs. 3.12–3.15.

TABLE 3.10 Dimensionless frequencies Ω of free vibration of an HH FGM tube beam with $R_i/R_o = 0.5$.

Mode	$L/R_o = 20$			$L/R_o = 1$			$L/R_o = 5$		
	$N = -8$	$N = 1$	$N = 10$	$N = -8$	$N = 1$	$N = 10$	$N = -8$	$N = 1$	$N = 10$
1	9.7531	9.7057	9.2928	9.2119	9.1405	8.7444	7.7583	7.6461	7.3008
2	36.8477	36.5620	34.9775	31.0333	30.5845	29.2031	21.5040	21.0083	20.0108
3	76.5159	75.6586	72.3084	57.7987	56.6664	54.0286	35.7030	34.7390	33.0530

TABLE 3.11 Dimensionless frequencies Ω of free vibration of a CF FGM tube beam with $R_i/R_o = 0.5$.

	$L/R_o = 20$			$L/R_o = 1$			$L/R_o = 5$		
Mode	$N = -8$	$N = 1$	$N = 10$	$N = -8$	$N = 1$	$N = 10$	$N = -8$	$N = 1$	$N = 10$
1	3.5165	3.5007	3.3524	3.4267	3.4049	3.2582	3.1334	3.0945	2.9551
2	20.8817	20.7216	19.8216	17.9628	17.6874	16.8792	12.5326	12.2091	11.6165
3	54.2321	53.6195	51.2363	41.7888	40.9380	39.0136	26.2912	25.5707	24.325

TABLE 3.12 Dimensionless frequencies Ω of free vibration of a CC FGM tube beam with $R_i/R_o = 0.5$.

	$L/R_o = 20$			$L/R_o = 1$			$L/R_o = 5$		
Mode	$N = -8$	$N = 1$	$N = 10$	$N = -8$	$N = 1$	$N = 10$	$N = -8$	$N = 1$	$N = 10$
1	20.8600	20.6660	19.7574	17.4272	17.0964	16.2911	11.7787	11.4105	10.8287
2	52.6620	51.9404	49.5876	38.8686	37.9044	36.0596	23.1163	22.3890	21.2639
3	93.8265	92.2070	87.9352	63.7228	62.0131	58.9723	36.4386	35.3381	33.5850

TABLE 3.13 Dimensionless frequencies Ω of free vibration of a CH FGM tube beam with $R_i/R_o = 0.5$.

	$L/R_o = 20$			$L/R_o = 1$			$L/R_o = 5$		
Mode	$N = -8$	$N = 1$	$N = 10$	$N = -8$	$N = 1$	$N = 10$	$N = -8$	$N = 1$	$N = 10$
1	14.8346	14.7320	14.0953	13.1655	12.9854	12.3969	9.7172	9.4761	9.0149
2	44.6938	44.2163	42.2567	35.1596	34.4556	32.8342	22.4601	21.8354	20.7635
3	85.2838	84.0624	80.2516	60.8712	59.4384	56.5913	40.3317	34.7805	33.0668

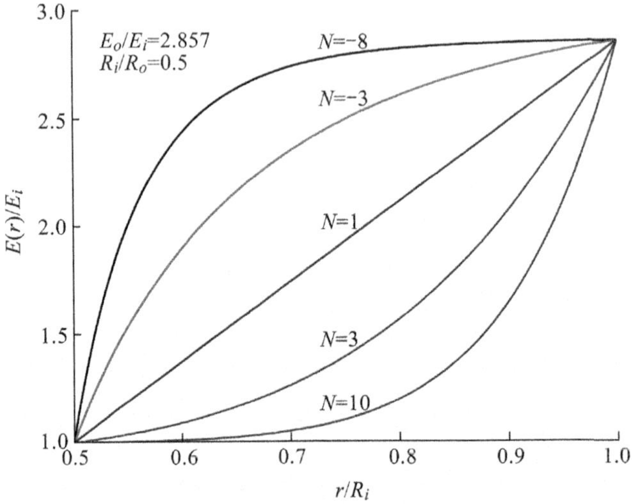

FIGURE 3.11 Variation of the radius-dependent Young's modulus.

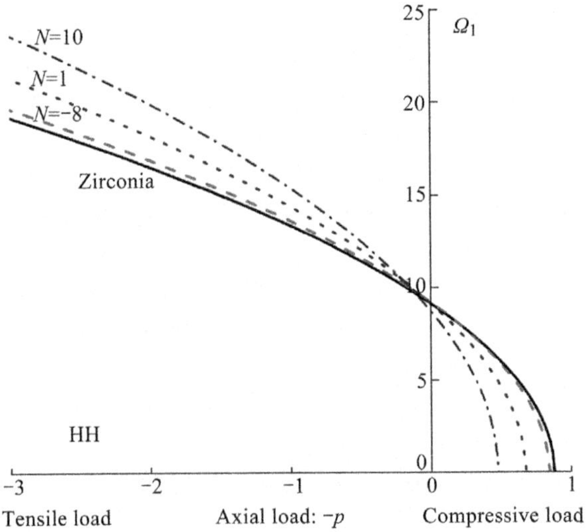

FIGURE 3.12 The influence of axial load P on the nondimensional fundamental frequency Ω_1 for a graded HH tube with $R_i/R_o = 0.5$, $L/R_o = 10$.

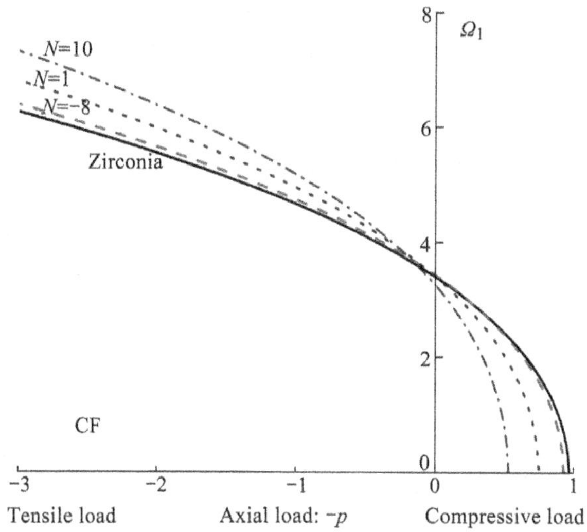

FIGURE 3.13 The influence of axial load P on the nondimensional fundamental frequency Ω_1 for a graded CF tube with $R_i/R_o = 0.5$, $L/R_o = 10$.

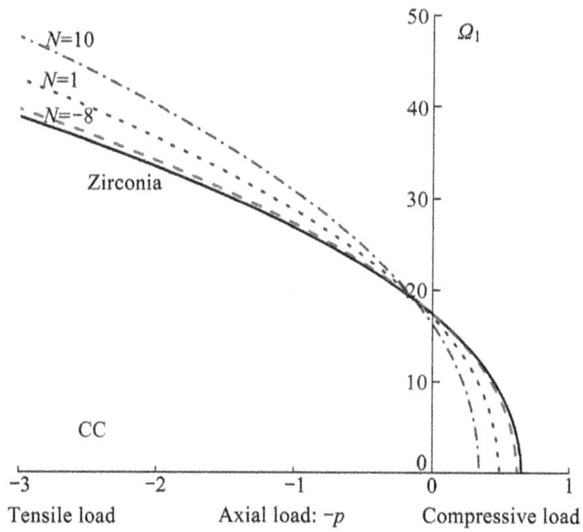

FIGURE 3.14 The influence of axial load P on the nondimensional fundamental frequency Ω_1 for a graded CC tube with $R_i/R_o = 0.5$, $L/R_o = 10$.

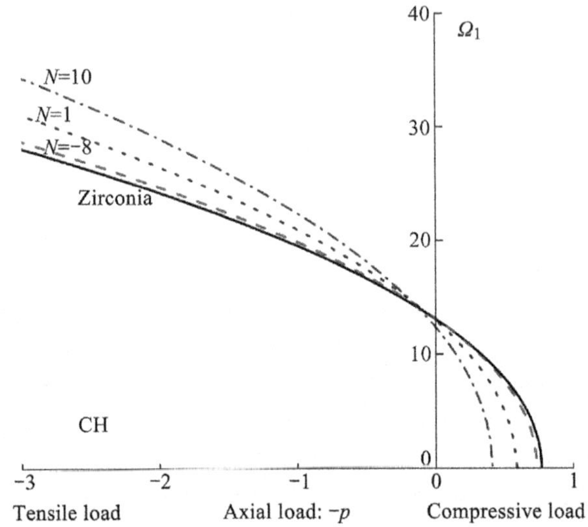

FIGURE 3.15 The influence of axial load P on the nondimensional fundamental frequency Ω_1 for a graded CH tube with $R_i/R_o = 0.5$, $L/R_o = 10$.

References

[1] Li XF. A unified approach for analyzing static and dynamic behaviors of functionally graded Timoshenko and Euler–Bernoulli beams. J Sound Vib 2008;318(4–5):1210–29.

[2] Levinson M. A new rectangular beam theory. J Sound Vib 1981;74(1):81–7.

[3] Thai HT, Vo TP. Bending and free vibration of functionally graded beams using various higher-order shear deformation beam theories. Int J Mech Sci 2012;62(1):57–66.

[4] Xie K, Wang Y, Fan X, et al. Nonlinear free vibration analysis of functionally graded beams by using different shear deformation theories. Appl Math Model 2020;77:1860–80.

[5] Huang Y, Li XF. Buckling of functionally graded circular columns including shear deformation. Mater Des 2010;31(7):3159–66.

[6] Huang Y, Li XF. Bending and vibration of circular cylindrical beams with arbitrary radial nonhomogeneity. Int J Mech Sci 2010;52(4):595–601.

[7] Love AEH. A treatise on the mathematical theory of elasticity. Courier Corporation; 1944.

[8] Timoshenko SP, Gere JM. Theory of elastic stability. Courier Corporation; 2012.

[9] Weaver JW, Timoshenko SP, Young DH. Vibration problems in engineering. John Wiley and Sons; 1991.

[10] Ma WL, Li XF, Yong Lee K. Third-order shear deformation beam model for flexural waves and free vibration of pipes. J Acoust Soc Am 2020;147(3):1634–47.

[11] Ma WL, Jiang ZC, Lee KY, et al. A refined beam theory for bending and vibration of functionally graded tube-beams. Compos Struct 2020;236:111878.

[12] Jiang ZC, Ma WL, Li XF. Stability of cantilever on elastic foundation under a subtangential follower force via shear deformation beam theories. Thin-Walled Struct 2020;154:106853.

[13] Ma WL, Jiang ZC, Li XF. Effect of warping shape on buckling of circular and rectangular columns under axial compression. Appl Math Model 2021;89:1475–90.

[14] Blevins RD. Formulas for dynamics, acoustics and vibration. John Wiley and Sons; 2015.

[15] Bokaian A. Natural frequencies of beams under compressive axial loads. J Sound Vib 1988;126(1):49–65.
[16] Bokaian A. Natural frequencies of beams under tensile axial loads. J Sound Vib 1990;142 (3):481–98.
[17] Pilkey WD. Formulas for stress, strain, and structural matrices. Hoboken, NJ, USA: John Wiley and Sons; 2005.

Further reading

[1] Elishakoff I, Kaplunov J, Nolde E. Celebrating the centenary of Timoshenko's study of effects of shear deformation and rotary inertia. Appl Mech Rev 2015;67(6):060802.

Chapter 4

Stability of higher-order functionally graded circular cylindrical columns

4.1 Introduction

The buckling of elastic columns is a fundamental and significant topic of structural safety in civil, mechanical, and marine engineering, and it plays a crucial role in the design of tall buildings, towers, pipes, etc. since their collapse results in a catastrophic failure. There is a long history of the study of the buckling of elastic columns. In this area, the earliest work can be traced back to Leonhard Euler, who first studied the stability of flexible compressed columns and described the critical buckling loads under various axial compressed forces. In contrast to the assumption of uniform shear stress at the cross section adopted in classical Timoshenko beam theory, the higher-order beam model proposed in this study provides a new approach to solving the stability problem. Based on the traction-free surface condition, two coupled governing equations for the deflection and rotation are derived, and a single governing equation is further obtained. A comparison between buckling loads derived from the proposed circular column model and those obtained from the Timoshenko and Euler–Bernoulli theories of beams is made. Moreover, the effects of the radial gradient on buckling loads of elastic columns with a circular cross section made of functionally graded materials are elucidated. The stability of double-walled carbon nanotubes is considered and the critical stress is determined and compared with existing results. In addition, a transition parameter is introduced to describe the direction of axial load from Engesser's hypothesis to Haringx's one, and then a unified method is presented to determine the critical load. The Haringx buckling load is greater than

Higher-Order Theories of Functionally Graded Circular Cylindrical Beams and Columns
https://doi.org/10.1016/B978-0-443-44718-1.00008-6

the Engesser buckling load which gives a conservative estimate of the critical load. A comparison of these buckling loads with Euler loads is provided.

4.2 Assumption and governing equations

Consider a solid functionally graded cylindrical column with uniform circular cross section of radius R and subjected to an axial compressive force P (Fig. 4.1). Functionally graded materials (FGMs) have spatially continuously varying material properties along certain direction(s). Usually, for rectangular and circular cylindrical columns made of FGMs, the material properties may vary along the thickness, radial, or axial directions [1–8]. In general, spatially varying material property Y, including Young's modulus, shear modulus, mass density, etc. may be expressed as:

$$Y = \sum_{j=1}^{k} Y_j V_j \tag{4.1}$$

where Y_j and V_j are the material property and volume fraction of the j-th constituent phase, respectively. For all V_j, they satisfy

$$\sum_{j=1}^{k} V_j = 1, 0 \leq V_j \leq 1. \tag{4.2}$$

For functionally graded circular cylindrical columns, we choose a Cartesian coordinate system such that the x-axis is oriented along the central axis of the cylindrical columns, and the positive z-axis is directed upward and perpendicular to the x-axis. The elastic displacement components in the x, y, and z directions are denoted as u, v, and w, respectively. Meanwhile, we also use polar cylindrical coordinates (x, r, θ) as well as the corresponding displacement vector (u, u_r, u_θ), which have the following relations (Fig. 4.1):

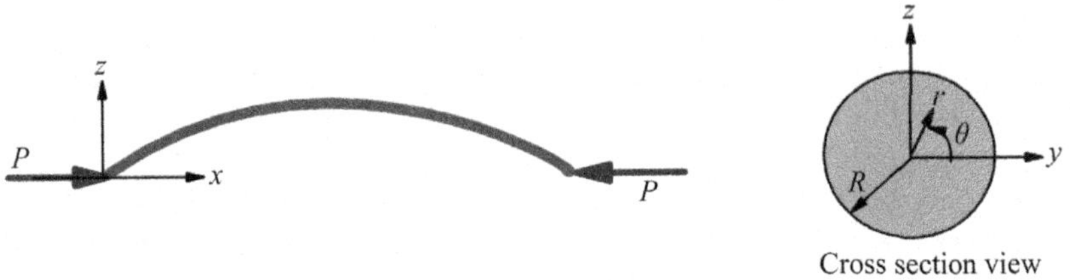

FIGURE 4.1 Schematic of a functionally graded circular cylindrical column along with the corresponding coordinates.

$$\begin{cases} y = r\cos\theta \\ z = r\sin\theta \\ u_r = v\cos\theta + w\sin\theta \end{cases} \tag{4.3}$$

When the columns are restricted to transverse deformation in the xoz-plane, the elastic displacement component v along the y-axis can be assumed to be independent of x, and the deflection w is only dependent on x, i.e., $w = w(x)$. We also introduce the rotation, ψ ($\psi > 0$ means rotation in a clockwise direction), of the cross section about the centroidal axis. w and ψ are chosen as two independent quantities. We start with the shear strain in terms of the displacement vector in polar coordinates

$$\gamma_{xr} = \frac{\partial u_r}{\partial x} + \frac{\partial u}{\partial r} \tag{4.4}$$

Using the coordinate transformations in Eq. (4.3), we have

$$\begin{aligned} \gamma_{xr} &= \cos\theta\frac{\partial v}{\partial x} + \sin\theta\frac{\mathrm{d}w}{\mathrm{d}x} + \cos\theta\frac{\partial u}{\partial y} + \sin\theta\frac{\partial u}{\partial z} \\ &= \frac{1}{r}\left(y\frac{\partial v}{\partial x} + z\frac{\mathrm{d}w}{\mathrm{d}x} + y\frac{\partial u}{\partial y} + z\frac{\partial u}{\partial z}\right) \end{aligned} \tag{4.5}$$

For convenience, we restrict the beam's deformation or buckling to the xoz-plane. In other words, the displacement along the y-direction of the beams is assumed to vanish, i.e., $v = 0$. With this in mind, if the shear strain γ_{xr} takes the form

$$y\frac{\partial u}{\partial y} + z\left(\frac{\mathrm{d}w}{\mathrm{d}x} + \frac{\partial u}{\partial z}\right) = z\left(y^2 + z^2 - R^2\right)\zeta(x) \tag{4.6}$$

where $\zeta(x)$ is a function to be determined, we find that the shear-stress-free boundary condition at the surface of a circular cylinder is satisfied completely. To obtain $\zeta(x)$, we divide u into two parts

$$u(x, y, z) = u_1(x, z) + \frac{zy^2}{3}\zeta(x) \tag{4.7}$$

which is substituted back into Eq. (4.6), leading to

$$\frac{\partial u_1}{\partial z} = -\frac{\mathrm{d}w}{\mathrm{d}x} + \left(z^2 - R^2\right)\zeta(x) \tag{4.8}$$

By integrating both sides of Eq. (4.8) with respect to z, one gets

$$u_1 = \frac{1}{3}z^3\zeta(x) - z\left[R^2\zeta(x) + \frac{\mathrm{d}w}{\mathrm{d}x}\right] \tag{4.9}$$

Now, if introducing a new function, i.e., rotation angle of the cross section ψ, as

$$\psi(x) = -R^2\zeta(x) - \frac{\mathrm{d}w}{\mathrm{d}x} \tag{4.10}$$

we eliminate $\zeta(x)$ from Eqs. (4.10) and (4.7), and obtain the displacement u expressed as

$$u = z\psi - \frac{z^3 + zy^2}{3R^2}\left(\frac{\mathrm{d}w}{\mathrm{d}x} + \psi\right) \tag{4.11}$$

by using kinematic equations

$$\begin{cases} \varepsilon_{xx} = \dfrac{\partial u}{\partial x}, \gamma_{xy} = \dfrac{\partial u}{\partial y} + \dfrac{\partial v}{\partial x} \\ \gamma_{xz} = \dfrac{\partial u}{\partial z} + \dfrac{\partial w}{\partial x}, \gamma_{xr} = \dfrac{\partial u_r}{\partial x} + \dfrac{\partial u}{\partial r} \end{cases} \tag{4.12}$$

we can obtain the normal strain ε_{xx} and shear strains $\gamma_{xy}, \gamma_{xz}, \gamma_{xr}$ in terms of w and ψ.

$$\varepsilon_{xx} = z\frac{\mathrm{d}\psi}{\mathrm{d}x} - \frac{z^3 + zy^2}{3R^2}\left(\frac{\mathrm{d}^2w}{\mathrm{d}x^2} + \frac{\mathrm{d}\psi}{\mathrm{d}x}\right) \tag{4.13}$$

$$\gamma_{xy} = -\frac{2zy}{3R^2}\left(\frac{\mathrm{d}w}{\mathrm{d}x} + \psi\right) \tag{4.14}$$

$$\gamma_{xz} = \frac{3R^2 - 3z^2 - y^2}{3R^2}\left(\frac{\mathrm{d}w}{\mathrm{d}x} + \psi\right) \tag{4.15}$$

$$\gamma_{xr} = \frac{z}{rR^2}\left(R^2 - z^2 - y^2\right)\left(\psi + \frac{\mathrm{d}w}{\mathrm{d}x}\right) \tag{4.16}$$

As can be seen from Eq. (4.16), at the surface of columns we have

$$\gamma_{xr} = 0, \text{as} \quad y^2 + z^2 = R^2 \tag{4.17}$$

which indicates that the shear stress free surface condition is satisfied completely owing to $\tau_{xr} = G(r)\gamma_{xr}$. Therefore, the above results allow us to analyze the stability and flexural behaviors of circular cylindrical columns with the shear stress at the cross section taken into account. Note that the bending and free vibration of an elastic beam with a circular cross section have been considered in Ref. [9].

For beam-column structures, as usual, we employ Saint-Venant's principle and have $\sigma_{yy} = \sigma_{zz} = \tau_{yz} = 0$. Consequently, keeping Eqs. (4.13)–(4.16) in mind and making use of the constitutive equations, one gets

$$\sigma_{xx} = E(r)\left[z\frac{\mathrm{d}\psi}{\mathrm{d}x} - \frac{z^3 + zy^2}{3R^2}\left(\frac{\mathrm{d}^2 w}{\mathrm{d}x^2} + \frac{\mathrm{d}\psi}{\mathrm{d}x}\right)\right] \tag{4.18}$$

$$\tau_{xy} = -\frac{2G(r)zy}{3R^2}\left(\frac{\mathrm{d}w}{\mathrm{d}x} + \psi\right) \tag{4.19}$$

$$\tau_{xz} = G(r)\frac{3R^2 - 3z^2 - y^2}{3R^2}\left(\frac{\mathrm{d}w}{\mathrm{d}x} + \psi\right) \tag{4.20}$$

where σ_{xx} is the normal stress, τ_{xy}, τ_{xz} are the shear stresses; and $E(r)$ and $G(r)$ are Young's modulus and shear modulus, respectively.

4.3 The formulation of stability

4.3.1 Functionally graded circular column

Furthermore, when the columns are subjected to an axial compressive load P along the centroidal axis, it is reasonable to assume that it is uniformly distributed at an arbitrary cross section, where $P > 0$ corresponds to compressive loading and $P < 0$ to tensile loading, respectively. That is, uniform stress $\sigma_{xx}^0 = P/A$ can be regarded as an initial axial stress. When subjected to distributed lateral loading q along the centroidal line, as well as axial compressive force P, the columns should be deflected in the xoz-plane due to the absence of applied loading in the y-direction. By neglecting the equilibrium equation along the y-direction, the following equilibrium equations based on the three dimensional elasticity theory are obtained:

$$\frac{\partial \sigma_{xx}}{\partial x} + \frac{\partial \tau_{xy}}{\partial y} + \frac{\partial \tau_{xz}}{\partial z} = 0 \tag{4.21}$$

$$\frac{\partial \tau_{xz}}{\partial x} + \frac{\partial \tau_{yz}}{\partial y} + \frac{\partial \sigma_{zz}}{\partial z} - \sigma_{xx}^0\frac{\partial^2 w}{\partial x^2} = 0 \tag{4.22}$$

In the above, the axial shortening under the compressive load P has been neglected due to the small deflection deformation, according to the standard method [10], the bending moment M and shear force Q at any position x of the columns can be evaluated through the following integrals

$$\begin{cases} M = \displaystyle\int_A \sigma_{xx} z \mathrm{d}A - Pw \\ Q = \displaystyle\int_A \tau_{xz} \mathrm{d}A - P\frac{\mathrm{d}w}{\mathrm{d}x} \end{cases} \tag{4.23}$$

where $A = \pi R^2$ is the cross sectional area, and we then obtain

$$M=\widetilde{E}_2\frac{\mathrm{d}\psi}{\mathrm{d}x}-E_4\frac{\mathrm{d}^2w}{\mathrm{d}x^2}-Pw \tag{4.24}$$

$$Q=\widetilde{G}_0\left(\frac{\mathrm{d}w}{\mathrm{d}x}+\psi\right)-P\frac{\mathrm{d}w}{\mathrm{d}x} \tag{4.25}$$

where

$$\begin{cases}\widetilde{E}_2=E_2-E_4\\ \widetilde{G}_0=G_0-G_2\\ E_2=\pi\int_0^R E(r)r^3\mathrm{d}r\\ E_4=\dfrac{\pi}{3R^2}\int_0^R E(r)r^5\mathrm{d}r\\ G_0=2\pi\int_0^R G(r)r\mathrm{d}r\\ G_2=\dfrac{4\pi}{3R^2}\int_0^R G(r)r^3\mathrm{d}r\end{cases} \tag{4.26}$$

Now both sides of Eq. (4.21) are multiplied by z and then integrated over the cross section. After some manipulations, we obtain

$$\frac{\mathrm{d}M}{\mathrm{d}x}-Q=0 \tag{4.27}$$

In addition, a straightforward integration over the cross section leads to

$$\frac{\mathrm{d}Q}{\mathrm{d}x}+q-P\frac{\mathrm{d}^2w}{\mathrm{d}x^2}=0 \tag{4.28}$$

Using Eqs. (4.24) and (4.25), the above-derived two equilibrium equations [Eqs. (4.27) and (4.28)] can be rewritten as

$$\widetilde{E}_2\frac{\mathrm{d}^2\psi}{\mathrm{d}x^2}-E_4\frac{\mathrm{d}^3w}{\mathrm{d}x^3}-\widetilde{G}_0\left(\frac{\mathrm{d}w}{\mathrm{d}x}+\psi\right)=0 \tag{4.29}$$

$$\widetilde{G}_0\left(\frac{\mathrm{d}^2w}{\mathrm{d}x^2}+\frac{\mathrm{d}\psi}{\mathrm{d}x}\right)-P\frac{\mathrm{d}^2w}{\mathrm{d}x^2}+q=0 \tag{4.30}$$

Hence, two coupled governing differential equations in terms of the transverse deflection w and rotation ψ for functionally graded circular cylindrical columns have been derived. In order to further simplify the above equations, we introduce a new unknown function F in place of two unknown quantities. To this end, we express w and ψ as follows:

$$w = F - \frac{\widetilde{E}_2}{\widetilde{G}_0}\frac{\mathrm{d}^2F}{\mathrm{d}x^2} \tag{4.31}$$

$$\psi = -\frac{E_4}{\widetilde{G}_0}\frac{\mathrm{d}^3F}{\mathrm{d}x^3} - \frac{\mathrm{d}F}{\mathrm{d}x} \tag{4.32}$$

which further provide the bending moment M and shear force Q in terms of F below:

$$\begin{cases} M = -E_2\dfrac{\mathrm{d}^2F}{\mathrm{d}x^2} - Pw \\ Q = -E_2\dfrac{\mathrm{d}^3F}{\mathrm{d}x^3} - P\dfrac{\mathrm{d}w}{\mathrm{d}x} \end{cases} \tag{4.33}$$

It is readily checked that when substituting Eqs. (4.31) and (4.32) into Eq. (4.29), it is completely fulfilled. On the other hand, when substituting Eqs. (4.31) and (4.32) into Eq. (4.30), we get

$$\left(E_2 - P\frac{\widetilde{E}_2}{\widetilde{G}_0}\right)\frac{\mathrm{d}^4F}{\mathrm{d}x^4} + P\frac{\mathrm{d}^2F}{\mathrm{d}x^2} = q \tag{4.34}$$

Once an admissible solution F is found, the critical buckling load, denoted as P_{cr}, can be determined by the following expression

$$P_{cr} = -\frac{E_2F^{IV} - q}{F'' - \widetilde{E}_2\widetilde{G}_0^{-1}F^{IV}} \tag{4.35}$$

which immediately follows from Eq. (4.34). It is noted that F can be determined by appropriate end conditions.

In the following, we consider classical Euler–Bernoulli and Timoshenko beams with a circular cross section made of radially functionally graded materials, and then compare their critical buckling loads. For Euler–Bernoulli beams, it is sufficient to set $\widetilde{G}_0 \to \infty$ in the above results, implying that $\psi = -\mathrm{d}w/\mathrm{d}x$ or shear locking occurs. If the critical buckling load of functionally graded Euler–Bernoulli beams is denoted by P_{cr}^E, we have

$$P_{cr}^E = -\frac{E_2F^{IV} - q}{F''} \tag{4.36}$$

From Eq. (4.34), we find that in addition to a polynomial of degree one, all admissible solutions of F are composed of sine and cosine functions, which means that F'' and F^{IV} always have opposite signs. This indicates that the critical load P_{cr}, which takes different values depending on various end supports, is always lower than P_{cr}^E. In other words, the critical buckling load of Euler–Bernoulli beams is indeed overestimated.

One of the most important cases in practice is the case of $q = 0$. For this case,

$$P_{cr} = -\frac{E_2 F^{IV}}{F'' - \widetilde{E}_2 \widetilde{G}_0^{-1} F^{IV}} = \frac{P_{cr}^E}{1 + \widetilde{E}_2 \widetilde{G}_0^{-1} E_2 P_{cr}^E} \tag{4.37}$$

or

$$\frac{1}{P_{cr}} = \frac{1}{P_{cr}^E} + \frac{\widetilde{E}_2}{E_2 \widetilde{G}_0} \tag{4.38}$$

following from Eq. (4.34). The above result has a marked advantage, that is, one does not need to solve discrete problems to obtain the critical compressive loading. Instead, with knowledge of the critical buckling load of functionally graded Euler–Bernoulli beams, the desired critical buckling load may be directly determined using Eq. (4.38). In other words, when neglecting the contribution of shear deformation, the buckling loads corresponding to functionally graded Euler–Bernoulli columns are recovered, which further reduces to the classical ones if the columns are also homogeneous.

Next, let us turn our attention to Timoshenko beams with a circular cross-section made of radially functionally graded materials. For this problem, the longitudinal displacement should take the form of

$$u(x, z) = z\psi \tag{4.39}$$

instead of Eq. (4.11). Moreover, we have

$$\begin{cases} M = E_2 \dfrac{d\psi}{dx} - Pw \\ Q = \kappa G_0 \left(\dfrac{dw}{dx} + \psi \right) - P \dfrac{dw}{dx} \end{cases} \tag{4.40}$$

where κ is a shear correction factor, and E_2 and G_0 are defined by Eq. (4.26), respectively. Similarly, one can get two coupled equations [11]:

$$E_2 \frac{d^2\psi}{dx^2} - \kappa G_0 \left(\frac{dw}{dx} + \psi \right) = 0 \tag{4.41}$$

$$\kappa G_0 \left(\frac{d^2 w}{dx^2} + \frac{d\psi}{dx} \right) - P \frac{d^2 w}{dx^2} + q = 0 \tag{4.42}$$

If taking

$$\begin{cases} w = F - \dfrac{E_2}{\kappa G_0} \dfrac{d^2 F}{dx^2} \\ \psi = -\dfrac{dF}{dx} \end{cases} \tag{4.43}$$

we immediately find that the coupled governing Eqs. (4.41) and (4.42) which can be converted into a single governing equation as follows:

$$\left(E_2 - P\frac{E_2}{\kappa G_0}\right)\frac{\mathrm{d}^4 F}{\mathrm{d}x^4} + P\frac{\mathrm{d}^2 F}{\mathrm{d}x^2} = q \tag{4.44}$$

Obviously, if the shear correction factor is chosen as $E_2\widetilde{G}_0/\widetilde{E}_2 G_0$, we find that the governing Eqs. (4.34) and (4.44) are identical. Furthermore, the critical buckling load of functionally graded Timoshenko beams, denoted as P_{cr}^T, can be calculated by

$$P_{cr}^T = -\frac{E_2 F^{IV} - q}{F'' - E_2\kappa^{-1}G_0^{-1}F^{IV}} \tag{4.45}$$

Considering the case of $q = 0$, from Eqs. (4.36) and (4.45), we can find the relation between P_{cr}^T and P_{cr}^E.

$$P_{cr}^T = -\frac{E_2 F^{IV}}{F'' - E_2\kappa^{-1}G_0^{-1}F^{IV}} = \frac{P_{cr}^E}{1 + \kappa^{-1}G_0^{-1}P_{cr}^E} \tag{4.46}$$

or

$$\frac{1}{P_{cr}^T} = \frac{1}{P_{cr}^E} + \frac{1}{\kappa G_0} \tag{4.47}$$

4.3.2 Homogeneous columns with a circular cross section

For homogeneous columns with a circular cross section, the material properties take constants: $E(r) = E, G(r) = G$. Under such circumstances, we have

$$\begin{cases} E_2 = EI \\ E_4 = \frac{2}{9}EI \\ \widetilde{E}_2 = \frac{7}{9}EI \\ G_0 = GA \\ G_2 = \frac{1}{3}GA \\ \widetilde{G}_0 = \frac{2}{3}GA \end{cases} \tag{4.48}$$

where $I = \pi R^4/4$ is the moment of inertia of the cross sectional area. Substituting the above equations into Eq. (4.34) leads to the governing equation of homogeneous circular cylindrical columns

$$EI\left(1-\frac{7P}{6GA}\right)\frac{d^4F}{dx^4}+P\frac{d^2F}{dx^2}=q \tag{4.49}$$

Consequently, in the case of $q=0$, we have

$$\begin{cases} \dfrac{1}{P_{cr}}=\dfrac{1}{P_{cr}^E}+\dfrac{7}{6GA} \\ \dfrac{1}{P_{cr}}=\dfrac{1}{P_{cr}^T}+\dfrac{7}{6GA}\left(1-\dfrac{6}{7\kappa}\right) \end{cases} \tag{4.50}$$

which implies that the critical loads for homogeneous circular columns are identical to those obtained from the Timoshenko beams with $\kappa=6/7$.

4.3.3 Cylindrical tube beams subject to axial force

In the preceding chapter, we investigated the vibration behavior of radially graded hollow cylinders subjected to transverse and axial loadings (see Section 3.2.2). Static stability is a special case of dynamic behavior of an axially loaded beam-column. By removing all the time-dependent terms in the previous relationships, that is, from Eq. (4.36), the governing equation on F for buckling of a straight tube becomes

$$\left[(EI)_0-\frac{P(EI)^*}{(GA)^*}\right]\frac{d^4F}{dx^4}+P\frac{d^2F}{dx^2}=0 \tag{4.51}$$

As representative examples, we consider several typical end conditions. Since the ordinary differential Eq. (4.51) with constant coefficients is easily solved, a general solution can be given as:

$$F(x)=C_1\cos(\lambda x)+C_2\sin(\lambda x)+C_3x+C_4 \tag{4.52}$$

with

$$\lambda=\sqrt{\frac{P(GA)^*}{(EI)_0(GA)^*-P(EI)^*}} \tag{4.53}$$

where $C_j (j=1, \ldots, 4)$ are unknown constants. Substituting the above expression for F into Eqs. (4.38), (4.39), (4.41), and (4.42), and after removing all of the time-dependent terms, one gets

$$w=\beta_1[C_1\cos(\lambda x)+C_2\sin(\lambda x)]+C_3x+C_4 \tag{4.54}$$

$$\psi=\lambda\beta_2[C_1\sin(\lambda x)-C_2\cos(\lambda x)]-C_3 \tag{4.55}$$

$$M=\lambda^2(EI)_0[C_1\cos(\lambda x)+C_2\sin(\lambda x)] \tag{4.56}$$

$$V=-PC_3 \tag{4.57}$$

where

$$\begin{cases} \beta_1 = 1 + \lambda^2 \dfrac{(EI)^*}{(GA)^*} \\ \beta_2 = 1 + \lambda^2 \dfrac{(EI)^* - (EI)_0}{(GA)^*} \end{cases} \tag{4.58}$$

Next, we consider cylindrical tubes with four typical boundary conditions: hinged–hinged (HH), clamped–free (CF), clamped–clamped (CC), and clamped–hinged (CH) ends, respectively. For the above four differential ends, we have

$$\begin{cases} \text{HH}: w(0) = 0, M(0) = 0, w(L) = 0, M(L) = 0 \\ \text{CF}: w(0) = 0, \psi(0) = 0, M(L) = 0, V(L) = 0 \\ \text{CC}: w(0) = 0, \psi(0) = 0, w(L) = 0, \psi(L) = 0 \\ \text{CH}: w(0) = 0, \psi(0) = 0, w(L) = 0, M(L) = 0 \end{cases} \tag{4.59}$$

By applying four given boundary conditions for each case, a system of algebraic equations can be obtained. The necessary condition for the existence of nonzero solutions to the above system of algebraic equations is that the determinant of the coefficient matrix must be zero. Omitting the details, the characteristic equations of critical buckling loads of the corresponding columns are derived:

$$\begin{cases} \text{HH}: \sin(\lambda L) = 0 \\ \text{CF}: \cos(\lambda L) = 0 \\ \text{CC}: \sin\left(\dfrac{\lambda L}{2}\right) = 0 \\ \text{CH}: \tan(\lambda L) = \dfrac{\lambda \beta_2 L}{\beta_1} \end{cases} \tag{4.60}$$

For the first three cases, the first positive root of the characteristic equations gives the buckling load, which can be expressed by

$$\begin{cases} P_{cr} = \dfrac{\dfrac{\pi^2}{(\mu L)^2}(EI)_0}{1 + \dfrac{\pi^2}{(\mu L)^2}\dfrac{(EI)^*}{(GA)^*}} \\ \mu = \begin{cases} 1, \text{HH} \\ 2, \text{CF} \\ 0.5, \text{CC} \end{cases} \end{cases} \tag{4.61}$$

For the last case, we employ numerical methods to determine the buckling load. Moreover, if shear deformation of the cross section is locked, meaning

$(GA)^* \to \infty$, we also find $\beta_1 = \beta_2 = 1$, and the above-derived characteristic Eq. (4.60) or (4.61) reduces to the classical ones [12]. These results are in agreement with those derived in Ref. [13].

4.4 The buckling analysis

In this section, four illustrative examples are provided. The first example is introduced to make a comparison of buckling loads among several available beam models. The second is devoted to the effect of the radial inhomogeneity on critical buckling loads, the third pertains to the buckling of double-walled carbon nanotubes under an axial compressive load, and the last studies the bucking behaviors of a tube beam with force.

4.4.1 Buckling loads of homogeneous columns

For the first example, we consider simply supported homogeneous columns using the theories of Euler–Bernoulli beams, Timoshenko beams, and Reddy–Bickford beams and compare our results with theirs. The latter two theories have been considered with the effect of shear deformation taken into account. When using the Timoshenko beam theory, the shear correction factor must be given. Up to now, several different values have been suggested, and in the present work, we take two possible values of κ: $\kappa_1 = 6(1+\nu)/(7+6\nu)$ [14,15] and $\kappa_2 = 6(1+\nu)^2/(7+12\nu+4\nu^2)$ [16]. Using the Reddy–Bickford beam theory, the corresponding critical load, denoted as P^R_{cr}, is related to the Euler critical load P^E_{cr} through the following relationship for columns with a circular cross section [17]:

$$P^R_{cr} = P^E_{cr}\left(1+\frac{P^E_{cr}}{90GA}\right)\left(1+\frac{101P^E_{cr}}{90GA}\right)^{-1} \tag{4.62}$$

Nondimensional critical buckling loads $\overline{P}_{cr} = P_{cr}l^2/EI$ are calculated, and the numerical results are presented in Table 4.1 and compared. From Table 4.1 we see that the critical loads calculated by the Euler–Bernoulli beam theory are larger than those calculated by the present approach and the other two beam theories, especially for small length-to-radius ratios. This is due to the fact that the Euler–Bernoulli assumption neglects the effect of shear deformation. However, the numerical results calculated by the present approach are very close to those obtained from the Timoshenko beam and Reddy–Bickford beam theories, which indicates that the present model is very efficient.

4.4.2 Effects of the gradient index on buckling loads

Here we consider a two-phase composite as a mixture of metal and ceramic, denoted as phases 1 and 2, respectively, where the volume fraction V takes the following form

TABLE 4.1 Nondimensional critical buckling loads $\overline{P}_{cr}$ for homogeneous columns with $\nu = 0.3$

		Timoshenko			
L/R	Euler–Bernoulli	κ_1	κ_2	Reddy–Bickford	Present
5	9.8696	7.6538	7.7266	7.6847	7.5956
10	9.8696	9.2035	9.2296	9.2133	9.1824
15	9.8696	9.5620	9.5745	9.5666	9.5519
20	9.8696	9.6942	9.7014	9.6968	9.6883
25	9.8696	9.7566	9.7613	9.7583	9.7528

$$\begin{cases} V_m = 1 - \left(\frac{r}{R}\right)^n \\ V_c = \left(\frac{r}{R}\right)^n \end{cases} \tag{4.63}$$

For simplicity, the material property Y at radius r is assumed to obey the Voigt rule:

$$Y(r) = Y_m\left[1 - \left(\frac{r}{R}\right)^n\right] + Y_c\left(\frac{r}{R}\right)^n \tag{4.64}$$

where subscripts m and c denote metal and ceramic phases, respectively. The center is rich in metal and the surface is rich in ceramic. Two materials, aluminum and Al_2O_3, are selected, respectively, and their material properties are [18]: $E_m = 70$ GPa, $\nu_m = 0.3$ for Al, and $E_c = 380$ GPa, $\nu_c = 0.3$ for Al_2O_3.

First, we examine the effects of the gradient index n on the ratio λ, defined as the ratio of the critical buckling load P_{cr} to its counterpart $P_0 = \pi^2 E_m I/L^2$ for the classical Euler–Bernoulli beam theory. Fig. 4.2 shows the variation of the ratio λ against *L/R* for simply supported columns subjected to a compressive load. Here, for the Timoshenko beams, we have taken $\kappa_1 = 6(1+\nu)/(7+6\nu)$ in this example. From Fig. 4.2, it can be seen that λ^E for the Euler–Bernoulli beam is independent of *L/R*, whereas the obtained ratios λ are very close to λ^T for the Timoshenko beams, and increase with *L/R* increasing. This shows the influence of the shear deformation on buckling loads. The effects become stronger for smaller *L/R*, meaning that the shorter the circular columns, the more pronounced the contribution of shear deformation, as expected. This trend is similar to that observed in the Timoshenko beams. Another interesting observation is that the critical buckling loads rise when the gradient parameter n declines, inferring that whether the shear deformation is included or not, graded columns with smaller gradient parameters n are capable of bearing higher critical loads when Young's modulus at the surface

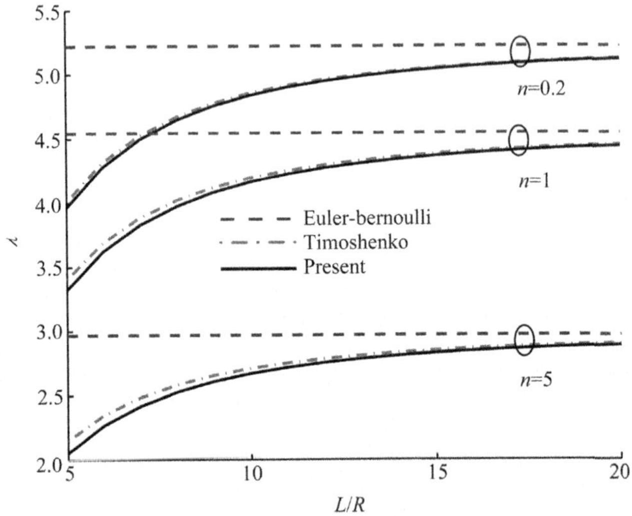

FIGURE 4.2 Variation of the ratio λ against L/R with different gradient indexes n for simply supported beams.

is greater than that at the center. Although the obtained ratios λ are close to λ^T, the distribution of shear stress at the cross section is completely different.

Figs. 4.3 and 4.4 show a representative distribution of shear stress $\tau_{xz}(L/2, 0, z)/E_m$ for simply supported functionally graded columns with different gradient indexes n when using the present model and Timoshenko

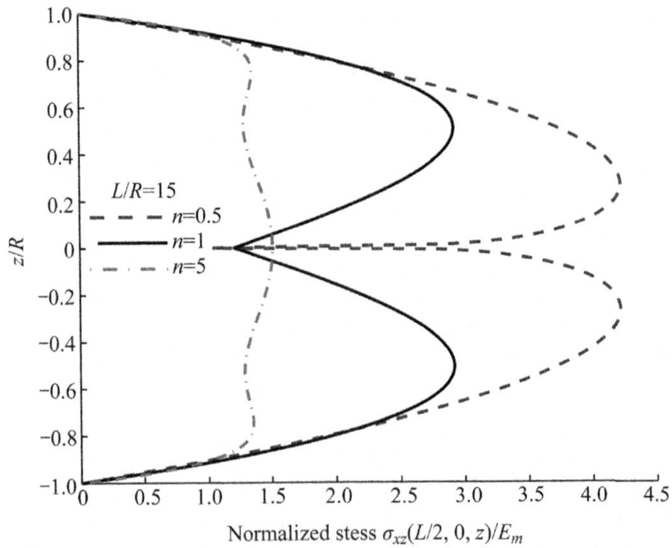

FIGURE 4.3 The distribution of the nondimensional critical shear stress $\tau_{xz}(L/2, 0, z)/E_m$ in simply supported functionally graded circular columns at buckling: the present model.

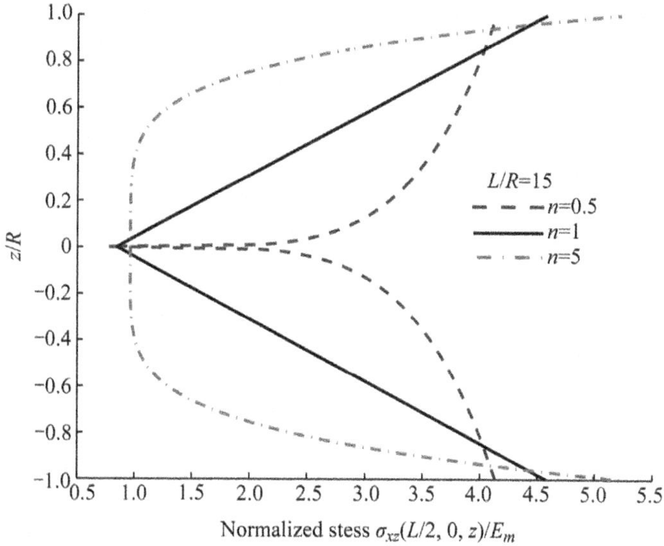

FIGURE 4.4 The distribution of the nondimensional critical shear stress $\tau_{xz}(L/2,\ 0,\ z)/E_m$ in simply supported functionally graded circular columns at buckling: the Timoshenko beam model.

beam theory at buckling, respectively, where a constant factor has been neglected. An apparent difference is that the shear stress at the top and bottom surfaces vanishes when adopting the present model, as expected, whereas it behaves differently in the Timoshenko beam theory, in particular reaching its maximum at the top or bottom surfaces for the gradient considered. It is worth noting that for the homogeneous Timoshenko beams, the shear stress is uniformly distributed according to the Timoshenko beam assumption.

4.4.3 Buckling loads of double-walled carbon nanotubes

As the last example, let us consider the elastic stability of double-walled carbon nanotubes (DWCNTs) based on a continuum mechanics approach. Previous analysis on this topic can be found in Ref. [19], where a DWCNT was simulated as two interacting Euler–Bernoulli beams. Based on the above analysis, we have the following governing equations

$$\begin{cases} EI_1\left(1-\dfrac{7P_1}{6GA_1}\right)\dfrac{\mathrm{d}^4F_1}{\mathrm{d}x^4}+P_1\dfrac{\mathrm{d}^2F_1}{\mathrm{d}x^2}=-c(w_1-w_2) \\ EI_2\left(1-\dfrac{7P_2}{6GA_2}\right)\dfrac{\mathrm{d}^4F_2}{\mathrm{d}x^4}+P_2\dfrac{\mathrm{d}^2F_2}{\mathrm{d}x^2}=-kw_2-c(w_2-w_1) \end{cases} \tag{4.65}$$

where the subscripts $j = 1, 2$ are used to denote the quantities associated with the inner and outer walls, respectively, P_j denotes the axial compressive load of the j-th tube where the negative sign $(-P_j)$ indicates compression [19], c denotes the intertube interaction coefficient per unit length which describes the van der Waals forces between two adjacent tubes, and k denotes a spring coefficient which accounts for the interaction of the outer tube and a surrounding elastic medium [20].

Assume that the inner and outer tubes have the same effective material constants. Following the disposal described in Ref. [19] and recalling Eq. (4.31) for w, we get the governing equations

$$EI_1\left(1-\frac{7\sigma_{xx}^0}{6G}\right)\frac{\mathrm{d}^4F_1}{\mathrm{d}x^4}+\left(\sigma_{xx}^0A_1-\frac{7cEI_1}{6GA_1}\right)\frac{\mathrm{d}^2F_1}{\mathrm{d}x^2}+cF_1=c\left(F_2-\frac{7EI_2}{6GA_2}\frac{\mathrm{d}^2F_2}{\mathrm{d}x^2}\right) \tag{4.66}$$

$$\begin{aligned}EI_2\left(1-\frac{7\sigma_{xx}^0}{6G}\right)\frac{\mathrm{d}^4F_2}{\mathrm{d}x^4}+\left[\sigma_{xx}^0A_2-\frac{7(c+k)EI_2}{6GA_2}\right]\frac{\mathrm{d}^2F_2}{\mathrm{d}x^2}\\+(c+k)F_2=c\left(F_1-\frac{7EI_1}{6GA_1}\frac{\mathrm{d}^2F_1}{\mathrm{d}x^2}\right)\end{aligned} \tag{4.67}$$

where the axial stress σ_{xx}^0 is uniform over the entire cross section. If disregarding the effect of shear deformation and letting $k=0$, Eqs. (4.66) and (4.67) are identical to those derived from the Euler–Bernoulli beam theory [19].

In what follows, let us consider a simply supported DWCNT with the following boundary conditions

$$w_j(0)=w_j(L)=0, M_j(0)=M_j(L)=0, j=1,2 \tag{4.68}$$

For such boundary conditions, it is appropriate to take a solution to Eqs. (4.66) and (4.67) of the form

$$F_1=f_1\sin\frac{\xi\pi}{L}x, F_2=f_2\sin\frac{\xi\pi}{L}x \tag{4.69}$$

where f_j ($j=1,2$) are constants, and ξ is an integer. After substituting Eq. (4.69) into Eqs. (4.66) and (4.67), we further simplify two equations and then get a system of linear equations in f_j ($j=1,2$). To obtain a nontrivial solution, the determinant of the coefficient matrix of the resulting system has to vanish, namely

$$\left(\frac{\xi\pi}{L}\right)^4\left(\sigma_{xx}^0\right)^2+a_1\left(\frac{\xi\pi}{L}\right)^2\sigma_{xx}^0+a_2=0 \tag{4.70}$$

where

$$\begin{cases}a_1=-\left[\dfrac{\beta_1+c\alpha_1}{A_1\alpha_1}+\dfrac{\beta_2+(c+k)\alpha_2}{A_2\alpha_2}\right]\\ a_2=\dfrac{\beta_1+c\alpha_1}{A_1\alpha_1}\cdot\dfrac{\beta_2+(c+k)\alpha_2}{A_2\alpha_2}-\dfrac{c^2}{A_1A_2}\\ \alpha_j=1+\dfrac{7EI_j}{6GA_j}\left(\dfrac{\xi\pi}{L}\right)^2, \beta_j=EI_j\left(\dfrac{\xi\pi}{L}\right)^4, j=1,2\end{cases} \tag{4.71}$$

Eq. (4.70) is clearly a quadratic equation in σ_{xx}^0. Prior to solving this equation, let us make some observations for some special cases. If neglecting the effect of shear deformation, which means $G \to \infty$, we have $\alpha_j = 1$, and Eq. (4.70) then simplifies to

$$\left(\frac{\sigma_{xx}^0}{E}\right)^2 + b_1\frac{\sigma_{xx}^0}{E} + b_2 = 0 \tag{4.72}$$

where

$$\begin{cases} b_1 = -\left(\frac{\xi\pi}{L}\right)^{-2}\left[\left(\frac{I_1}{A_1}+\frac{I_2}{A_2}\right)\left(\frac{\xi\pi}{L}\right)^4 + \frac{1}{E}\left(\frac{c}{A_1}+\frac{c+k}{A_2}\right)\right] \\ b_2 = \frac{I_1 I_2}{A_1 A_2}\left(\frac{\xi\pi}{L}\right)^4 + \frac{(c+k)I_1 + cI_2}{A_1 A_2 E} + \frac{ck}{A_1 A_2 E^2}\left(\frac{\xi\pi}{L}\right)^{-4} \end{cases} \tag{4.73}$$

Furthermore, if considering a DWCNT in a free space, i.e., $k = 0$ in Eq. (4.72), we find that the equation obtained is in exact agreement with that derived in Ref. [19]. If the interlayer radial relative displacements are prohibited, the critical axial stress is reduced to

$$\begin{cases} \sigma_{xx}^0 A = EI\left(\frac{\xi\pi}{L}\right)^2 \\ I = I_1 + I_2 \\ A = A_1 + A_2 \end{cases} \tag{4.74}$$

in exact agreement with the classical formula [11]. Therefore, when shear deformation is considered, the critical axial buckling stress of a DWCNT embedded in a surrounding elastic medium is determined by the smallest root of Eq. (4.70) with $\xi = 1$, i.e.,

$$\sigma_{xx}^0 = \frac{1}{2}\left(\frac{\pi}{L}\right)^{-2}\left\{\left[\frac{\beta_1 + c\alpha_1}{A_1\alpha_1} + \frac{\beta_2 + (c+k)\alpha_2}{A_2\alpha_2}\right] - \sqrt{\left[\frac{\beta_1 + c\alpha_1}{A_1\alpha_1} - \frac{\beta_2 + (c+k)\alpha_2}{A_2\alpha_2}\right]^2 + \frac{4c^2}{A_1 A_2}}\right\} \tag{4.75}$$

Next, the numerical results are evaluated for a DWCNT with inner radius $R_1 = 0.35$ nm, outer radius $R_2 = 0.69$ nm, $E = 1.2$ TPa, $\nu = 0.28$, and $c = 0.0694$ TPa [21]. For a DWCNT in a free space, the variations of the critical axial strain $\varepsilon_{cr} = \sigma_{xx}^0/E$ against the ratio L/R_2 are plotted in Fig. 4.5. By comparing our results with those in Ref. [19], it can be seen that the two results are very close for sufficiently large values of L/R_2. However, for small values of L/R_2, two critical buckling strains have a notable difference, which may reach or exceed 8.8% when $L/R_2 < 10$. As a result, the critical strains

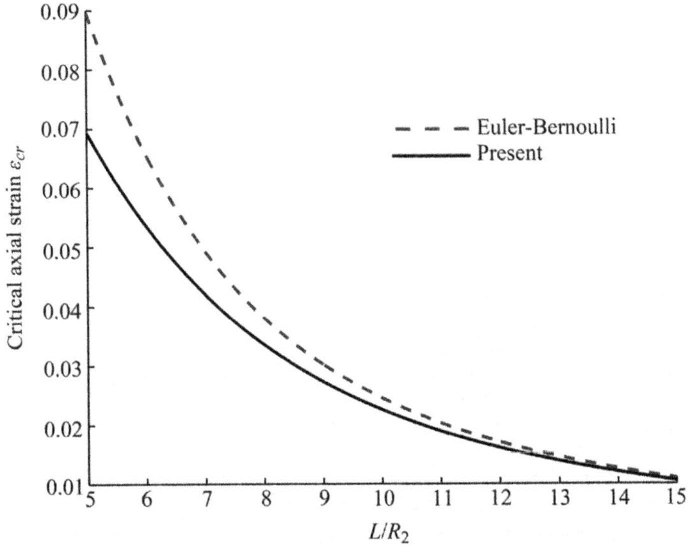

FIGURE 4.5 The variations of the critical axial strain $\varepsilon_{cr} = \sigma_{xx}^0/E$ for a DWCNT against the ratio L/R_2.

evaluated by the Euler–Bernoulli beam theory are overestimated. Furthermore, the effect of the spring coefficient k on the critical axial buckling strain is examined. Evaluated values of the critical axial strain are listed in Table 4.2. Due to the interaction between a DWCNT and the surrounding elastic medium, the corresponding critical strains are obviously intensified as compared to a DWCNT in a free space.

4.4.4 Cylindrical tube beams with axial force

In this section, numerical examples are given to show the effects of the gradient and the warping shapes on the buckling loads and natural frequencies for radially graded tubes. For radially graded tubes, we assume that the material properties vary according to the following power-law form:

$$Y(r) = Y_i + (Y_o - Y)_i \frac{r^N - R_i^N}{R_o^N - R_i^N} \tag{4.76}$$

where Y_i and Y_o represent the material properties such as Young's modulus E, shear modulus G, and mass density ρ. Here the subscript i indicates the inner surface, while o describes the outer surface. We denote gradient index as N to present the volume fraction of two kinds of constituents. Two different materials, aluminum and zirconia, are chosen as the internal and outside components of the gradient, respectively, and their material properties are as follows:

TABLE 4.2 Critical axial strains ε_{cr} with $\nu = 0.28$.

	$L/R_2 = 5$		$L/R_2 = 10$		$L/R_2 = 15$	
k/E	Euler–Bernoulli	Present	Euler–Bernoulli	Present	Euler–Bernoulli	Present
0	0.0896	0.0693	0.0246	0.0226	0.0110	0.0105
0.001	0.0902	0.0700	0.0277	0.0258	0.0182	0.0178
0.005	0.0926	0.0727	0.0404	0.0384	0.0469	0.0465
0.01	0.0954	0.0759	0.0560	0.0541	0.0825	0.0821
0.05	0.1158	0.0995	0.1731	0.1714	0.3497	0.3493
0.1	0.1355	0.1229	0.2971	0.2957	0.6341	0.6338

$$\begin{cases} \text{Al}: E_i = 70\ \text{GPa}, G_i = 27\ \text{GPa}, \rho_i = 2702\ \text{kg/m}^3 & (4.77) \\ \text{ZrO}_2: E_o = 200\ \text{GPa}, G_o = 77\ \text{GPa}, \rho_o = 5700\ \text{kg/m}^3 & (4.78) \end{cases}$$

Using the above material properties, the variation of the radius-dependent Young's modulus is shown in Fig. 4.6 for $R_i/R_o = 0.5$. When the value of the gradient index N is lower, $E(r)$ is approximately equal to E_o, and the tube can be considered as full of the outer constituent. When N increases, the constituent of the tube changes quickly. Taking $N = 1$, this means a linear material constituent from the inner surface to the outer surface. When N becomes larger, the tube constituent or the proportion of inner surface material aluminum varies. For other varying material properties, a similar trend is seen.

For simplicity, in the subsequent numerical calculations, we only consider the warping shape of rational function [Eq. (4.25)]. In this case, we have

$$\frac{\partial f}{\partial z} = 1 + \frac{\left(3R_i^2R_o^2 + r^4\right)\left(y^2 - z^2\right) - 2r^6}{3r^4\left(R_i^2 + R_o^2\right)} \tag{4.79}$$

Moreover, we can calculate the following integral

$$\int_{R_i}^{R_o} Y(r) r^k \mathrm{d}r = \frac{R_o^{k+1} - R_i^{k+1}}{k+1} Y_i + \frac{Y_o - Y_i}{R_o^N - R_i^N}\left(\frac{R_o^{N+k+1} - R_i^{N+k+1}}{N+k+1} - \frac{R_i^N\left(R_o^{k+1} - R_i^{k+1}\right)}{k+1}\right) \tag{4.80}$$

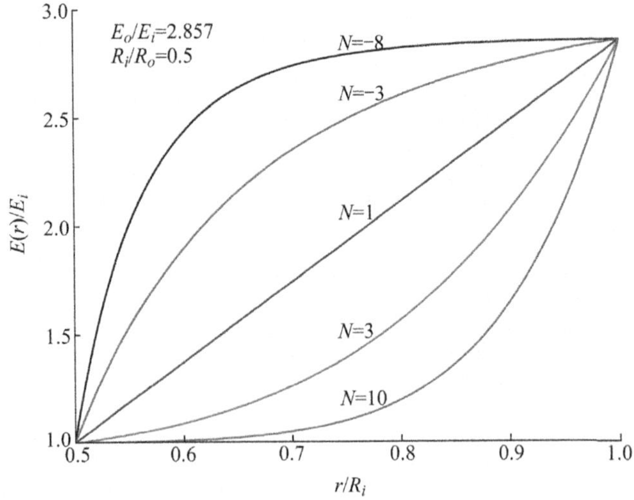

FIGURE 4.6 The variation of the radius-dependent Young's modulus.

With the above integral result, the parameters involved can be readily determined, and the buckling loads can then be evaluated.

To verify the effectiveness of the presented method, we introduce a nondimensional buckling load $\overline{P}_{cr} = P_{cr}L^2/\pi^2 E_o I$. Tables 4.3 and 4.4 provide a comparison of the nondimensional buckling load with the FEM results for hinged–hinged and clamped–free homogeneous cylindrical tubes made of zirconia. For comparison, the classical Euler buckling loads, P^E_{cr}, the Timoshenko buckling loads, P^T_{cr}, and Reddy–Bickford buckling loads, P^R_{cr}, are listed in Tables 4.3 and 4.4, according to the following relationships [13,17,22]:

$$\begin{cases} P^T_{cr} = P^E_{cr}\left(1 + \dfrac{P^E_{cr}}{\kappa GA}\right)^{-1} \\ P^R_{cr} = P^E_{cr}\left(1 + \dfrac{P^E_{cr}}{90GA}\right)\left(1 + \dfrac{101P^E_{cr}}{90GA}\right)^{-1} \end{cases} \tag{4.81}$$

where the shear correction coefficient [23] $\kappa = \left[(7+6\nu)\left(1+j^2\right)^2 + (20+12\nu)j^2\right] / 6(1+\nu)\left(1+j^2\right)^2$, ν being Poisson's ratio and $j = R_i/R_o$, is selected when using the Timoshenko beam theory.

From Tables 4.3 and 4.4, one can readily see that the numerical results agree with the calculated results based on the Timoshenko beam and Reddy–Bickford beam theories as well as the FEM results, irrespective of the length of the tubes. However, it is interesting to mention that there are some errors between the Reddy–Bickford buckling loads and other buckling loads. A possible cause for these may be that the Reddy–Bickford buckling load formula used here is based on columns with a solid circular cross section, rather than an annular cross section [17]. Here we directly apply it to hollow tubes. In addition, for short tubes, the Euler buckling loads are overestimated, as expected, because the effect of shear deformation is neglected. For sufficiently long tubes, the dimensionless buckling loads reduce to 1 and 0.25 for hinged–hinged tubes and clamped–free tubes, respectively, in accordance with what is anticipated.

The effects of the aspect ratio L/R_o on the nondimensional buckling load $\overline{P}_{cr}$ have been examined. For a thick tube with $R_i/R_o = 0.5$, Table 4.5 lists a comparison of $\overline{P}_{cr}$ of functionally graded tubes with the gradient index $N = 1$. It is interesting to note that because of the variations in the constituents of the graded tube, the dimensionless buckling loads for sufficiently long tubes are no longer 1, 0.25, 4, and 2 for hinged–hinged tubes, clamped–free tubes, clamped–clamped tubes, and clamped–hinged tubes, respectively. However, the numerical relationships between the dimensionless buckling loads themselves for sufficiently long tubes with the above end constraints remain unchanged. This is mainly due to the fact that the influence of shear deformation decreases as the tube becomes longer.

TABLE 4.3 Nondimensional buckling loads $\overline{P}_{cr}$ of hinged–hinged homogeneous columns with $R_i/R_o = 0.5$

L/R_o	Present	FEM	Euler–Bernoulli [11]	Timoshenko [11]	Reddy–Bickford [17]
5	0.6474	0.6308	1	0.6593	0.7381
10	0.8801	0.8775	1	0.8858	0.9183
15	0.9429	0.9423	1	0.9457	0.9620
20	0.9671	0.9668	1	0.9687	0.9782
50	0.9946	0.9946	1	0.9949	0.9965
100	0.9986	0.9986	1	0.9987	0.9991

TABLE 4.4 Nondimensional buckling loads $\overline{P}_{cr}$ of clamped–free homogeneous columns with $R_i/R_o = 0.5$

L/R_o	Present	FEM	Euler–Bernoulli [11]	Timoshenko [11]	Reddy–Bickford [17]
5	0.2200	0.2194	0.2500	0.2214	0.2296
10	0.2418	0.2417	0.2500	0.2422	0.2446
15	0.2463	0.2463	0.2500	0.2465	0.2476
20	0.2479	0.2479	0.2500	0.2480	0.2486
50	0.2497	0.2497	0.2500	0.2497	0.2498
100	0.2499	0.2499	0.2500	0.2499	0.2499

TABLE 4.5 Nondimensional buckling loads $\overline{P}_{cr}$ of FGM tube columns with $R_i/R_o = 0.5$, $N = 1$.

L/R_o	HH	CF	CC	CH
5	0.4841	0.1684	0.9112	0.6816
10	0.6736	0.1867	1.9365	1.1882
15	0.7262	0.1905	2.4462	1.3798
20	0.7467	0.1919	2.6944	1.4625
50	0.7700	0.1934	3.0260	1.5795
100	0.7735	0.1936	3.0802	1.5638

Fig. 4.7 displays the variations in the ratio of the buckling loads to the corresponding Euler loads for homogeneous circular cylindrical tube columns with the same radius, i.e., P_{cr}/P_{cr}^{Eu}, where $P_{cr}^{Eu} = \pi^2 E_o I/(\mu L)^2$, and $\mu = 1, 2, 0.5, 0.7$ for HH, CF, CC, and CH ends, respectively. One can see that P_{cr}/P_{cr}^{Eu} is greatly affected by the gradient index, and decreases significantly as the gradient index increases. Moreover, all the curves gradually climb to a steady value when the length of the radially graded tube increases. This also means that the influence of the shear deformation along with warping shapes on the

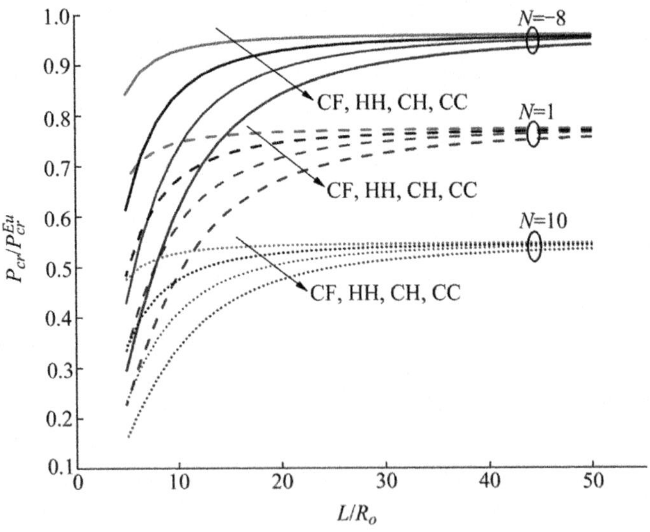

FIGURE 4.7 The variation of the ratio of the buckling loads to the corresponding Euler loads for a homogeneous circular cylindrical tube with the same radius.

buckling loads immediately disappears when the aspect ratio increases. This is easily understood since the results of all of the beam theories for long beams finally tend to those of the Euler–Bernoulli beam theory. Usually, for short beams, the shear deformation and warping shapes of the cross section play important roles in determining the buckling loads. It is particularly noted from Fig. 4.7 that we do not conclude that a longer column could bear higher loads since the Euler buckling loads contain the length factor. For clarity, Fig. 4.8 demonstrates the variation of P_{cr}/E_oA against L/R_o, the longer the tube beam, the lower the buckling loads, as expected. Also, the influence of the gradient index on the buckling loads is viewed again. A negative gradient index results in an increase in the load-carrying capacity of tubes, which is in agreement with that observed in Fig. 4.8. In fact, it is can be explained from Fig. 4.6 that since most of the material constituent of the radially graded tubes is made of zirconia with a larger Young'' modulus, although the inner and outer surface materials do not change, the volume fraction of both materials is adjusted in the inside.

4.5 Effects of Engesser's and Haringx's hypotheses

Consider buckling of a uniform prismatic column of length L subjected to an axially compressive force P, as shown in Fig. 4.1, where its cross section is either a rectangle or circle. For a rectangular cross section, we denote its width and height as b and h, respectively, while for a circular cross section, we denote the radius as a. If taking a Cartesian coordinate system with the x-axis along the centroidal axis of the longitudinal direction, and the z-axis

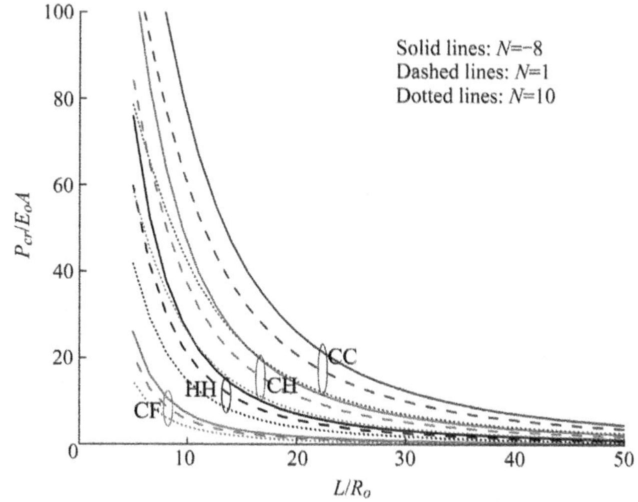

FIGURE 4.8 The variation of P_{cr}/E_oA against L/R_o.

perpendicular to the x-axis and upward, the column lies in domain: $0 \le x \le L$, $-b/2 \le y \le b/2$, $-h/2 \le z \le h/2$ for rectangular columns and $0 \le x \le L$, $0 \le r \le a$ for circular cylindrical columns, where $r = \sqrt{y^2 + z^2}$. For rectangular columns, without a loss of generality, to that ensure buckling occurs in the x–z plane, it is required that $h \le b$. First, we choose an infinitesimal element of arc length ds, as shown in Fig. 4.9. In this element, by using the equilibrium of transverse force, we obtain

$$\frac{\partial V}{\partial x} = 0 \tag{4.82}$$

with $V = Q - P\theta_0$, where V specifies the shear force perpendicular to the undeformed centroidal line, θ_0 is the rotation angle, $\theta_0 = \mathrm{d}w/\mathrm{d}x$ for Engesser's assumption, and $\theta_0 = -\psi$ for Haringx's assumption. To treat them in a unified way, we introduce a transition parameter α and denote [22]

$$\theta_0 = (1-\alpha)\frac{\partial w}{\partial x} - \alpha\psi \tag{4.83}$$

Clearly, Engesser columns correspond to $\alpha = 0$ and Haringx columns correspond to $\alpha = 1$. Then one gets

$$V = Q - P\left[(1-\alpha)w' - \alpha\psi\right] \tag{4.84}$$

In addition, according to the equilibrium of bending moment on arc length ds and the assumption d$s \approx$ dx, by neglecting the higher-order infinitesimal, we get

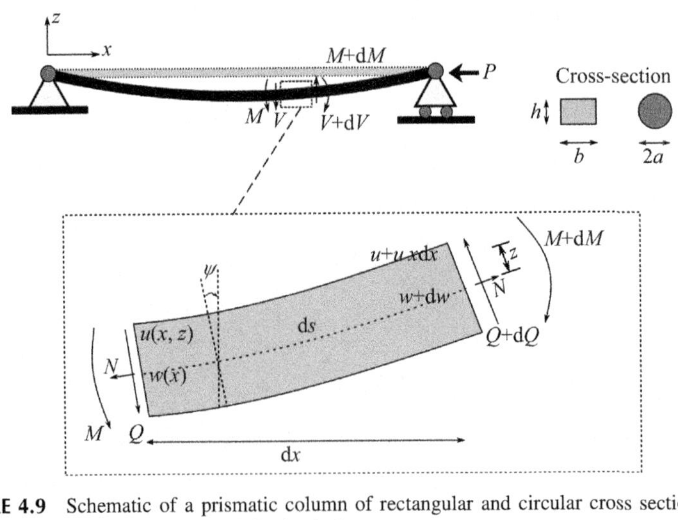

FIGURE 4.9 Schematic of a prismatic column of rectangular and circular cross section under axial compression along with an infinitesimal element.

$$\frac{\partial M}{\partial x} - V - P\frac{\partial w}{\partial x} = 0 \tag{4.85}$$

Next, we establish the relationships of the shear force and bending moment with the deflection and rotation. Since plane buckling is restricted, we neglect the corresponding displacement component v in the y direction. Then there are only axial displacement and transverse displacement, which are denoted as u and w.

4.5.1 Formulation of the problem

Now we turn our attention to the buckling of a uniform cylindrical column with a circular cross section under axial compression. For a circular cylindrical column, the axial displacement component u takes the following form:

$$u(x,y,z) = u_0(x) - zw'(x) + f(y,z)\left[\psi(x) + w'(x)\right] \tag{4.86}$$

where $f(y,z)$ is a function of both y and z:

$$y = r\cos\theta, z = r\sin\theta, r < a \tag{4.87}$$

Using the above expression for u and neglecting the displacement component in the y direction, we obtain the shear strain

$$\gamma_{xr} = \frac{\partial u}{\partial r} + \frac{\partial(w\sin\theta)}{\partial x} = \left(\frac{\partial f}{\partial y}\cos\theta + \frac{\partial f}{\partial r}\sin\theta\right)\left[\psi(x) + w'(x)\right] \tag{4.88}$$

A polynomial warping shape has been presented in Ref. [24]. Here we further propose two new trigonometric and exponential warping shapes for circular columns:

$$\begin{cases} f(y,z) = \dfrac{2za}{\pi r}\sin\dfrac{\pi r}{2a}, \text{WSC1} \\ f(y,z) = ze^{-r^2/(2a^2)}, \text{WSC2} \end{cases} \tag{4.89}$$

All of these are listed in Table 4.6. After performing derivative of these warping shape functions, we easily obtain the shear strain

$$\begin{cases} \gamma_{xr} = \dfrac{z}{\sqrt{y^2+z^2}}\cos\dfrac{\pi\sqrt{y^2+z^2}}{2a}(\psi + w'), \text{WSC1} \\ \gamma_{xr} = \dfrac{z}{\sqrt{y^2+z^2}}\left(1 - \dfrac{y^2+z^2}{a^2}\right)e^{-(y^2+z^2)/(2a^2)}(\psi + w'), \text{WSC2} \end{cases} \tag{4.90}$$

The shear strain γ_{xr} and shear stress σ_{xr} then vanish on the surface of a column. Consequently, no shear correction factor is required to relax the shear stress free condition on the surface. Furthermore, with the above expression

TABLE 4.6 The warping function $f(y, z)$ and related quantities for Levinson rectangular and circular cylindrical columns.

Ref.	$f(y,z)$	$\frac{\partial f}{\partial z}$	A_f/A	I_f/I	AI_f/IA_f
[25]	$z-\frac{4z^3}{3h^2}$	$1-\frac{4z^2}{h^2}$	2/3	4/5	1.2
[24]	$z\left(1-\frac{y^2+z^2}{3a^2}\right)$	$1-\frac{y^2+3z^2}{3a^2}$	2/3	7/9	1.1667
WSC1	$\frac{2za}{\pi r}\sin\frac{\pi r}{2a}$	$\frac{2ay^2}{\pi r^3}\sin\frac{\pi r}{2a}+\frac{z^2}{r^2}\cos\frac{\pi r}{2a}$	$2/\pi$	$\frac{64(\pi-2)}{\pi^4}$	1.1782
WSC2	$ze^{-r^2/2a^2}$	$\left(1-\frac{z^2}{a^2}\right)e^{-r^2/2a^2}$	$\frac{1}{\sqrt{e}}$	$8-\frac{12}{\sqrt{e}}$	1.1898

for u, we immediately find the normal strain ε_{xx} and shear strains γ_{xz} as follows:

$$\begin{cases} \varepsilon_{xx} = u_0' - zw'' + f(y,z)\left(\psi' + w''\right) \\ \gamma_{xz} = \dfrac{\partial f}{\partial z}\left(\psi + w'\right) \end{cases} \tag{4.91}$$

With the above shear strain γ_{xz}, we can give the distribution of shear stress σ_{xz} over the circular cross section. Clearly, the distribution is solely related to $\partial f/\partial z$, and the distribution of the normal shear stress is shown in Fig. 4.10. The use of the constitutive equations $\sigma_{xx} = E\varepsilon_{xx}$, $\sigma_{xz} = G\gamma_{xz}$ allows one to obtain

$$\begin{cases} M = -EIw'' + EI_f\left(\psi' + w''\right) \\ Q = GA_f\left(\psi + w'\right) \end{cases} \tag{4.92}$$

where

$$\begin{cases} I_f = \displaystyle\int_A zf(y,z)\mathrm{d}A \\ A_f = \displaystyle\int_A \frac{\partial f}{\partial z}\mathrm{d}A \end{cases} \tag{4.93}$$

Now we substitute Eq. (4.92) for circular cylindrical columns and Eq. (4.84) into the equilibrium Eqs. (4.82) and (4.85), then obtain

$$\begin{cases} \left(GA_f + \alpha P\right)\left(\psi' + w''\right) - Pw'' = 0 \\ EIw''' - EI_f\left(\psi'' + w'''\right) + \left(GA_f + \alpha P\right)\left(\psi + w'\right) = 0 \end{cases} \tag{4.94}$$

Thus, we have derived a system of coupled differential equations governing the buckling of Levinson rectangular and circular prismatic columns under an axially compressive force.

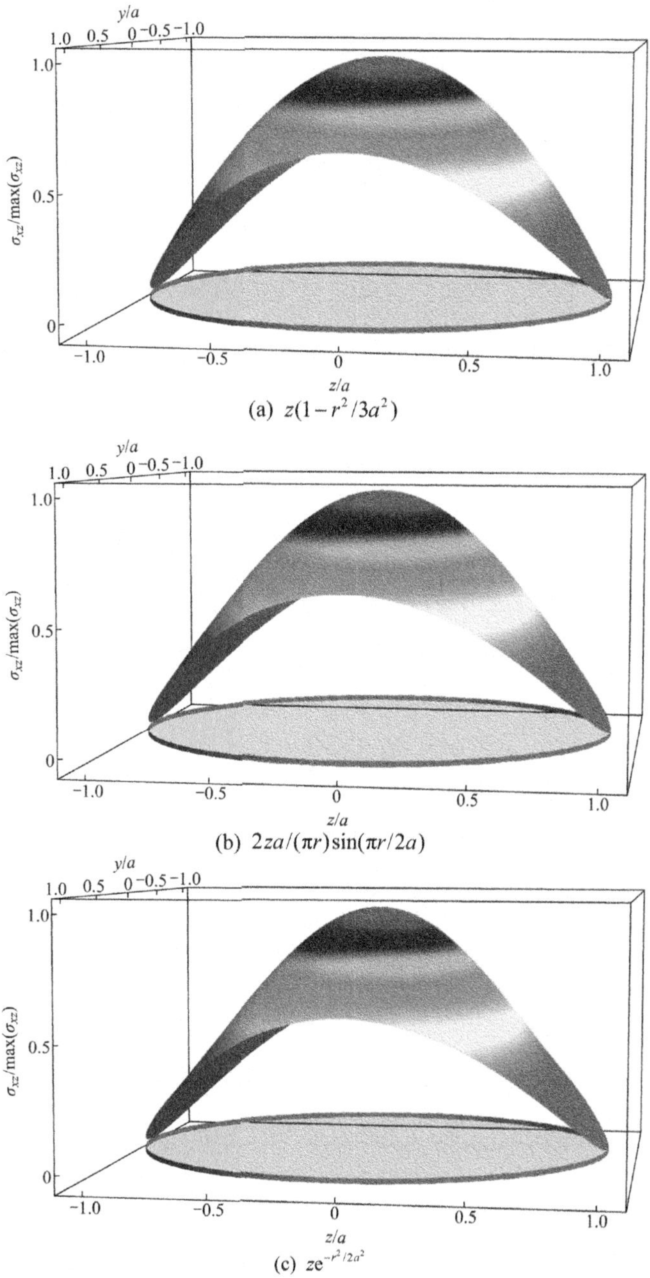

FIGURE 4.10 Distribution of normalized shear stress $\sigma_{xz}(x, 0, z)/\sigma_{xz}(x, 0, 0)$ for a circular cylindrical column, where the warping function $f(y, z)$ takes (a)$z(1-r^2/3a^2)$; (b)$2za/(\pi r)\sin(\pi r/2a)$; and (c)$ze^{-r^2/2a^2}$.

4.5.2 Buckling load

This section presents the determination of the buckling loads of a prismatic column under axial compression. For convenience, we introduce the dimensionless quantities

$$\begin{cases} \xi = \dfrac{x}{L}, W(\xi) = \dfrac{w(x)}{L,} \quad \Psi(\xi) = \psi(x) \\ p_b = \dfrac{PL^2}{EI}, p_s = \dfrac{P}{GA_f}, k = \sqrt{\dfrac{(1+\alpha p_s)I}{I+(\alpha I - I_f)p_s}\dfrac{PL^2}{EI}} \end{cases} \tag{4.95}$$

Then Eq. (4.94) can be rewritten as

$$(1+\alpha p_s)(\Psi' + W'') - p_s W'' = 0 \tag{4.96}$$

$$\frac{p_s}{p_b}W''' - \frac{p_s I_f}{p_b I}(\Psi'' + W''') + (1+\alpha p_s)(\Psi + W') = 0 \tag{4.97}$$

Following Ref. [22], we construct a new unknown function $F(\xi)$ such that it satisfies the following governing equation

$$F^{IV} + k^2 F'' = 0 \tag{4.98}$$

If expressing the deflection w and rotation ψ in dimensionless form in terms of the introduced function $F(\xi)$ as follows:

$$\begin{aligned} W &= (1+\alpha p_s)F - \frac{p_s I_f}{p_b I}F'' \\ \Psi &= -\left[(1+\alpha p_s)F' + \frac{p_s}{p_b}\left(1 - \frac{I_f}{I}\right)F'''\right] \end{aligned} \tag{4.99}$$

we find that the system of coupled ordinary differential Eqs. (4.96) and (4.97) is automatically fulfilled. As a result, it is sufficient to determine a nontrivial solution of the governing Eq. (4.98). With the above expressions for the deflection and rotation, we obtain the shear strain below

$$\Psi + W' = -\frac{p_s}{p_b}F''' \tag{4.100}$$

furthermore, from Eqs. (4.92) and (4.84), the bending moment and shear force can take the following dimensionless form

$$\begin{aligned} M &= -\frac{EI}{L}(1+\alpha p_s)F'' \\ V &= -\frac{EI}{L^2}\left(1 + \frac{\alpha I - I_f}{I}p_s\right)(F''' + k^2 F') \end{aligned} \tag{4.101}$$

The remaining task is to determine the nontrivial solution by solving the governing Eq. (4.98) under appropriate boundary conditions. For a prismatic column with elastically restrained ends connected to translational and rotational springs, the characteristic equation can be derived, however the details are not provided here.

The general solution of the governing Eq. (4.98) takes

$$F(\xi) = C_1 \cos k\xi + C_2 \sin k\xi + C_3 k\xi + C_4 \tag{4.102}$$

where $C_j (j = 1, \ldots, 4)$ are constants that are not all zero. With the general solution, from Eqs. (4.99) and (4.101) we easily obtain

$$W = \left(1 + \alpha p_s + \frac{k^2 p_s I_f}{p_b I}\right)(C_1 \cos k\xi + C_2 \sin k\xi) + (1 + \alpha p_s)(C_3 k\xi + C_4) \tag{4.103}$$

$$\Psi = \left[1 + \alpha p_s - \frac{k^2 p_s}{p_b}\left(1 - \frac{I_f}{I}\right)\right] k(C_1 \sin k\xi - C_2 \cos k\xi) - (1 + \alpha p_s)kC_3 \tag{4.104}$$

$$M = \frac{k^2 EI}{L}(1 + \alpha p_s)(C_1 \cos k\xi + C_2 \sin k\xi) \tag{4.105}$$

$$V = -\frac{EI}{L^2}\left(1 + \frac{\alpha I - I_f}{I} p_s\right) k^3 C_3 \tag{4.106}$$

4.5.3 Pinned–pinned columns

For the first case, the boundary conditions of pinned–pinned columns can be stated as:

$$\begin{aligned} &w = 0, M = 0, x = 0 \\ &w = 0, M = 0, x = L \end{aligned} \tag{4.107}$$

Using Eqs. (4.103) and (4.105), we get

$$\begin{cases} \left(1 + \alpha p_s + \dfrac{k^2 p_s I_f}{p_b I}\right) C_1 + (1 - \alpha p_s) C_4 = 0 \\ \left(1 + \alpha p_s + \dfrac{k^2 p_s I_f}{p_b I}\right)(C_1 \cos k + C_2 \sin k) + (1 + \alpha p_s)(C_3 k + C_4) = 0 \\ C_1 = 0 \\ C_1 \cos k + C_2 \sin k = 0 \end{cases} \tag{4.108}$$

The existence condition of a nontrivial solution of the system of linear algebraic equations requires that the determinant of the coefficient matrix of the system must be zero, which allows us to obtain the characteristic equation

$$\sin k = 0 \tag{4.109}$$

From the above characteristic equation, one gets the lowest positive value as $k = \pi$, which gives the critical buckling load meeting the relationship

$$\frac{p_b}{\pi^2} = \frac{I + (\alpha I - I_f)p_s}{(1 + \alpha p_s)I} \tag{4.110}$$

After simplification and collection, one finds that the critical load of a simply supported column should satisfy

$$\frac{1}{P_{cr}} = \frac{L^2}{\pi^2 EI} + \left(\frac{I_f}{I} - \alpha\right)\frac{1}{GA_f} + \frac{P_{cr}L^2}{\pi^2 EI}\frac{\alpha}{GA_f} \tag{4.111}$$

This is a quadratic equation, and its solution or the critical load can be exactly determined, i.e.,

$$\frac{P_{cr}}{P_{cr}^E} = \frac{2}{\sqrt{\left[1 + \left(\frac{I_f}{I} - \alpha\right)\frac{\pi^2 EI}{GA_f L^2}\right]^2 + 4\frac{\alpha\pi^2 EI}{GA_f L^2}} + \left[1 + \left(\frac{I_f}{I} - \alpha\right)\frac{\pi^2 EI}{GA_f L^2}\right]} \tag{4.112}$$

where P_{cr}^E specifies the Euler buckling load of the corresponding end conditions. Another root is negative and discarded since it has no physical meanings.

4.5.4 Double-clamped column

Secondly, we consider a double-clamped column with the boundary conditions as follows:

$$\begin{cases} w = 0, \psi = 0, x = 0 \\ w = 0, \psi = 0, x = L \end{cases} \tag{4.113}$$

Using Eqs. (4.103) and (4.104), we get

$$\begin{cases} \left(1+\alpha p_s+\dfrac{k^2 p_s I_f}{p_b I}\right)C_1+(1+\alpha p_s)C_4=0 \\ \left(1+\alpha p_s+\dfrac{k^2 p_s I_f}{p_b I}\right)(C_1\cos k+C_2\sin k)+(1+\alpha p_s)(C_3 k+C_4)=0 \\ \left[1+\alpha p_s-\dfrac{k^2 p_s}{p_b}\left(1-\dfrac{I_f}{I}\right)\right]C_2+(1+\alpha p_s)C_3=0 \\ \left[1+\alpha p_s-\dfrac{k^2 p_s}{p_b}\left(1-\dfrac{I_f}{I}\right)\right](C_1\sin k-C_2\cos k)-(1+\alpha p_s)C_3=0 \end{cases} \tag{4.114}$$

The existence condition of a nontrivial solution for the system of linear algebraic equations requires that the determinant of the coefficient matrix vanishes, which allows us to obtain the characteristic equation. In a similar manner, one obtains

$$\sin\frac{k}{2}\left(\sin\frac{k}{2}-\frac{\zeta k}{2}\cos\frac{k}{2}\right)=0 \tag{4.115}$$

where

$$\zeta=1-\frac{k^2 p_s I}{(1+\alpha p_s)p_b I+k^2 p_s I_f} \tag{4.116}$$

It is clear that $0<\zeta<1$, and we easily conclude that the lowest positive root is $k=2\pi$. This gives the critical buckling load meeting the relationship

$$\frac{p_b}{4\pi^2}=\frac{I+(\alpha I-I_f)p_s}{(1+\alpha p_s)I} \tag{4.117}$$

After simplification and collection, one finds that the critical load of a clamped–clamped column is

$$\frac{P_{cr}}{P_{cr}^E}=\frac{2}{\sqrt{\left[1+\left(\dfrac{I_f}{I}-\alpha\right)\dfrac{4\pi^2 EI}{GA_f L^2}\right]^2+\dfrac{16\alpha\pi^2 EI}{GA_f L^2}}+\left[1+\left(\dfrac{I_f}{I}-\alpha\right)\dfrac{4\pi^2 EI}{GA_f L^2}\right]} \tag{4.118}$$

4.5.5 Clamped–free column

Finally, we consider a clamped–free column with the boundary conditions as follows:

$$\begin{cases} w = 0, \psi = 0, x = 0 \\ M = 0, V = 0, x = L \end{cases} \tag{4.119}$$

Using Eqs. (4.105) and (4.106), we get

$$\begin{cases} \left(1 + \alpha p_s + \dfrac{k^2 p_s I_f}{p_b I}\right) C_1 + (1 + \alpha p_s) C_4 = 0 \\ \left[1 + \alpha p_s - \dfrac{k^2 p_s}{p_b}\left(1 - \dfrac{I_f}{I}\right)\right] C_2 + (1 + \alpha p_s) C_3 = 0 \\ C_1 \cos k + C_2 \sin k = 0 \\ C_3 = 0 \end{cases} \tag{4.120}$$

From the above system, making use of the existence condition of a nontrivial solution for the system of linear algebraic equations, we obtain

$$\cos k = 0 \tag{4.121}$$

It is clear that the lowest positive root is $k = \pi/2$. This gives the critical buckling load meeting the relationship

$$\frac{4p_b}{\pi^2} = \frac{I + (\alpha I - I_f) p_s}{(1 + \alpha p_s) I} \tag{4.122}$$

After simplification and collection, one finds that the critical load of a clamped–free column has the following expression

$$\frac{P_{cr}}{P_{cr}^E} = \frac{2}{\sqrt{\left[1 + \left(\frac{I_f}{I} - \alpha\right)\frac{\pi^2 EI}{4GA_f L^2}\right]^2 + \frac{\alpha\pi^2 EI}{GA_f L^2}} + \left[1 + \left(\frac{I_f}{I} - \alpha\right)\frac{\pi^2 EI}{4GA_f L^2}\right]} \tag{4.123}$$

4.5.6 Effect of the transition parameter α

In the previous section, based on the shear deformation beam theory, we derived the critical loads for buckling of prismatic columns with typical end constraints both for Engesser's and Haringx's hypotheses. Based on the obtained results in the foregoing section, we rewrite the critical load in alternative form

$$P_{cr} = \frac{2P_{cr}^{E}}{\sqrt{\left[1+\left(\frac{I_f}{I}-\alpha\right)\frac{P_{cr}^{E}}{GA_f}\right]^2 + 4\alpha\frac{P_{cr}^{E}}{GA_f}} + 1 + \left(\frac{I_f}{I}-\alpha\right)\frac{P_{cr}^{E}}{GA_f}} \tag{4.124}$$

and

$$P_{cr} = \frac{GA_f}{2\alpha}\left(\sqrt{\left[1+\left(\frac{I_f}{I}-\alpha\right)\frac{P_{cr}^{E}}{GA_f}\right]^2 + 4\alpha\frac{P_{cr}^{E}}{GA_f}} - \left[1+\left(\frac{I_f}{I}-\alpha\right)\frac{P_{cr}^{E}}{GA_f}\right]\right) \tag{4.125}$$

where P_{cr}^{E} denotes the Euler buckling load, i.e., $P_{cr}^{E} = \pi^2 EI/\left(\mu^2 L^2\right)$, μ denotes the end constraint coefficient that takes 1, 0.5, and 2 corresponding to pinned–pinned, clamped–clamped, and clamped–free ends, respectively. From Eq. (4.124), the buckling load is always less than the Euler buckling load. Furthermore, if setting the transition parameter $\alpha = 0$, we find

$$\begin{cases} P_{cr} = \dfrac{P_{cr}^{E}}{1+\beta\dfrac{P_{cr}^{E}}{GA}} \\ \beta = \dfrac{AI_f}{IA_f} \end{cases} \tag{4.126}$$

The Engesser buckling load [11] is recovered. In the present study, we provide an effective approach to calculate the factor β, whose value is approximately 1.2, as shown in Table 4.6. Moreover, it is seen that the arbitrariness of $f'(0)$ does not affect the buckling load since the value of β is invariant. On the other hand, if setting $I_f/I = A_f/A = \alpha = 1$, from Eq. (4.125) we deduce

$$P_{cr} = \frac{GA}{2\alpha}\left(\sqrt{1+4\frac{P_{cr}^{E}}{GA}} - 1\right) \tag{4.127}$$

which is identical to the Haringx buckling load [11]. The discrepancy between these two results, regarded as a paradox, can be well explained as two extreme cases through the transition parameter α. Strictly speaking, Haringx's assumption does not give $I_f/I = A_f/A = 1$ directly because of the results in Table 4.6 for higher-order analysis. In the case of $I_f/I = A_f/A = 1$, Fig. 4.11 shows the difference of the present buckling load for two different values of the transition parameter $\alpha = 0, 1$, where $\alpha = 0$ gives the Engesser buckling load and $\alpha = 1$ gives the Haringx buckling load. For prismatic columns, it clearly has $P_{cr}^{E}/GA < 1$. From Fig. 4.11, one finds that when $GA/P_{cr}^{E} > 10$, the maximum relative error of Engesser and Haringx buckling loads is within 0.77%. For $GA/P_{cr}^{E} < 10$, with P_{cr}^{E} rising, the Haringx buckling load is

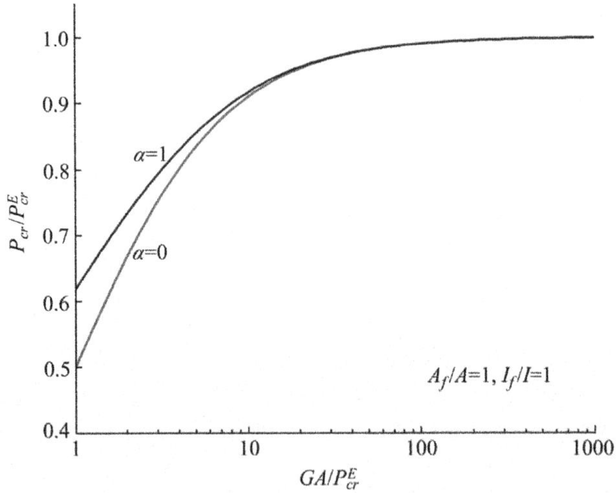

FIGURE 4.11 Critical load ratio P_{cr}/P_{cr}^E against GA/P_{cr}^E in the case of $I_f/I = A_f/A = 1$ for $\alpha = 0$. (Engesser's assumption) and $\alpha = 1$ (Haringx's assumption).

gradually overestimated compared to the Engesser buckling load. The above comparison is for $I_f/I = A_f/A = 1$. However, when higher-order beam theories are employed, they do not hold, as is shown in Table 4.6.

Fig. 4.12 shows a comparison of the buckling loads using the Levinson beam theory according to Engesser's ($\alpha = 0$) and Haringx's ($\alpha = 1$) assumptions, respectively. In the Levinson beam theory, the shear deformation

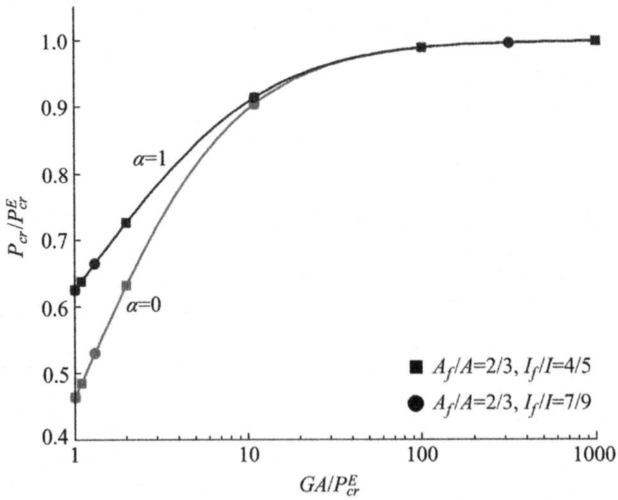

FIGURE 4.12 Critical load ratio P_{cr}/P_{cr}^E against GA/P_{cr}^E based on the Levinson beam theory for $\alpha = 0$ (Engesser's assumption) and $\alpha = 1$ (Haringx's assumption).

and warping of the cross section have been considered. By comparing Fig. 4.11, it can be seen from Fig. 4.12 that the buckling load has a slight decline owing to the influence of shear deformation and warping of the cross-section. The buckling load shows negligible differences between rectangular columns and circular cylindrical columns. Moreover, the Engesser buckling load is again seen to be less than the Haringx buckling load.

4.5.7 Effect of shear rigidity

To examine the effect of shear rigidity on the buckling load, based on the Levinson beam theory, we plot the variation of P_{cr}/P_{cr}^E as a function of κ, where κ is the ratio of shear rigidity to bending stiffness, GAL^2/π^2EI, in Figs. 4.13 and 4.14 for a column of rectangular and circular cross section with three typical end conditions. Clearly, as κ becomes large, the buckling load tends to the Euler buckling load, while for small values of κ (i.e., columns with low shear rigidity where shear deformation readily occurs), the buckling load is significantly lower than the Euler load. However, the buckling load is almost the same for rectangular and circular cross sections. This is readily understood since they have the same $A_f = 2/3$, and the values of I_f have only a slight difference, $I_f = 0.8$ for a rectangular cross section and $I_f = 0.78$ for a circular cross section.

Similarly, Figs. 4.15 and 4.16 plot P_{cr}/GA as a function of κ for columns with both rectangular and circular cross sections, under three typical end

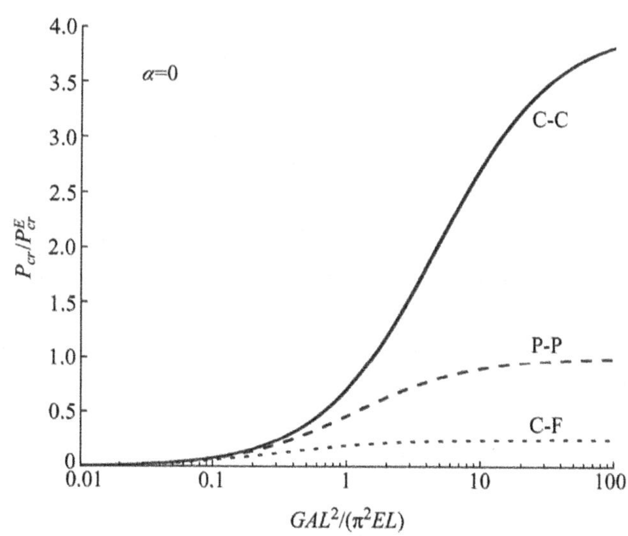

FIGURE 4.13 The variation of P_{cr}/P_{cr}^E against κ, ($\kappa = GAL^2/\pi^2EI$) for columns of rectangular and circular cross sections with three typical end conditions, $\alpha = 0$.

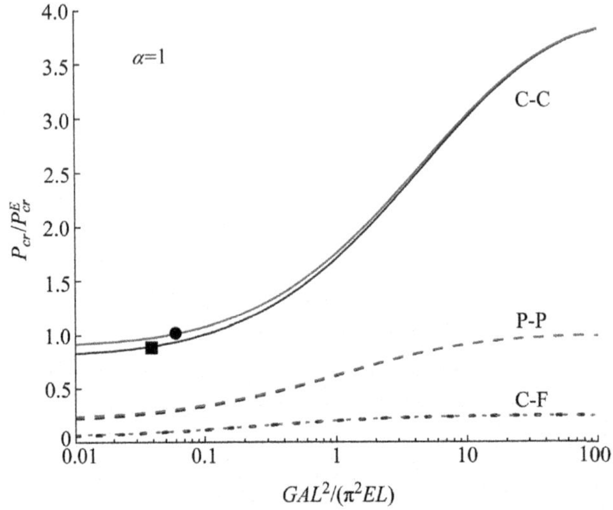

FIGURE 4.14 The variation of P_{cr}/P_{cr}^E against κ, ($\kappa = GAL^2/\pi^2 EI$) for columns of rectangular and circular cross sections with three typical end conditions, $\alpha = 1$.

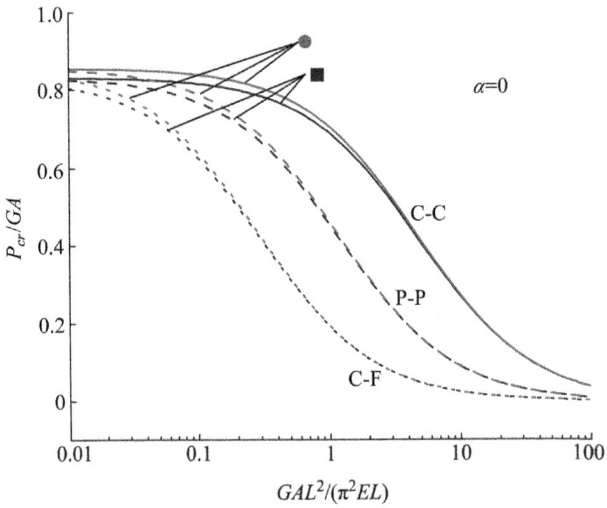

FIGURE 4.15 The variation of P_{cr}/GA against κ, ($\kappa = GAL^2/\pi^2 EI$) for columns of rectangular and circular cross sections with three typical end conditions, $\alpha = 0$.

conditions. When shear deformation easily occurs, meaning that the column has low shear rigidity, one finds that κ becomes smaller, P_{cr}/GA progressively approaches a steady value for $\alpha = 0$ and infinity for $\alpha = 1$, regardless of the column cross section. This is due to the fact that for various warping shape functions, we have $I_f/I < 1$, then $I_f/I - \alpha$ is negative if $\alpha = 1$, which results in a vanishing denominator in Eq. (4.124).

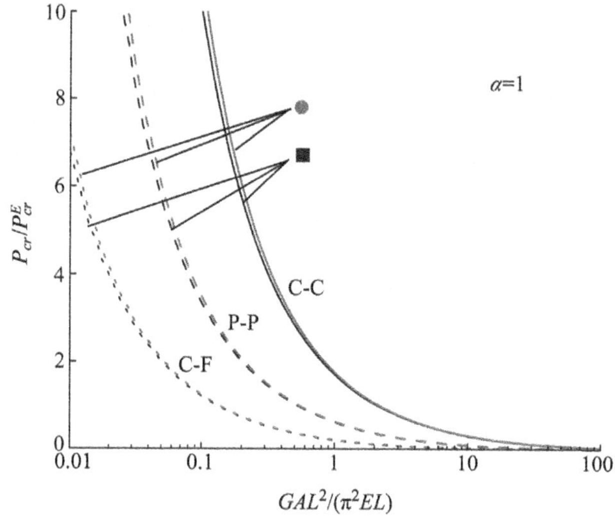

FIGURE 4.16 The variation of $\frac{P_{cr}}{GA}$ against κ, ($\kappa = \frac{GAL^2}{\pi^2 EI}$) for columns of rectangular and circular cross sections with three typical end conditions, $\alpha = 1$.

For homogeneous isotropic prismatic columns, Young's modulus is related to shear modulus by $E = 2(1 + \nu)G$, ν is the Poisson's ratio. Then we rewrite the buckling load [Eq. (4.124)] as:

$$P_{cr} = \frac{\pi^2 EI}{\mu^2 L^2} \frac{2}{\sqrt{\left[1 + \eta\left(\beta - \alpha\frac{A}{A_f}\right)\right]^2 + 4\alpha\eta\frac{A}{A_f}} + 1 + \eta\left(\beta - \alpha\frac{A}{A_f}\right)} \tag{4.128}$$

or

$$P_{cr} = EA\frac{A_f/A}{(1+\nu)\left[\sqrt{\left(\frac{I_f}{I} + \frac{A_f}{\eta A} - \alpha\right)^2 + 4\alpha\frac{A_f}{\eta A}} + \frac{I_f}{I} + \frac{A_f}{\eta A} - \alpha\right]} \tag{4.129}$$

where

$$\eta = \frac{2(1+\nu)\pi^2 I}{\mu^2 L^2 A} \tag{4.130}$$

In the case of $\nu = 0.3$, we calculate the buckling load ratio $P_{cr}/\ P_{cr}^E$ for various warping shape functions. If taking the slenderness ratio $L^2 A/\ I = 50$, we find that all buckling load ratios take the same values of 0.96, 0.99, and 0.997 for clamped–clamped, pinned–pinned, and clamped–free columns, irrespective of whether Engesser's or Haringx's assumption is used. This means that for a slender column, the Engesser and Haringx buckling loads are almost equal to the Euler loads. Nonetheless, if taking a lower value of $L^2A/I = 10$, the calculated buckling load ratios P_{cr}/P_{cr}^E are listed in Table 4.7. It is readily found that the Engesser and Haringx buckling loads significantly decrease, in particular for clamped–clamped end constraints. On the other hand, despite various warping shape functions, almost the same Engesser (or Haringx) buckling load is predicted, where the difference between the Engesser and Haringx buckling loads is highest among the three end constraints.

4.5.8 Effect of warping shapes

Finally, the influence of the warping shape of the cross section is discussed. Here we compare our results for $\alpha = 0$ with previous ones for cantilevers. If using the Timoshenko–Engesser beam theory, the corresponding buckling load is given by:

$$P_{cr}^T = \frac{P_{cr}^E}{1 + \dfrac{P_{cr}^E}{\kappa GA}} \tag{4.131}$$

where κ is the shear correction factor. Additionally, if using the Reddy–Bickford beam theory, the buckling load is given by [17]:

$$\begin{cases} P_{cr}^R = \dfrac{P_{cr}^E\left(1 + \dfrac{P_{cr}^E}{70GA}\right)}{1 + \dfrac{17P_{cr}^E}{14GA}} \quad \text{(for square cross – section)} \\ P_{cr}^R = \dfrac{P_{cr}^E\left(1 + \dfrac{P_{cr}^E}{90GA}\right)}{1 + \dfrac{101P_{cr}^E}{90GA}} \quad \text{(for circular cross – section)} \end{cases} \tag{4.132}$$

TABLE 4.7 Buckling load ratio P_{cr}/P_{cr}^E for columns with various warping shapes with $L^2A/I = 0$ and $\nu = 0.3$.

$f(z)$	C–C		P–P		C–F	
	$\alpha = 0$	$\alpha = 1$	$\alpha = 0$	$\alpha = 1$	$\alpha = 0$	$\alpha = 1$
$z - \frac{4z^3}{3h^2}$	0.4481	0.6119	0.7646	0.8099	0.9285	0.9340
$z\left(1 - \frac{y^2+z^2}{3a^2}\right)$	0.4551	0.6201	0.7696	0.8144	0.9304	0.9357
$\frac{2za}{\pi r}\sin\frac{\pi r}{2a}$	0.4526	0.6239	0.7679	0.8146	0.9297	0.9354
$ze^{-r^2/2a^2}$	0.4502	0.6282	0.7661	0.8150	0.9291	0.9351

TABLE 4.8 Buckling load $P_{cr}L^2/EI$ for a cantilever with a rectangular cross-section.

EI/GAL^2	Timoshenko	Reddy	$f(z)$ (Levinson, 1981)
0.0	2.4674	2.4674	2.4674
0.15	1.7086	1.7113	1.7086
0.3	1.3067	1.3132	1.3067

TABLE 4.9 Buckling load $P_{cr}L^2/EI$ for a cantilever with a circular cross-section.

			$f(y, z)$		
EI/GAL^2	Timoshenko	Reddy	[24]	WSC1	WSC2
0.0	2.4674	2.4674	2.4674	2.4674	2.4674
0.15	1.7484	1.7505	1.7233	1.7182	1.7131
0.3	1.3539	1.3589	1.3240	1.3180	1.3120

A comparison of the buckling load is provided in Tables 4.8 and 4.9 using various beam theories, where $\kappa = 5/6$ for a square column and 9/10 for a circular column are chosen in the Timoshenko beam theory. From Tables 4.8 and 4.9 it is seen that when $EI/GAL^2 = 0$, all the buckling loads are identical, irrespective of the column cross section with any warp shapes. This is easily understood since no warping occurs because of the sufficiently high shear rigidity. Nonetheless, for a finite shear rigidity such as $EI/GAL^2 = 0.15, 0.3$, the effect of shear deformation with various warping shapes on the buckling load is seen in Tables 4.4 and 4.5. For circular columns with the results given in Table 4.9, the buckling load based on the Reddy–Bickford beam theory is the highest, whereas the model corresponding to $f(z) = ze^{-r^2/(2a^2)}$ yields the lowest value.

References

[1] Aydogdu M, Taskin V. Free vibration analysis of functionally graded beams with simply supported edges. Mater Des 2007;28(5):1651–6.

[2] Elishakoff I, Endres J. Extension of Euler's problem to axially graded columns: two hundred and sixty years later. J Intell Mater Syst Struct 2005;16(1):77–83.

[3] Kang YA, Li XF. Bending of functionally graded cantilever beam with power-law non-linearity subjected to an end force. Int J Non Lin Mech 2009;44(6):696–703.
[4] Kapuria S, Bhattacharyya M, Kumar AN. Bending and free vibration response of layered functionally graded beams: a theoretical model and its experimental validation. Compos Struct 2008;82(3):390–402.
[5] Li XF. A unified approach for analyzing static and dynamic behaviors of functionally graded Timoshenko and Euler-Bernoulli beams. J Sound Vib 2008;318(4–5):1210–29.
[6] Oh SY, Librescu L, Song O. Vibration and instability of functionally graded circular cylindrical spinning thin-walled beams. J Sound Vib 2005;285(4–5):1071–91.
[7] Sankar BV. An elasticity solution for functionally graded beams. Compos Sci Technol 2001;61(5):689–96.
[8] Sina SA, Navazi HM, Haddadpour H. An analytical method for free vibration analysis of functionally graded beams. Mater Des 2009;30(3):741–7.
[9] Huang Y, Li XF. Bending and vibration of circular cylindrical beams with arbitrary radial nonhomogeneity. Int J Mech Sci 2010;52(4):595–601.
[10] Blaauwendraad J. Timoshenko beam–column buckling. Does Dario stand the test? Eng Struct 2008;30(11):3389–93.
[11] Timoshenko SP, Gere JM. Theory of elastic stability. Courier Corporation; 2012.
[12] Wiggers S, Pedersen P. Structural stability and vibration. Berlin: Springer Tracts in Mechanical Engineering. Springer; 2018.
[13] Li SR, Batra RC. Relations between buckling loads of functionally graded Timoshenko and homogeneous Euler–Bernoulli beams. Compos Struct 2013;95:5–9.
[14] Cowper GR. The shear coefficient in Timoshenko's beam theory. J Appl Mech 1966;33 (2):335–40.
[15] Wang CM, Wang CY. Exact solutions for buckling of structural members. CRC Press; 2004.
[16] Hutchinson JR. Shear coefficients for Timoshenko beam theory. J Appl Mech 2001;68 (1):87–92.
[17] Shear deformable beams and plates: relationships with classical solutions. Elsevier; 2000.
[18] Matsunaga H. Free vibration and stability of functionally graded circular cylindrical shells according to a 2D higher-order deformation theory. Compos Struct 2009;88(4):519–31.
[19] Ru CQ. Column buckling of multiwalled carbon nanotubes with interlayer radial displacements. Phys Rev B 2000;62(24):16962.
[20] Li XF, Wang BL, Mai YW. Effects of a surrounding elastic medium on flexural waves propagating in carbon nanotubes via nonlocal elasticity. J Appl Phys 2008;103(7).
[21] Sudak LJ. Column buckling of multiwalled carbon nanotubes using nonlocal continuum mechanics. J Appl Phys 2003;94(11):7281–7.
[22] Li XF, Lee KY. Effects of Engesser's and Haringx's hypotheses on buckling of Timoshenko and higher-order shear-deformable columns. J Eng Mech 2018;144(1):04017150.
[23] Pilkey WD. Formulas for stress, strain, and structural matrices. Hoboken, NJ, USA: John Wiley and Sons; 2005.
[24] Huang Y, Li XF. Buckling of functionally graded circular columns including shear deformation. Mater Des 2010;31(7):3159–66.
[25] Levinson M. A new rectangular beam theory. J Sound Vib 1981;74(1):81–7.

Further reading

[1] Thai HT, Vo TP. Bending and free vibration of functionally graded beams using various higher-order shear deformation beam theories. Int J Mech Sci 2012;62(1):57–66.

[2] Xie K, Wang Y, Fan X, et al. Nonlinear free vibration analysis of functionally graded beams by using different shear deformation theories. Appl Math Model 2020;77:1860–80.

[3] Love AEH. A treatise on the mathematical theory of elasticity. Courier Corporation; 1944.

[4] Weaver JW, Timoshenko SP, Young DH. Vibration problems in engineering. John Wiley and Sons; 1991.

[5] Ma WL, Li XF, Yong Lee K. Third-order shear deformation beam model for flexural waves and free vibration of pipes. J Acoust Soc Am 2020;147(3):1634–47.

[6] Ma WL, Jiang ZC, Lee KY, et al. A refined beam theory for bending and vibration of functionally graded tube-beams. Compos Struct 2020;236:111878.

[7] Jiang ZC, Ma WL, Li XF. Stability of cantilever on elastic foundation under a subtangential follower force via shear deformation beam theories. Thin-Walled Struct 2020;154:106853.

[8] Ma WL, Jiang ZC, Li XF. Effect of warping shape on buckling of circular and rectangular columns under axial compression. Appl Math Model 2021;89:1475–90.

Chapter 5

Buckling and stability with various elastic foundations

5.1 Introduction

Foundations are important parts of physical systems due to their wide range of applications in different science and engineering areas including the mechanical, civil, electrical, nanotechnology, and even biotechnology sectors. They are of great interest to researchers due to their critical applications in various systems. Foundations play a pivotal role in preserving structural systems under oscillations. The demand for solutions to various problems regarding beams, plates, and shells continuously supported by elastic or viscoelastic media has become increasingly pressing. In this chapter, the structural stability of clamped–free beam-columns resting on elastic foundations under a nonconservative force is investigated. Several recently suggested foundation models (Winkler model, Pasternak model, Vlasov model, Kerr model, and several viscoelastic models) are discussed briefly. A higher-order beam theory is developed by considering the warping of the cross section, rotary inertia, and axial force. As an example, the characteristic equation of dynamic stability of rectangular/circular cantilevers on a Winkler–Pasternak foundation is obtained for a generalized follower force. Critical loads for divergence instability and flutter instability are evaluated. The influences of foundation stiffness, cross sectional warping shape, and the tangency coefficient on the critical loads are analyzed. The load frequency interaction curves are provided for different directions of follower force. The normal/shear stiffness of foundation on divergence and flutter loads has different effects. The normal stiffness increases the first divergence load but decreases the flutter load, reducing to the Herrmann–Smith paradox for Euler columns. The rise of shear stiffness enlarges the first divergence load and lowers the flutter loads. These effects are more sensitive for short beam-columns.

A coordinate system is established for a beam/column of length L as follows: the x-axis coincides with the centroidal axis, and the z-axis is perpendicular to the x-axis with its positive direction upward (see Fig. 5.1). Meanwhile, we also use the polar coordinates for columns with circular cross

Higher-Order Theories of Functionally Graded Circular Cylindrical Beams and Columns
https://doi.org/10.1016/B978-0-443-44718-1.00004-9

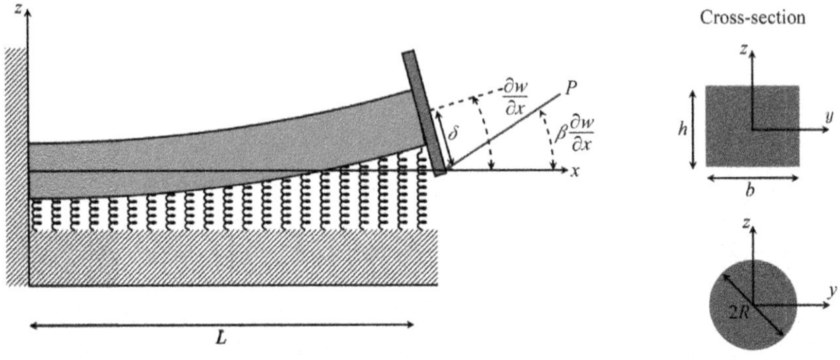

FIGURE 5.1 Schematic diagrams of a generalized Beck–Rent's column subjected to a nonconservative force on a massless rigid attachment mounted to the column at its free end.

section, and their relationship with the Cartesian coordinates is governed by: $y = r\cos\theta$, $z = r\sin\theta$.

Usually, there are two basic approaches to derive the governing equation for dynamic behaviors of axially loaded columns/beams. One is to appeal to Hamilton's principle from the viewpoint of energy conservation, and the other is to apply equilibrium equations for forces and moments. Here we adopt the latter. First, for an infinitesimal element of arc length ds, utilization of equilibrium of transverse force allows one to obtain [1]:

$$\frac{\partial V}{\partial x} + q = \rho A \frac{\partial^2 w}{\partial t^2} \tag{5.1}$$

with

$$V = Q - P\theta_0 \tag{5.2}$$

where ρ is the mass density, A is the cross sectional area, q is applied transverse loading, V specifies the shearing force perpendicular to the undeformed centroidal axis, Q denotes the shearing force on cross section, and θ_0 is the rotation angle that takes $\theta_0 = \partial w/\partial x$ if Engesser's hypothesis is adopted and $\theta_0 = -\psi$ if Haringx's hypothesis is adopted [$w(x, t)$: the deflection at the neutral surface, $\psi(x, t)$: the rotation of the cross section at the neutral surface]. Note that $\partial w/\partial x$ is the slope angle of the neutral surface that is positive along the anticlockwise direction, whereas the rotation angle ψ of the cross section is positive along the clockwise direction. Similar to the treatment in Ref. [2], we introduce a transition parameter α in the following expression

$$\theta_0 = \frac{\partial w}{\partial x} - \alpha\left(\psi + \frac{\partial w}{\partial x}\right) \tag{5.3}$$

We find that Engesser's and Haringx's hypotheses are two extreme cases of $\alpha = 0, \alpha = 1$, respectively. Thus, we come up with a unified approach to analyze the dynamic stability of cantilevered columns under nonconservative force simultaneously, irrespective of Engesser's or Haringx's hypothesis. Substituting Eqs. (5.2) and (5.3) into Eq. (5.1) leads to

$$\frac{\partial Q}{\partial x}-(1-\alpha)P\frac{\partial^2 w}{\partial x^2}+\alpha P\frac{\partial \psi}{\partial x}+q=\rho A\frac{\partial^2 w}{\partial t^2} \tag{5.4}$$

Considering the bending moment equilibrium on arc length ds and noting that d$s\approx$ dx when higher-order infinitesimal quantities are neglected, we get [1]

$$\frac{\partial M}{\partial x}-V-P\frac{\partial w}{\partial x}=\rho\int_A z\frac{\partial^2 u}{\partial t^2}\mathrm{d}A \tag{5.5}$$

where u denotes the axial displacement of any position of cross section, M denotes the bending moment. We have derived two equilibrium equations for forces and moments: Eqs. (5.4) and (5.5).

To further solve the problem, we need to relate the shearing force and bending moment to the deflection and rotation. To this end, let us express the axial displacement component in terms of the transverse deflection $w(x,t)$ [3]

$$u=-z\frac{\partial w}{\partial x}+f(y,z)\left(\psi+\frac{\partial w}{\partial x}\right) \tag{5.6}$$

where the extension and shrinkage of the neutral surface are neglected, from which the shear strains γ_{xz} and γ_{xr} are expressed by

$$\gamma_{xz}=\left(\psi+\frac{\partial w}{\partial x}\right)\frac{\partial f}{\partial z},\gamma_{xr}=\left(\psi+\frac{\partial w}{\partial x}\right)\frac{\partial(rf/z)}{\partial r} \tag{5.7}$$

where $f(y,z)$ or $f(r\cos\theta, r\sin\theta)$ describes the warping shape of the cross section during bending deformation. If setting the warping shape function $f(z)=0$, the Euler–Bernoulli hypothesis is reduced and the corresponding shear strain vanishes, while if setting $f(z)=z$, the Timoshenko hypothesis is retrieved and uniform shear strain can be found. Incidentally, in the case of $f(z)=z$, the uniform shear stress on the surface of circular beams violates traction-free conditions on the surface; so the shear correction factor in the Timoshenko beam theory is needed to remedy this shortage.

One of our tasks is to seek suitable warping shape functions so that the shear stress satisfies traction-free conditions on the surface of the beam/column. Three typical warping shape functions are listed in Table 5.1. It is interesting to note that for rectangular beams/columns, there have been many warping shape functions available and some have been collected in Ref. [3]. Here we consider three typical cross section's warping shape functions: polynomial, exponential, and trigonometric functions, which are listed in Table 5.1 (h denotes the height of the cross section of the rectangular beams/columns, and R denotes the radius of the cross section of the circular beams/

TABLE 5.1 Typical warping shape functions for beams with rectangular and circular cross sections.

Cross section	Warping shape	$f(y, z)$	$\frac{\partial f}{\partial z}$	$\frac{\partial (rf/z)}{\partial r}$
Rectangular	Polynomial [4]	$z - \frac{4z^3}{3h^2}$	$1 - \frac{4z^2}{h^2}$	–
	Exponential [5]	$z \exp\left(-\frac{2z^2}{h^2}\right)$	$\left(1 - \frac{4z^2}{h^2}\right)\exp\left(-\frac{2z^2}{h^2}\right)$	–
	Trigonometric [6]	$\frac{h}{\pi}\sin\frac{\pi z}{h}$	$\cos\frac{\pi z}{h}$	–
Circular	Polynomial [7]	$z\left(1 - \frac{r^2}{3R^2}\right)$	$1 - \frac{r^2+2z^2}{3R^2}$	$1 - \frac{r^2}{R^2}$
	Exponential [3]	$z \exp\left(-\frac{r^2}{2R^2}\right)$	$\left(1 - \frac{z^2}{R^2}\right)\exp\left(-\frac{r^2}{2R^2}\right)$	$\left(1 - \frac{r^2}{R^2}\right)\exp\left(-\frac{r^2}{2R^2}\right)$
	Trigonometric [3]	$\frac{2zR}{\pi r}\sin\frac{\pi r}{2R}$	$\frac{2y^2R}{\pi r^3}\sin\frac{\pi r}{2R} + \frac{z^2}{r^2}\cos\frac{\pi r}{2R}$	$\cos\frac{\pi r}{2R}$

Remark: $r^2 = y^2 + z^2$.

columns, respectively). It is worth mentioning that for circular cantilevers, only polynomial warping shape was reported [7,8], and the other warping shapes available are very limited. Here we propose two new cross sectional warping shapes.

Using the above-provided warping shape functions and relations, Eq. (5.7), along with the Eq. (5.6) for u, one applies Hooke's law to obtain the normal stress σ_{xx} and shear stress σ_{xz} as follows:

$$\sigma_{xx} = -Ez\frac{\partial^2 w}{\partial x^2} + f(y,z)E\left(\frac{\partial \psi}{\partial x} + \frac{\partial^2 w}{\partial x^2}\right) \tag{5.8}$$

$$\sigma_{xz} = G\frac{\partial f}{\partial z}\left(\psi + \frac{\partial w}{\partial x}\right) \tag{5.9}$$

and

$$\sigma_{xz} = G\frac{\partial (rf/z)}{\partial r}\left(\psi + \frac{\partial w}{\partial x}\right) \tag{5.10}$$

where E, G are Young's modulus and shear modulus, respectively. From Table 5.1, one finds $\sigma_{xz} = 0$ as $z = \pm h/2$ and $\sigma_{xr} = 0$ as $y^2 + z^2 = R^2$. This indicates that if choosing the above-mentioned three warping functions, the shear stresses vanish on the surface of the beams. Consequently, we can obtain the bending moment and shearing force expressed in terms of w and ψ:

$$M = -EI\frac{\partial^2 w}{\partial x^2} + EI_f\left(\frac{\partial \psi}{\partial x} + \frac{\partial^2 w}{\partial x^2}\right) \tag{5.11}$$

$$Q = GA_f\left(\psi + \frac{\partial w}{\partial x}\right) \tag{5.12}$$

with

$$I_f = \int_A zf(y,z)\mathrm{d}A, A_f = \int_A \frac{\partial f}{\partial z}\mathrm{d}A \tag{5.13}$$

The substitution of Eq. (5.11) into Eq. (5.5) yields

$$\frac{\partial M}{\partial x} - V - P\frac{\partial w}{\partial x} = -\rho I\frac{\partial^3 w}{\partial t^2 \partial x} + \rho I_f\left(\frac{\partial^2 \psi}{\partial t^2} + \frac{\partial^3 w}{\partial t^2 \partial x}\right) \tag{5.14}$$

On the other hand, from Eqs. (5.2), (5.3), and (5.12), one gets

$$V = (GA_f + \alpha P)\left(\psi + \frac{\partial w}{\partial x}\right) - P\frac{\partial w}{\partial x} \tag{5.15}$$

In order to consider the effect of foundations, the transverse loading q in Eq. (5.4) can be replaced by the foundation reaction.

The foundation models presented in the following sections are developed based on the Winkler model. In fact, researchers have started with the Winkler theory and inserted more parameters and elements into the Winkler model in order to present a more accurate reaction behavior for a structural foundation. The Pasternak model, the most well-known two parameter foundation model, is also described. In the Pasternak model, the foundation elements are assumed to be connected by a layer, which acts as a horizontal linkage of vertical elements. The Vlasov, viscoelastic, and Kerr models are also introduced.

5.2 Winkler–Pasternak foundation

5.2.1 Winkler model

The simplest representation of a continuous elastic foundation was proposed by Winkler, who assumed that the foundation consists of closely spaced, independent linear springs (see Fig. 5.2). Such a foundation is equivalent to that of a liquid base. The relationship between the pressure and the deflection of the foundation surface, both parallel to the z-axis, is expressed as follows:

$$p(x,y) = kw(x,y) \tag{5.16}$$

where k is the foundation modulus or specific weight of the liquid base; and $w(x,y)$ denotes the vertical displacement.

As a result, the foundation's reaction force is proportional to displacement, regardless of its direction. This type of mechanical behavior is referred to as bilateral (bilateral foundation). In mathematical modeling, the mechanical behavior of a foundation is idealized. The linear one parameter model of Winkler is the simplest; however, because of the various properties of different materials and soils (granular soils, clay soils, wet soils, pressured soils, etc.), more parameters should be involved to represent the behavior of materials used as a foundation.

5.2.2 Pasternak model

To develop a physically accurate and mathematically simple representation of these materials in the contact area, we begin with the Winkler foundation and introduce assumptions regarding interactions between spring elements.

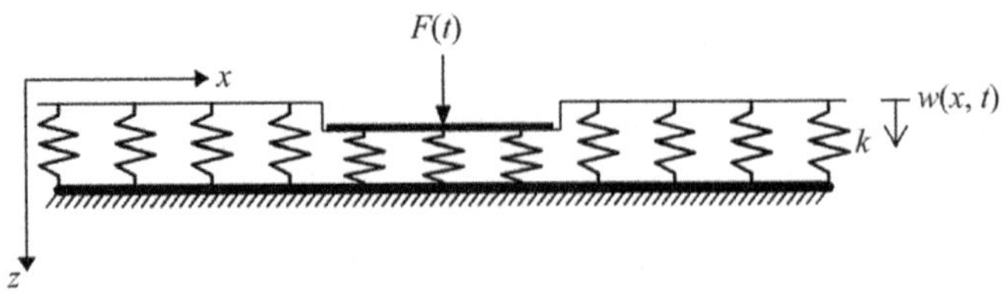

FIGURE 5.2 Winkler model.

Pasternak postulated the existence of shear interactions between the spring elements. This may be accomplished by connecting the ends of the springs to a beam or plate consisting of incompressible vertical elements, as shown in Fig. 5.3, which deforms only by transverse shear.

To derive the load–deflection relationship, the vertical equilibrium of a "shear layer" element bounded by the surfaces x, $x + \mathrm{d}x$, y and $y + \mathrm{d}y$ is considered, as shown in Fig. 5.4. Assuming that the foundation material is homogeneous and isotropic in the x–y plane, hence $G_x = G_y = G$, then

$$\begin{aligned}\tau_{xz} &= G\gamma_{xz} = G\frac{\partial w}{\partial x}\\ \tau_{yz} &= G\gamma_{yz} = G\frac{\partial w}{\partial y}\end{aligned} \tag{5.17}$$

Noting that the deflections of the contact area are $w = w(x, y)$, the shear forces per unit length of the shear layer are

$$\begin{aligned}N_x &= \int_0^1 \tau_{xz}\mathrm{d}z = G\frac{\partial w}{\partial x}\\ N_y &= \int_0^1 \tau_{yz}\mathrm{d}z = G\frac{\partial w}{\partial y}\end{aligned} \tag{5.18}$$

According to Fig. 5.4, the equilibrium equation is given by

$$\frac{\partial N_x}{\partial x} + \frac{\partial N_y}{\partial y} + p - q_s = 0 \tag{5.19}$$

Substituting Eq. (5.18) into Eq. (5.19), with $q_s = kw$, we obtain,

$$p = kw - G\nabla^2 w \tag{5.20}$$

The second term on the right-hand side of Eq. (5.20) is the effect of the shear interactions of the vertical elements. Zhaohua and Cook [9] introduced the use of two parameter foundations under beam elements. They related the second parameter to shear deformations or bending properties and presented the following general equation for the restoring force of two parameter foundations:

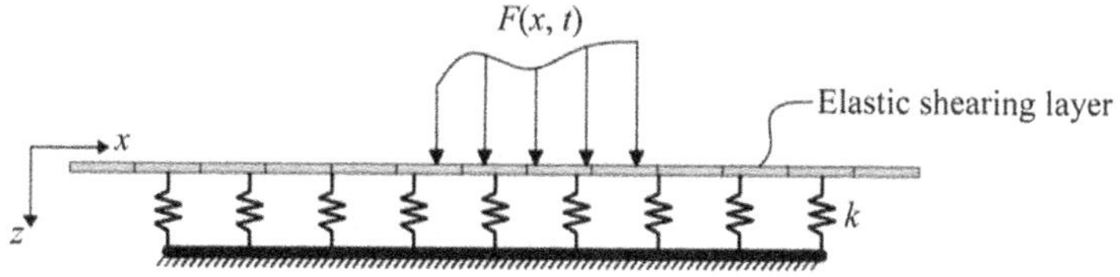

FIGURE 5.3 Pasternak model.

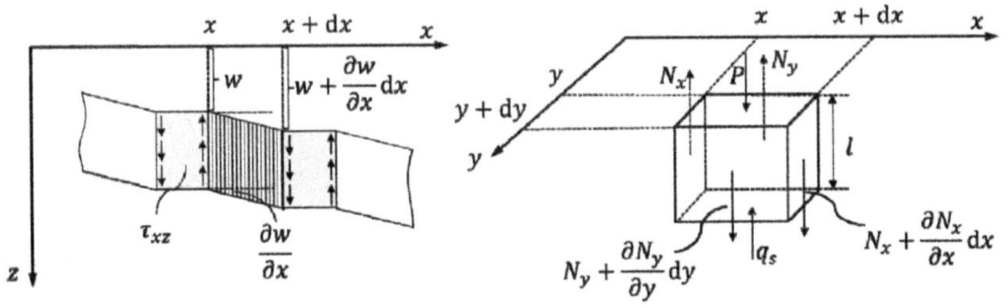

FIGURE 5.4 Element of shearing layer.

$$p(x,y,t) = K_1 w(x,y,t) - K_2 \frac{\partial^2 w(x,y,t)}{\partial x^2} \tag{5.21}$$

where K_1 is the Winkler's modulus and K_2 is the second parameter related to the interactions between springs. Eq. (5.21) is the presentation of the Pasternak model. The study concluded that when K_2 is not large, the Winkler foundation can provide a reliable result. In contrast, when K_2 is large and approaches K_1, use of the Winkler model leads to considerable error [9]. Since the introduction of two parameter foundation models, the Pasternak model has attracted the most attention among the different types of two parameter foundations, and it is considered to be the most generalized two parameter foundation [10–13]. Wang and Stephens investigated the effect of the shear layer by comparing the natural frequencies of a beam resting on the Winkler and Pasternak foundations and demonstrated that the shear layer increases the frequencies [14]. Shen [15] studied the buckling and postbuckling phenomena for orthotropic rectangular plates resting on a Pasternak foundation. They demonstrated that the postbuckling behavior is significantly influenced by the foundation parameters.

5.2.3 Vlasov model

V. Z. Vlasov approached the problem from a continuum mechanics perspective. He formulated the problem using a variational method. Imposing certain restrictions upon the possible deformations of an elastic layer, he obtained a load–deflection relation under a specific set of assumptions, which is also considered to be a two parameter foundation and identical to Eq. (5.21) after the constants are redefined. Here, a summary for the case of Vlasov model is presented [16]. For the plane strain condition of the elastic layer in the x–z plane (Fig. 5.5), the displacement components are

$$u(x,z) = 0, w(x,z) = w(x)h(z) \tag{5.22}$$

where the function $h(z)$ provides the vertical variations of displacement. For a linear variation case, the function can be written as:

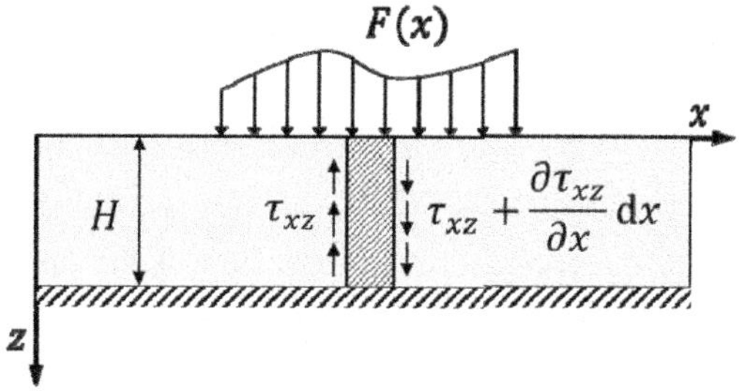

FIGURE 5.5 Vlasov model.

$$h(z) = 1 - \frac{z}{H} \tag{5.23}$$

in which H is the vertical thickness of elastic medium. By using stress–strain relation and the Lagrange's principle of virtual work, the force–displacement relation of the Vlasov foundation can be expressed as:

$$p(x) = kw(x) - k_1 \frac{d^2 w(x)}{dx^2} \tag{5.24}$$

in which

$$\begin{cases} k = \dfrac{E_0}{(1 - \nu_0^2)} \displaystyle\int_0^H \left(\frac{dh(z)}{dz}\right)^2 dz \\ k_1 = \dfrac{E_0}{2(1 + \nu_0)} \displaystyle\int_0^H h(z)^2 \, dz \\ E_0 = \dfrac{E_s}{(1 - \nu_s^2)} \\ \nu_0 = \nu_s (1 - \nu_s) \end{cases} \tag{5.25}$$

where E_s is the elastic modulus and ν_s is the Poisson's ratio of the elastic media.

The procedure for determining the first and second parameters of the Vlasov model also provides insight into the nature of the shear and elastic moduli used in the Pasternak and other two parameter models. Mullapudi and Ayoub demonstrated the importance of including the second parameter in the mathematical modeling of soil reaction forces. They also presented an iterative algorithm in order to calculate the parameters of a soil medium based on the Vlasov and Pasternak models [17]. Ozgan and Daloglu used finite element analysis to study the vibrations of a plate on a Vlasov foundation [18]. In their vibration system, the Vlasov model was used to present the interactions between soil and the plate.

5.3 Kerr foundation

The Kerr foundation is shown in Fig. 5.6. For derivation of the plate equation, we consider the plate deflection w consisting of two parts

$$w(x,y) = w_1(x,y) + w_2(x,y) \tag{5.26}$$

where w_1 is the deflection due to the contraction or extension of the upper spring layer and w_2 is due to the rest of the foundation. According to Fig. 5.6 (b) and Eqs. (5.16) and (5.20), we obtain

$$p(x,y) = cw_1 \tag{5.27}$$

and

$$p(x,y) = kw_2 - G\nabla^2 w_2 \tag{5.28}$$

Performing k/c and $G\nabla^2/c$ on Eq. (5.27) and subtracting the resulting equations yields

$$\frac{k}{c}p - \frac{G}{c}\nabla^2 p = kw_1 - G\nabla^2 w_1 \tag{5.29}$$

Adding Eqs. (5.28) and (5.29), we obtain the relation between the interface pressure and the deflection of the plate w as

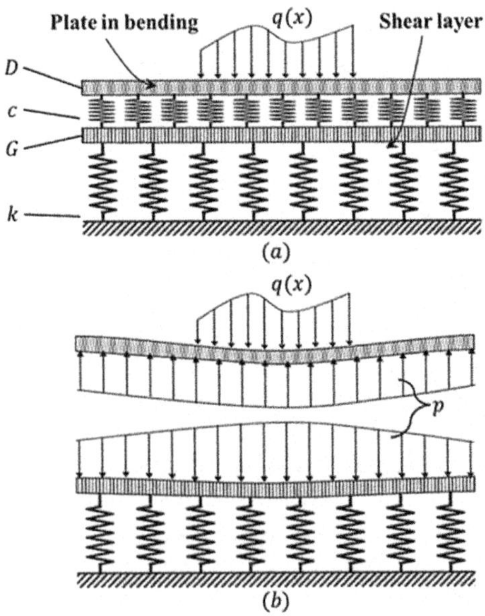

FIGURE 5.6 Kerr model.

$$\left(1+\frac{k}{c}\right)p - \frac{G}{c}\nabla^2 p = kw - G\nabla^2 w$$

or (5.30)

$$p - \frac{G}{k+c}\nabla^2 p = \frac{ck}{k+c} \cdot - \frac{cG}{k+c}\nabla^2 w$$

This foundation model is primarily used to simulate the behavior of flexible soil and to represent the boundary conditions between a structure and the soil. Avramidis and Morfidis made a comparison among the responses of one-, two-, and three parameter foundation models and showed that the Kerr-type foundation provides more realistic results for the considered case [19]. Hetenyi [20] presented the general procedure for analyzing beams on a Kerr-type foundation using an analytical scheme. Hetenyi's procedure enables the determination of the dynamic response of a beam on a Kerr foundation under arbitrary loads and boundary conditions. Based on the exact solution of the governing differential equation, Morfidis formulated the problem of a Timoshenko beam on the three parameter foundation in matrix form and investigated the natural frequency of the structure's vibrations [21,22]. Limkatanyu et al. [23] derived the foundation's stiffness matrix using the virtual force principle. Wang and Zhang employed the three parameter foundation to model the adhesive layer between curved beam structures and fiber-reinforced polymer (FRP) plates [24]. They also used the three parameter model to satisfy the boundary condition of zero shear stress at the edge of adhesive joints [25]. Wang et al. studied the parametric instability of an inextensional beam, considering Winkler, Pasternak, and Kerr foundations [26].

5.4 Viscoelastic foundation

5.4.1 Kelvin–Voigt model

A weakness of the Winkler model is that it does not include viscoelastic elements. The Kelvin–Voigt model includes viscoelastic behavior and presents the behavior of a medium with a series of discrete springs parallel with dampers (as shown in Fig. 5.7). In other words, the Kelvin–Voigt model introduces the effects of viscoelasticity into the elastic Winkler model. The relationship between the restoring force and displacement reads as follows [27]:

$$p(x,t) = \left(k + c\frac{\partial}{\partial t}\right)w(x,t) \tag{5.31}$$

where c is the damping of viscous elements. The Kelvin–Voigt model has been used in several studies, including nonlinear frequency analysis of beams on Kelvin–Voigt foundations, finite element formulation of infinite beams on Kelvin–Voigt foundations, and wave propagation studies of infinite beams on such foundations [28–30].

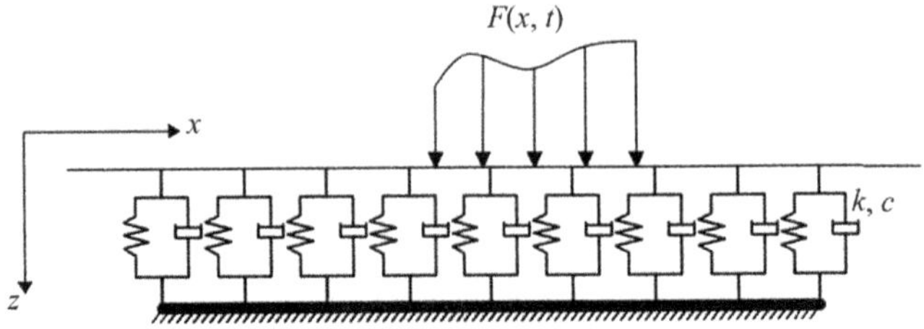

FIGURE 5.7 Kelvin–Voigt model.

5.4.2 Maxwell model

In this model, a series of springs along with dampers are placed as shown in Fig. 5.8. The reaction force of the foundation can be calculated from the following equation [27]:

$$\left.\begin{array}{l} p(x,t) = k\Delta_1(x,t) = c\Delta_2(x,t) \\ w(x,t) = \Delta_{\text{tot}}(x,t) = \Delta(x,t) + \Delta_2(x,t) \end{array}\right\} \Rightarrow \frac{\partial}{\partial t} w(x,t) = \left(\frac{1}{k}\frac{\partial}{\partial t} + \frac{1}{c}\right) p(x,t) \tag{5.32}$$

where $p(x, t)$ is the foundation reaction, and $\Delta_1(x,t)$ and $\Delta_2(x,t)$ are the displacement changes of spring elements and damper elements, respectively (presented in Fig. 5.8). Hatada et al. modeled the assembly of braces and viscous dampers for tall buildings using the Maxwell model. In their analysis, the Maxwell model was selected because of the arrangement of the braces and dampers [31]. Soukup and Volek used the Maxwell model to describe the viscoelastic behavior of plates, and then conducted stress analysis for a thin plate [32]. Other viscoelastic models (Zener model, Poynting–Thomson model, Burger model, etc.) can be found in literature [33].

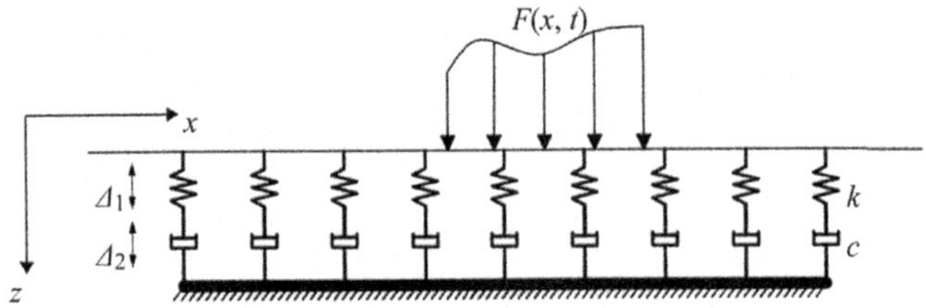

FIGURE 5.8 Maxwell model.

5.5 The formulation of stability and results

For a higher-order shear deformation beam on a Winkler–Pasternak foundation, eliminating the bending moment M and shearing force V from Eqs. (5.4), (5.11), (5.14), and (5.15), we obtain the following coupled governing differential equations

$$\left(GA_f+\alpha P\right)\left(\frac{\partial\psi}{\partial x}+\frac{\partial^2 w}{\partial x^2}\right)-P\frac{\partial^2 w}{\partial x^2}-K_1 w+K_2\frac{\partial^2 w}{\partial x^2}=\rho A\frac{\partial^2 w}{\partial t^2} \tag{5.33}$$

$$EI_f\left(\frac{\partial^2\psi}{\partial x^2}+\frac{\partial^3 w}{\partial x^3}\right)-EI\frac{\partial^3 w}{\partial x^3}-\left(GA_f+\alpha P\right)\left(\psi+\frac{\partial w}{\partial x}\right)$$
$$=-\rho I\frac{\partial^3 w}{\partial t^2\partial x}+\rho I_f\left(\frac{\partial^2\psi}{\partial t^2}+\frac{\partial^3 w}{\partial t^2\partial x}\right) \tag{5.34}$$

In order to further simplify the coupled partial differential equations, we introduce a new unknown function, $F(x, t)$, to express the deflection w and rotation ψ as follows:

$$w=F-\frac{EI_f}{GA_f+\alpha P}\frac{\partial^2 F}{\partial x^2}+\frac{\rho I_f}{GA_f+\alpha P}\frac{\partial^2 F}{\partial t^2} \tag{5.35}$$

$$\psi=-\left(\frac{\partial F}{\partial x}+\frac{EI-EI_f}{GA_f+\alpha P}\frac{\partial^3 F}{\partial x^3}+\frac{\rho I_f-\rho I}{GA_f+\alpha P}\frac{\partial^3 F}{\partial t^2\partial x}\right) \tag{5.36}$$

from which

$$\psi+\frac{\partial w}{\partial x}=\frac{\rho I}{GA_f+\alpha P}\frac{\partial^3 F}{\partial t^2\partial x}-\frac{EI}{GA_f+\alpha P}\frac{\partial^3 F}{\partial x^3} \tag{5.37}$$

follows. Using the above expressions, it is easily seen that the partial differential equations [Eqs. (5.33) and (5.34)] are identically satisfied if the function F satisfies the following governing equation

$$EI\left[1-\frac{(P+K_2)I_f}{(GA_f+\alpha P)I}\right]\frac{\partial^4 F}{\partial x^4}+\left(P+K_2-\frac{K_1 EI_f}{GA_f+\alpha P}\right)\frac{\partial^2 F}{\partial x^2}+K_1 F-$$
$$\rho I\left[1-\frac{PI_f-(EA+K_2)I_f}{(GA_f+\alpha P)I}\right]\frac{\partial^4 F}{\partial t^2\partial x^2}+\rho A\left[1+\frac{K_1 I_f}{(GA_f+\alpha P)A}\right]\frac{\partial^2 F}{\partial t^2}+\frac{\rho A\rho I_f}{GA_f}+\alpha P\frac{\partial^4 F}{\partial t^4}=0 \tag{5.38}$$

As a consequence, based on the higher-order shear deformation beam theory, we have derived a single governing differential equation for the dynamic behavior of an axially loaded beam/column on a two-parameter elastic foundation. This can be applied to solve free vibration and buckling problems. In the following, we employ it to deal with dynamic stability. It is noted that both shear deformation and rotary inertia are taken into account. Moreover, the shear stress automatically fulfills traction-free conditions on the surface of the beam/column. Note that the extreme case of $\alpha = 0$ gives the governing equation under the Engesser hypothesis, and the extreme case of $\alpha = 1$ gives the one under the Haringx hypothesis. For convenience, the two cases are named higher-order Engesser beams (HEB) and higher-order Haringx beams (HHB). Clearly, the transition parameter α comes into play only when shear deformation is considered. From Eq. (5.37), it can be seen that if shear rigidity is sufficiently large, $\psi = -\partial w/\partial x$ is deduced and it infers that the slope angle is the same as the rotation of the cross section. In this case, we find that the governing equation [Eq. (5.38)] reduces to

$$EI\frac{\partial^4 F}{\partial x^4} + (P + K_2)\frac{\partial^2 F}{\partial x^2} + K_1 F - \rho I\frac{\partial^4 F}{\partial t^2 \partial x^2} + \rho A\frac{\partial^2 F}{\partial t^2} = 0 \tag{5.39}$$

which gives the governing equation for dynamic behaviors of axially loaded Rayleigh beams resting on an elastic foundation. When Winkler foundation is considered, i.e., K_1 remains the same and $K_2 = 0$, the governing equation in Ref. [34] is recovered from Eq. (5.39). If both of the elastic stiffness coefficients are absent, the governing equation for axially loaded Rayleigh beams in a free space [35,36] is restored. Although shear deformation is neglected, rotary inertia of the cross section, $\rho I\partial^4 F/\partial t^2 \partial x^2$, is still considered. When further removing the stiffness constants K_1 and K_2 of elastic foundation and the rotary inertia of the cross section, $\rho I\partial^4 F/\partial t^2 \partial x^2$, the well-known differential equation governing the dynamic behavior of an axially loaded Euler–Bernoulli beam is retrieved [37].

Finally, in closing this section, we compare the resulting Eq. (5.38) with that for Timoshenko beams with $\alpha = 0$. To achieve it, let us consider a simple case of neither elastic foundation nor axial loading. The governing equation [Eq. (5.38)] in this case becomes

$$EI\frac{\partial^4 F}{\partial x^4} + \rho A\frac{\partial^2 F}{\partial t^2} - \rho A\frac{I}{A}\left(1 + \frac{EAI_f}{GA_f I}\right)\frac{\partial^4 F}{\partial t^2 \partial x^2} + \frac{\rho^2 A I_f}{GA_f}\frac{\partial^4 F}{\partial t^4} = 0 \tag{5.40}$$

By comparing it with the governing equation for Timoshenko beams [38], we find that the shear correction factor κ introduced in the Timoshenko beam theory is

$$\kappa = \frac{IA_f}{I_f A} \tag{5.41}$$

The result provides us with another way to determine the shear correction factor through the present higher-order beam theory. In this way, the values of the shear correction factor for given warping shape functions are tabulated in Table 5.2. Of course, it should be mentioned that it is not necessary to introduce the shear correction factor when a higher-order beam model is adopted. Moreover, it does not cause any superfluous difficulty in solving the governing equation because its order is the same as that of the Timoshenko beam theory. On the contrary, its solution based on the higher-order beam theory is more accurate than that based on the Timoshenko beam theory.

TABLE 5.2 Typical warping shape functions for beams with rectangular and circular cross sections.

Cross section	Warping shape	$f(y,z)$	A/A_f	κ
Rectangular	Polynomial	$z-\frac{4z^3}{3h^2}$	$\frac{3}{2}$	$\frac{5}{6}$ (or 0.8333)
	Trigonometric	$\frac{h}{\pi}\sin\frac{\pi z}{h}$	$\frac{\pi}{2}$	0.8225
	Exponential	$z\exp\left(-\frac{2z^2}{h^2}\right)$	$\sqrt{e}$	0.8117
Circular	Polynomial	$z\left(1-\frac{y^2+z^2}{3R^2}\right)$	$\frac{3}{2}$	$\frac{6}{7}$ (or 0.8571)
	Trigonometric	$\frac{2zR}{\pi\sqrt{y^2+z^2}}\sin\frac{\pi\sqrt{y^2+z^2}}{2R}$	$\frac{\pi}{2}$	0.8488
	Exponential	$z\exp\left(-\frac{y^2+z^2}{2R^2}\right)$	$\sqrt{e}$	0.8405

In the foregoing section, we have derived a single governing differential equation for dynamic behaviors of an axially loaded beam/column resting on a Winkler–Pasternak elastic foundation based on a higher-order shear deformation beam theory. This section is concerned with the analysis of the dynamic stability of a cantilevered beam/column under the action of a nonconservative force P that is applied at a massless rigid attachment mounted to the column at its free end, as shown in Fig. 5.1. The force is applied at an angle $\beta\partial w/\partial x$ to the undeformed axis of the column. This problem is related to the dynamic stability of a generalized Reut's column [39]. When the force is applied at the free tip, the generalized Beck's column [39] is reduced to the case of a subtangential follower force applied at the free tip. Thus, the corresponding boundary conditions can be stated as

$$w=0, \psi=0, \text{at } x=0 \tag{5.42}$$

$$M-Pw\delta=0, V+P\beta\frac{\partial w}{\partial x}=0, \text{at } x=L \tag{5.43}$$

where β is often termed as the tangency coefficient, and δ is a constant taking 0 or 1. When $\delta=0$, the term Pw in Eq. (5.43) vanishes and this case reduces to Beck's columns under a generalized follower force. In particular, the case with $\beta=1$ corresponds to a follower force, whereas $\beta=0$ corresponds to a conservative force (dead load). In the case of $\delta=1$, the classical Rent's column is recovered from $\beta=0$. Therefore, the case of $\delta\neq 0$ and $\beta\neq 0$ gives generalized Rent's and Beck's columns.

To solve the above problem, for the governing equation [Eq. (5.38)], one applies the method of separation of variables and assumes

$$\begin{cases} F(x,t) = y(\xi)\mathrm{e}^{\mathrm{i}\omega t} \\ \xi = \dfrac{x}{L} \end{cases} \tag{5.44}$$

where ω is angle frequency, and $y(\xi)$ is an unknown function. Substitution of Eq. (5.44) into Eq. (5.38) leads to the following ordinary differential equation

$$a_0 y^{IV} + a_1 y'' - a_2 y = 0 \tag{5.45}$$

with

$$\begin{cases} a_0 = 1 - \dfrac{(p+k_2)\eta}{\zeta + \alpha p} \\ a_1 = p + k_2 - \dfrac{k_1\eta}{\zeta + \alpha p} + \dfrac{\Omega^2}{s^2}\left[1 - \eta\dfrac{p - (s^2 + k_2)}{\zeta + \alpha p}\right] \\ a_2 = \Omega^2\left[1 + \dfrac{k_1\eta}{s^2(\zeta + \alpha p)}\right] - \dfrac{\eta\Omega^4}{s^2(\zeta + \alpha p)} - k_1 \end{cases} \tag{5.46}$$

For the convenience of later analysis, the dimensionless quantities:

$$\begin{cases} \xi = \dfrac{x}{L} \\ p = \dfrac{PL^2}{EI} \\ k_1 = \dfrac{K_1 L^4}{EI} \\ k_2 = \dfrac{K_2 L^2}{EI} \\ \Omega = \omega L^2\sqrt{\dfrac{\rho A}{EI}} \\ \eta = \dfrac{I_f}{I} \\ s = L\sqrt{\dfrac{A}{I}} \\ \zeta = \dfrac{A_f s^2}{2(1+\nu)A} \end{cases} \tag{5.47}$$

are defined.

The remaining task is to seek a function satisfying Eq. (5.45) along with appropriate boundary conditions. From Eqs. (5.7) and (5.37) we can represent the shear strain as:

$$\gamma_{xz} = -\frac{\Omega^2 y' + s^2 y'''}{s^2(\zeta + \alpha p)L}\frac{\partial f}{\partial z} \tag{5.48}$$

Also, Eqs. (5.35) and (5.36) allow one to derive the deflection and rotation:

$$w = \left[1 - \frac{\eta\Omega^2}{s^2(\zeta + \alpha p)}\right]y - \frac{\eta}{\zeta + \alpha p}y'' \tag{5.49}$$

$$\psi = -\frac{1}{L}\left\{\left[1 - \frac{1-\eta}{s^2(\zeta + \alpha p)}\Omega^2\right]y' + \frac{1-\eta}{\zeta + \alpha p}y'''\right\} \tag{5.50}$$

In view of Eqs. (5.11) and (5.12), the bending moment and shearing force can take the following dimensionless form:

$$M = -\frac{EI}{L^2}y'' \tag{5.51}$$

$$Q = -\frac{GA_f}{s^2(\zeta + \alpha p)L}\left(\Omega^2 y' + s^2 y'''\right) \tag{5.52}$$

Additionally, from Eq. (5.15) one has

$$V = -\frac{EI}{L^3}\left\{\left(1 - \frac{p\eta}{\zeta + \alpha p}\right)y''' + \left[p + \frac{1}{s^2}\left(1 - \frac{p\eta}{\zeta + \alpha p}\right)\Omega^2\right]y'\right\} \tag{5.53}$$

Next, we solve the governing equation [Eq. (5.45)] and acquire a general solution

$$y(\xi) = C_1\cos(\chi_1\xi) + C_2\sin(\chi_1\xi) + C_3\cosh(\chi_2\xi) + C_4\sinh(\chi_2\xi) \tag{5.54}$$

where $C_j(j = 1, \ldots, 4)$ are unknown constants and $\chi_j(j = 1, 2)$ are the roots of the following quadratic equation in χ^2.

$$a_0\chi^4 - a_1\chi^2 - a_2 = 0 \tag{5.55}$$

Solving this equation, its roots can be expressed as:

$$\chi_{1,2} = \sqrt{\frac{\sqrt{a_1^2 + 4a_0a_2} \pm a_1}{2a_0}} \tag{5.56}$$

After substituting Eq. (5.54) into the foregoing expressions for the deflection [Eq. (5.49)], rotation [Eq. (5.50)], bending moment [Eq. (5.51)] and vertical force [Eq. (5.53)], by applying the boundary conditions in Eqs. (5.42) and (5.43), one obtains

$$\begin{cases} b_1C_1 + b_3C_3 = 0 \\ b_2C_2 + b_4C_4 = 0 \\ b_5(C_1 \cos\chi_1 + C_2 \sin\chi_1) + b_7(C_3 \cosh\chi_2 + C_4 \sinh\chi_2) = 0 \\ b_6C_1 \sin\chi_1 - b_6C_2 \cos\chi_1 + b_8C_3 \sinh\chi_2 + b_8C_4 \cosh\chi_2 = 0 \end{cases} \tag{5.57}$$

where

$$\begin{cases} b_1 = 1 - \dfrac{\eta(\Omega^2 - s^2\chi_1^2)}{s^2(\zeta + \alpha p)} \\ b_2 = \left[1 - (1-\eta)\dfrac{\Omega^2 + s^2\chi_1^2}{s^2(\zeta + \alpha p)}\right]\chi_1 \\ b_3 = 1 - \dfrac{\eta(\Omega^2 + s^2\chi_2^2)}{s^2(\zeta + \alpha p)} \\ b_4 = \left[1 - (1-\eta)\dfrac{\Omega^2 - s^2\chi_2^2}{s^2(\zeta + \alpha p)}\right]\chi_2 \\ b_5 = p\delta\left(1 - \dfrac{\eta\Omega^2}{s^2(\zeta + \alpha p)}\right) - \chi_1^2\left(1 - \dfrac{p\eta\delta\Omega^2}{\zeta + \alpha p}\right) \\ b_6 = \chi_1^3\left[1 - \dfrac{(1-\beta)p\eta}{\zeta + \alpha p}\right] - \chi_1\left[(1-\beta)p + \dfrac{1}{s^2}\left(1 - \dfrac{(1-\beta)p\eta}{\zeta + \alpha p}\right)\Omega^2\right] \\ b_7 = p\delta\left(1 - \dfrac{\eta\Omega^2}{s^2(\zeta + \alpha p)}\right) + \chi_2^2\left(1 - \dfrac{p\eta\delta\Omega^2}{\zeta + \alpha p}\right) \\ b_8 = \chi_2^3\left[1 - \dfrac{(1-\beta)p\eta}{\zeta + \alpha\eta}\right] + \chi_2\left[(1-\beta)p + \dfrac{1}{s^2}\left(1 - \dfrac{(1-\beta)p\eta}{\zeta + \alpha p}\right)\Omega^2\right] \end{cases} \tag{5.58}$$

By applying the condition for the existence of a nontrivial solution to the linear system given by Eq. (5.57) and performing some simplification, the following characteristic equation is obtained

$$\frac{b_7b_8}{b_3b_4} - \frac{b_5b_6}{b_1b_2} + \left(\frac{b_6b_7}{b_2b_3} - \frac{b_5b_8}{b_1b_4}\right)\cos\chi_1 \cosh\chi_2 + \left(\frac{b_6b_7}{b_1b_4} + \frac{b_5b_8}{b_2b_3}\right)\sin\chi_1 \sinh\chi_2 = 0 \tag{5.59}$$

In what follows, we consider several special cases. First, let us restrict our attention to sufficiently long columns, i.e., the slenderness ratio $s \rightarrow \infty$, which is equivalent to the neglect of both shear deformation and rotary inertia of the cross section. To clearly elucidate this point, we may set G to be large enough, so that $s \rightarrow \infty$. In this case, one finds

$$\chi_{1,2} = \sqrt{\frac{\sqrt{\left(p + k_2 + \frac{\Omega^2}{s^2}\right)^2 + 4\left(\Omega^2 - k_1\right)} \pm \left(p + k_2 + \frac{\Omega^2}{s^2}\right)}{2}} \tag{5.60}$$

and

$$\begin{cases} b_1 = b_3 = 1 \\ b_2 = \chi_1 \\ b_4 = \chi_2 \\ b_5 = p\delta - \chi_1^2 \\ b_7 = p\delta + \chi_2^2 \\ b_6 = \chi_1^3 - \chi_1\left[(1-\beta)p + \frac{\Omega^2}{s^2}\right] \\ b_8 = \chi_2^3 + \chi_2\left[(1-\beta)p + \frac{\Omega^2}{s^2}\right] \end{cases} \tag{5.61}$$

Under such circumstances, with the help of the following results

$$\begin{cases} \chi_1\chi_2 = \sqrt{\Omega^2 - k_1} \\ \chi_2^2 - \chi_1^2 = -\left(p + k_2 + \frac{\Omega^2}{s^2}\right) \\ \chi_2^4 + \chi_1^4 = \left(p + k_2 + \frac{\Omega^2}{s^2}\right)^2 + 2\left(\Omega^2 - k_1\right) \end{cases} \tag{5.62}$$

one deduces the characteristic equation [Eq. (5.59)] below

$$\begin{aligned} & 2(\Omega^2 - k_1) + k_2\frac{\Omega^2}{s^2} + k_2^2 + p\left[(\beta+\delta)\left(p + \frac{\Omega^2}{s^2}\right) - 2p\beta\delta + k_2(1+\beta-\delta)\right] \\ & \quad + \left\{2(\Omega^2 - k_1) + k_2\frac{\Omega^2}{s^2} + p[p\beta\delta + (1+\delta-\beta)k_2] + \left[p(1-\delta) + \frac{\Omega^2}{s^2}\right]\left[p(1-\beta) + \frac{\Omega^2}{s^2}\right]\right\}\cos\chi_1\cosh\chi_2 \\ & \quad + \left\{\sqrt{\Omega^2 - k_1}\left[2p(\beta+\delta) + k_2 - \left(p + \frac{\Omega^2}{s^2}\right)\right] + \left(k_2 + p + \frac{\Omega^2}{s^2}\right)\frac{p\delta(k_2 + p\beta)}{\sqrt{\Omega^2 - k_1}}\right\}\sin\chi_1\sinh\chi_2 = 0 \end{aligned} \tag{5.63}$$

Clearly, in the final characteristic equation [Eq. (5.63)], although shear rigidity does not influence the dynamic stability of a column subjected to a nonconservative force on an eccentric position of a massless rigid attachment mounted to the free end, the other parameters including dimensionless elastic foundation stiffness, rotary inertia, the tangency coefficient, the load position, etc. alter the dynamic stability behavior. As mentioned earlier, since the shear rigidity is large enough that it prevents the cross section from deforming, the characteristic equation [Eq. (5.63)] is seen to be independent of the cross section warping shape function $f(y, z)$. For this reason, Haringx's and Engesser's hypotheses coincide. In other words, Eq. (5.63) gives the corresponding characteristic equation for the dynamic behavior of Rayleigh columns resting on an elastic foundation based on Beck's and Reut's load assumptions. Furthermore, when the Pasternak elastic foundation stiffness constant k_2 is zero, i.e., only a Winkler foundation is considered, the characteristic equation [Eq. (5.63)] becomes

$$
\begin{aligned}
&2(\Omega^2 - k_1) + p\left[(\beta+\delta)\left(p + \frac{\Omega^2}{s^2}\right) - 2p\beta\delta\right] \\
&\quad + \left\{2(\Omega^2 - k_1) + p^2\beta\delta + \left[p(1-\delta) + \frac{\Omega^2}{s^2}\right]\left[p(1-\beta) + \frac{\Omega^2}{s^2}\right]\right\}\cos\chi_1 \cosh\chi_2 \\
&\quad + \left\{\sqrt{\Omega^2 - k_1}\left[2p(\beta+\delta) - \left(p + \frac{\Omega^2}{s^2}\right)\right] + \left(p + \frac{\Omega^2}{s^2}\right)\frac{p^2\beta\delta}{\sqrt{\Omega^2 - k_1}}\right\}\sin\chi_1 \sinh\chi_2 = 0
\end{aligned}
\tag{5.64}
$$

An interesting conclusion is drawn from the above. A careful observation of the characteristic equation shows that the role of two parameters, β and δ is the same, equivalent to say, the characteristic equation for $\beta = 1$, $\delta = 0$ is the same as that for $\delta = 1$, $\beta = 0$. The values of δ and β can be interchanged with each other. Consequently, the conclusion of dynamic stability of generalized Reut's columns ($\beta = 0$) is in exact agreement with that of generalized Beck's columns ($\delta = 0$). It is worth noting that when $\delta \neq 0, \beta \neq 0$, the synergic influences of two parameters cannot be obtained by linear superposition due to a nonlinear effect, as shown in Eq. (5.63). Of course, if $k_2 \neq 0$, the above conclusion does not hold.

If neglecting the shear deformation and rotary inertia or letting the slenderness ratio s be sufficiently large, $s \to \infty$, for slender beams/columns, the characteristic equation [Eq. (5.59)] reduces to

$$
\begin{aligned}
&2(\Omega^2 - k_1) + k_2^2 + p[(\beta+\delta)p - 2p\beta\delta + k_2(1+\beta-\delta)] \\
&\quad + \{2(\Omega^2 - k_1) + p[p\beta\delta + (1+\delta-\beta)k_2] + p^2(1-\delta)(1-\beta)\}\cos\chi_1 \cosh\chi_2 \\
&\quad + \left\{\sqrt{\Omega^2 - k_1}\,[2p(\beta+\delta) + k_2 - p] + (k_2 + p)\frac{p\delta(k_2 + p\beta)}{\sqrt{\Omega^2 - k_1}}\right\}\sin\chi_1 \sinh\chi_2 = 0
\end{aligned}
\tag{5.65}
$$

Additionally, if only Beck's columns are concerned, i.e., $\delta = 0$, meaning that a tangential follower force is applied at the free end, the characteristic equation [Eq. (5.65)] reduces to

$$
\begin{aligned}
&2(\Omega^2 - k_1) + k_2^2 + p[\beta(p + k_2) + k_2] + [2(\Omega^2 - k_1) + p(k_2 + p)(1-\beta)]\cos\chi_1 \cosh\chi_2 \\
&\quad + \sqrt{\Omega^2 - k_1}[(2\beta - 1)p + k_2]\sin\chi_1 \sinh\chi_2 = 0
\end{aligned}
\tag{5.66}
$$

If the elastic foundation is also absent, the above characteristic equation [Eq. (5.66)] becomes

$$2\Omega^2 + p^2\beta + \left[2\Omega^2 + p^2(1-\beta)\right]\cos\chi_1 \cosh\chi_2 + p\Omega(2\beta - 1)\sin\chi_1 \sinh\chi_2 = 0 \tag{5.67}$$

identical to that derived in Ref. [40]. Especially, in the case of $\beta = 1$, the characteristic equation for dynamic stability of a cantilevered column subjected to a follower force

$$2\Omega^2 + p^2 + 2\Omega^2 \cos\chi_1 \cosh\chi_2 + p\Omega \sin\chi_1 \sinh\chi_2 = 0 \tag{5.68}$$

is recovered [37], whereas, in the case of $\beta = 0$, the characteristic equation for free vibration of a cantilevered beam under compressive axial forces [41].

$$2\Omega^2 + \left(2\Omega^2 + p^2\right)\cos\chi_1 \cosh\chi_2 - p\Omega \sin\chi_1 \sinh\chi_2 = 0 \tag{5.69}$$

is retrieved.

If keeping k_1 and k_2 present and letting $\delta = 0$ and $\beta = 1$, from Eq. (5.65) one gains

$$2 + \left(\frac{p + k_2}{\sqrt{\Omega^2 - k_1}}\right)^2 + 2\cos\chi_1 \cosh\chi_2 + \frac{p + k_2}{\sqrt{\Omega^2 - k_1}} \sin\chi_1 \sinh\chi_2 = 0 \tag{5.70}$$

When $k_2 = 0$, the model reduces to that of Beck's column on a Winkler elastic foundation [42]. The classical Beck's column is recovered when $k_1 = k_2 = 0$ [37]. If considering Reut's and Beck's columns subjected to a follower force, $\beta = \delta = 1$, for slender columns, we find that the characteristic equation [Eq. (5.65)] collapses to

$$\begin{aligned} 2 + \frac{k_2(p + k_2)}{\Omega^2 - k_1} + \left[2 + \frac{p(p + k_2)}{\Omega^2 - k_1}\right] \cos\chi_1 \cosh\chi_2 \\ + \left[\frac{3p + k_2}{\sqrt{\Omega^2 - k_1}} + \frac{p(p + k_2)^2}{\left(\Omega^2 - k_1\right)^{3/2}}\right] \sin\chi_1 \sinh\chi_2 = 0 \end{aligned} \tag{5.71}$$

From Eq. (5.66), with $\delta = 1$ and $\beta = 0$, one has the characteristic equation

$$\begin{aligned} 2 + \frac{k_2^2 + p^2}{\Omega^2 - k_1} + 2\left(1 + \frac{k_2 p}{\Omega^2 - k_1}\right) \cos\chi_1 \cosh\chi_2 \\ + \frac{k_2 + p}{\sqrt{\Omega^2 - k_1}}\left(1 + \frac{p k_2}{\Omega^2 - k_1}\right) \sin\chi_1 \sinh\chi_2 = 0 \end{aligned} \tag{5.72}$$

In the foregoing section, we have obtained the characteristic equation [Eq. (5.59)] for a column resting on a two parameter elastic foundation and subjected to a compressive force according to the higher-order beam theory. This section is devoted to numerical computations to show the influence of key parameters on the dynamic stability of cantilevered columns. Since many dimensionless parameters are included in our model, several typical situations are considered in the following. For comparison, we also calculate the numerical results based on the theory of Euler–Bernoulli beams (EBB) using the characteristic equation Eq. (5.65). In the HEB and HHB theories, s takes a finite value, but $\alpha = 0,\ 1$ are required, respectively. As a representative example, we only give discussion for a cantilevered column with length-to-radius ratio $s = 10$ subjected to a tangential follower force at the free end unless otherwise stated. In addition, Poisson's ratio ν takes 0.3 in the following computations.

First, let us carry out some numerical calculations to verify the effectiveness and validity of the proposed model. For this purpose, we consider the free vibration and buckling of cantilevered homogeneous beams which correspond, respectively, to $P = 0$ and $\Omega = 0$. The first five frequency parameters for a cantilevered rectangular beam are tabulated in Tables 5.3–5.5, corresponding to polynomial, trigonometric, and exponential warping shapes, respectively. For comparison, we also list the dimensionless frequencies corresponding to the EBB theory [43] and Timoshenko beam theory (TB) [44], respectively. Since the numerical results in Ref. [44] correspond to rectangular beams, we only calculate them for rectangular beams for comparison, circular beams can be similarly treated and are therefore omitted here. From Tables 5.3–5.5, it can be seen that when the slenderness ratio s is large enough ($s = 100$), our results are in close agreement with the classical ones based on the EBB theory. This is easily understood since the characteristic equation for sufficiently long beams reduces to the one for the EBB theory, as indicated in the previous section. Thus, all the natural frequencies for infinitely long beams are the same in theory as those for the latter case. It is anticipated that when s becomes smaller, the difference among these natural frequencies is larger. For a relatively short beam, the values of the dimensionless frequencies have a remarkable difference, as shown in Tables 5.3–5.5. For example, for $s = 10$, the dimensionless frequencies are also different from those for the EBB [43] and TB theories [44] with the same shear correction factor 5/6 for rectangular beams. This is due to the fact that the cross section is no longer rigid and has a warping shape in the higher-order beam theory, whereas the TB theory adopts a plane cross section hypothesis. Moreover, our results are always lower than the corresponding dimensionless frequencies of the TB theory. In other words, the dimensionless frequencies calculated via the TB theory are overestimated. When $s = 5$, the dimensionless frequencies further deviate from the dimensionless frequencies of EBB and TB theories. By careful inspection, our results based on the HEB and HHB are identical in the absence of axial load. Nonetheless, Haringx's and Engesser's hypotheses give different frequencies, depending on the magnitude of the axial load. In particular, the natural frequencies obtained from the HEB theory are slightly lower than those obtained from the HHB theory. This conclusion is in agreement with that for the free vibration of Timoshenko beams [45].

On the other hand, to validate our results, let us calculate the divergence load of cantilever columns with rectangular and circular cross sections. In this

TABLE 5.3 First five Ω_n for the polynomial warping shape of cantilevered rectangular beams for different slenderness ratios with $\beta = 0,\ \delta = 0,\ k_1 = k_2 = 0.$

		HEB ($\alpha = 0$)			HHB ($\alpha = 1$)			EBB [WW14]	TB [H13]
p	n	5	10	100	5	10	100		10
0	1	2.5468	3.1688	3.5119	2.5468	3.1688	3.5119	3.516	3.2271
	2	8.4338	14.0322	21.8728	8.4338	14.0322	21.8728	22.035	14.4689
	3	16.1774	30.7796	60.6689	16.1774	30.7796	60.6689	61.697	31.5030
	4	39.3536	47.5935	117.3257	39.3536	47.5935	117.3257	120.902	48.0827
	5	41.8787	61.4688	190.7947	41.8787	61.4688	190.7947	199.860	62.6069
1	1	1.8310	2.4445	2.7501	1.9089	2.4529	2.7501	2.7536	–
	2	7.4174	13.2265	21.1239	7.8076	13.3540	21.1240	21.285	–
	3	15.3821	29.9313	60.0398	16.4075	30.2582	60.0401	61.068	–
	4	37.0257	46.6697	116.7319	39.8224	47.2669	116.7329	120.31	–
	5	41.1120	60.9029	190.2186	42.2342	61.9364	190.2211	199.29	–

Remark: WW14 = [43], H13 = [44].

TABLE 5.4 First five Ω_n for the trigonometric warping shape of cantilevered rectangular beams for different slenderness ratios with $\beta = 0$, $\delta = 0$, $k_1 = k_2 = 0$.

		HEB ($\alpha = 0$)			HHB ($\alpha = 1$)			EBB [WW14]	TB [H13]
p	n	5	10	100	5	10	100		10
0	1	2.5185	3.1560	3.5117	2.5185	3.1560	3.5117	3.516	3.2271
	2	8.3783	13.9208	21.8684	8.3783	13.9208	21.8684	22.035	14.4689
	3	15.8865	30.5518	60.6464	15.8865	30.5518	60.646	61.697	31.5030
	4	39.1010	47.3311	117.2588	39.1010	47.3311	117.2588	120.902	48.0827
	5	41.8213	60.9000	190.6443	41.8213	60.9000	190.6443	199.860	62.6069
1	1	1.8065	2.4345	2.7499	1.8899	2.4437	2.7499	2.7536	–
	2	7.3466	13.1115	21.1196	7.7468	13.2457	21.1197	21.285	–
	3	15.1146	29.6967	60.0175	16.1874	30.0374	60.0178	61.068	–
	4	36.7547	46.3954	116.6650	39.6553	47.0144	116.6661	120.31	–
	5	41.0217	60.3438	190.0681	42.2154	61.4257	190.0708	199.29	–

Remark: WW14 = [43], H13 = [44].

TABLE 5.5 First five Ω_n for the exponential warping shape of cantilevered rectangular beams for different slenderness ratios with $\beta = 0$, $\delta = 0$, $k_1 = k_2 = 0$.

		HEB ($\alpha = 0$)			HHB ($\alpha = 1$)			EBB [WW14]	TB [H13]
p	n	5	10	100	5	10	100		10
0	1	2.4881	3.1420	3.5115	2.4881	3.1420	3.5115	3.516	3.2271
	2	8.3209	13.8028	21.8636	8.3209	13.8028	21.8636	22.035	14.4689
	3	15.5470	30.3121	60.6221	15.5470	30.3121	60.6221	61.697	31.5030
	4	38.8523	47.0589	117.1868	38.8523	47.0589	117.1868	120.902	48.0827
	5	41.7568	60.2663	190.4832	41.7568	60.2663	190.4832	199.860	62.6069
1	1	1.7802	2.4236	2.7498	1.8696	2.4336	2.7498	2.7536	–
	2	7.2729	12.9895	21.1149	7.6834	13.1309	21.1150	21.285	–
	3	14.8102	29.4499	59.9933	15.9340	29.8055	59.9936	61.068	–
	4	36.4881	46.1101	116.5931	39.5063	46.7522	116.5943	120.31	–
	5	40.9194	59.7244	189.9071	42.1899	60.8606	189.9100	199.29	–

Remark: WW14 = [43], H13 = [44].

case, the natural frequency Ω is set to zero and the first positive root of the characteristic equation [Eq. (5.59)] gives the critical compressive force for columns at buckling. Here we compute the critical load parameter p_{cr} for rectangular and circular columns with three different ratios of the bending stiffness to shear rigidity, i.e., $\zeta^{-1} = 0.001, 0.01, 0.1$. In the determination of the critical force p_{cr} using the characteristic equation [Eq. (5.59)], one fails to directly set Ω to zero due to the singular coefficient matrix of the resulting system of algebraic equations. Instead, Ω is assigned a sufficiently small value, and the evaluated results are listed in Table 5.6. From Table 5.6 it can be seen that the critical load p_{cr} accurately approaches the results of the EBB theory when $\zeta = 0.001$, as expected. For shorter columns, the critical loads p_{cr} based on the HEB and HHB theories also have a noticeable declination compared to those based on the EBB and TB theories. Precisely, the critical load using the EBB theory is overestimated, and that using the TB theory is also overestimated, but less so. However, for the tangency coefficient $\beta = 0.4$, the critical divergence load does not exist for $\zeta = 0.1$ (in this case the flutter behavior occurs, this is discussed later) and moreover, it has a slight increase for $\zeta = 0.01$ as compared to that for $\zeta = 0.001$, which has an opposite trend for other values of $\beta = 0, 0.2$. By comparing the numerical results of the HEB and HHB theories, the critical loads using the HEB theory are slightly lower than those using the HHB theory. This conclusion is in exact agreement with that derived in Ref. [49]. It is particularly pointed out that in Table 5.6, the ζ^{-1} value, rather than the slenderness ratio, is assumed to be given for comparison, so the critical load p_{cr} corresponding to the exponential warping shape function takes the largest among the three evaluated results. Note that if taking the slenderness ratio s instead of the ζ^{-1} value as prescribed, we find an opposite conclusion that the critical load corresponding to the exponential warping shape function is the smallest. For instance, for $s = 10$, $\beta = 0$, we can obtain $p_{cr} = 2.291$, 2.2889, 2.2867 for rectangular columns and $p_{cr} = 2.2956, 2.294, 2.2924$ for circular columns corresponding to polynomial, trigonometric, and exponential warping shape functions based on the Engesser hypothesis ($\alpha = 0$), and $p_{cr} = 2.3046$, 2.3032, 2.3018 for rectangular columns and $p_{cr} = 2.3089$, 2.308, 2.3072 for circular columns corresponding to polynomial, trigonometric, and exponential warping shape functions based on the Haringx hypothesis ($\alpha = 0$), respectively. Finally, we also give a verification of our results with those through the finite element method. We chose aluminum (Al) columns with $E = 70$ GPa, $\nu = 0.3$, and $s = 10$, $\beta = 0$ corresponding to $L/h = 5/\sqrt{3}$ and $L/R = 5$ for rectangular and circular cross sections, the finite element results give dimensionless buckling loads $p_{cr} = 2.2888$ for rectangular columns and $p_{cr} = 2.2937$ for circular columns, respectively. These results agree well with the above-mentioned numerical results. Their buckling mode shapes are illustrated in Figs. 5.9 and 5.10,

TABLE 5.6 Divergence loads p_{cr} of cantilever columns with $k_1 = k_2 = 0$.

		HEB		HHB		TEB	THB	EBB	EBB
β	ζ^{-1}	Rect.	Circ.	Rect.	Circ.	[A08]		[R88]	[M88]
0	0	2.4674 (p, t, e)	2.4674 (p, t, e)	2.4674 (p, t, e)	2.4674 (p, t, e)	–	–	2.4674	2.4675
	0.001	2.4625 (p), 2.4627 (t), 2.4629 (e)	2.4627 (p), 2.4628 (t), 2.4630 (e)	2.4626 (p), 2.4627 (t), 2.4629 (e)	2.4627 (p), 2.4629 (t), 2.4630 (e)	–	–	2.4674	2.4675
	0.01	2.4196 (p), 2.4212 (t), 2.4227 (e)	2.4209 (p), 2.4226 (t), 2.4242 (e)	2.4207 (p), 2.4222 (t), 2.4238 (e)	2.4220 (p), 2.4236 (t), 2.4252 (e)	–	–	2.4674	2.4675
	0.1	2.0606 (p), 2.0717 (t), 2.0833 (e)	2.0701 (p), 2.0821 (t), 2.0945 (e)	2.1219 (p), 2.1318 (t), 2.1421 (e)	2.1304 (p), 2.1410 (t), 2.1521 (e)	–	–	2.4674	2.4675
0.2	0	3.3251 (p, t, e)	3.3251 (p, t, e)	3.3251 (p, t, e)	3.3251(p, t, e)	3.325	3.325	3.3251	3.3251
	0.001	3.3194 (p), 3.3197 (t), 3.3199 (e)	3.3196 (p), 3.3199 (t), 3.3202 (e)	3.3194 (p), 3.3197 (t), 3.3200 (e)	3.3196 (p), 3.3199 (t), 3.3202 (e)	3.317	3.314	3.3251	3.3251
	0.01	3.2692 (p), 3.2720 (t), 3.2749 (e)	3.2716 (p), 3.2746 (t), 3.2777 (e)	3.2709 (p), 3.2737 (t), 3.2765 (e)	3.2733 (p), 3.2762 (t), 3.2792 (e)	3.248	3.221	3.3251	3.3251
	0.1	2.8672 (p), 2.8917 (t), 2.9175 (e)	2.8882 (p), 2.9148 (t), 2.9427 (e)	2.9533 (p), 2.9731 (t), 2.9936 (e)	2.9702 (p), 2.9915 (t), 2.30136 (e)	2.694	2.633	3.3251	3.3251
0.4	0	5.2924 (p, t, e)	5.2924 (p, t, e)	5.2924 (p, t, e)	5.2924 (p, t, e)	5.292	5.292	5.2924	5.2925
	0.001	5.2919 (p), 5.2926 (t), 5.2934 (e)	5.2925 (p), 5.2933 (t), 5.2941 (e)	5.2919 (p), 5.2926 (t), 5.2934 (e)	5.2925 (p), 5.2933 (t), 5.2941 (e)	5.286	5.265	5.2924	5.2925
	0.01	5.2963 (p), 5.3040 (t), 5.3119 (e)	5.3029 (p), 5.3111 (t), 5.3195 (e)	5.2956 (p), 5.3028 (t), 5.3103 (e)	5.3018 (p), 5.3095 (t), 5.3175 (e)	5.238	5.039	5.2924	5.2925
	0.1	–	–	–	–	–	3.831	5.2924	5.2925

Remark: data with p, t, and e correspond to those using polynomial, trigonometric, and exponential warping shapes, respectively; A08 = [46], R88 = [47], M08 = [48].

FIGURE 5.9 Buckling mode shape of Al columns with $s = 10$, $\beta = 0$, rectangular cross section.

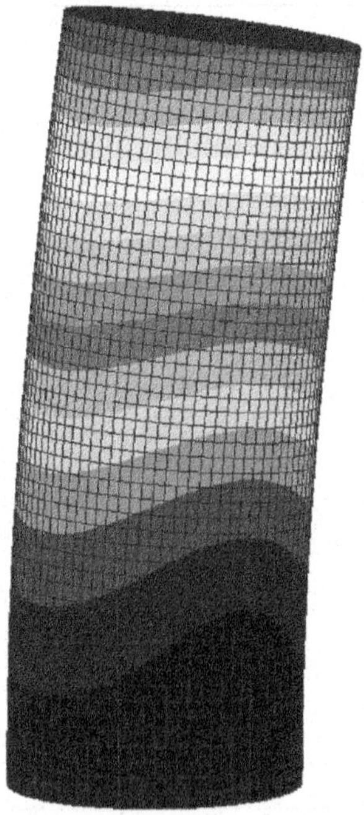

FIGURE 5.10 Buckling mode shape of Al columns with $s = 10$, $\beta = 0$, circular cross section.

respectively. The finite element results indicate the effectiveness of our method.

From the above, by comparing the dimensionless natural frequencies or buckling loads for three warping shape functions, we find that among these three warping shape functions, the exponential warping shape function produces the lowest natural frequencies and critical loads when the slenderness ratio is known, whereas the polynomial warping shape function produces the highest natural frequencies and critical loads. Based on this conclusion, in the following discussion, we only choose the exponential warping shape function $f_e(y,z)$ to calculate subsequent numerical results.

5.5.1 In a free space

In this subsection, the elastic foundation is assumed to be absent, i.e., $k_1 = k_2 = 0$. A circular beam-column with $s = 10$ and the exponential warping shape function $f_e(y,z)$ is considered and the load frequency interaction curves are displayed when $\alpha = 0, 1$ corresponding to HEB and HHB theories in Figs. 5.11 and 5.12, respectively. For comparison, we also plot the corresponding load frequency interaction curves in Fig. 5.13 using the EBB theory, which actually corresponds to an infinitely large s value, $s = 10^9$ is chosen in

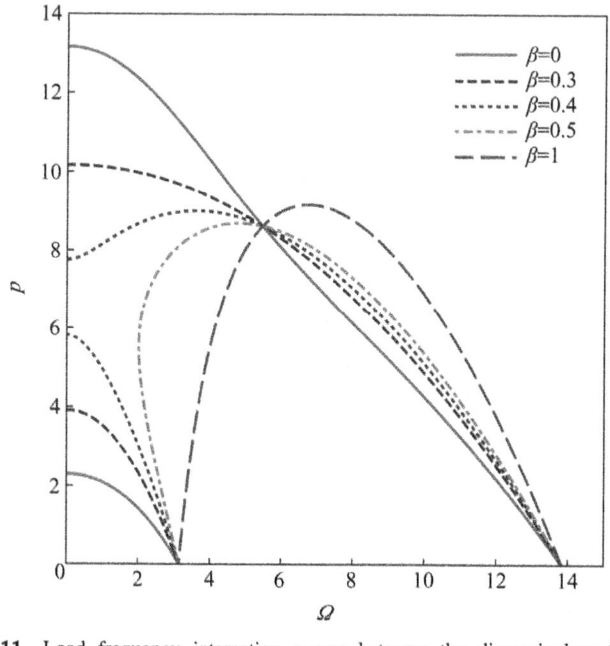

FIGURE 5.11 Load frequency interaction curves between the dimensionless force p and dimensionless frequency Ω for a generalized Beck's column in free space subjected to a subtangential follower force, HEB ($s = 10$, $\alpha = 0$).

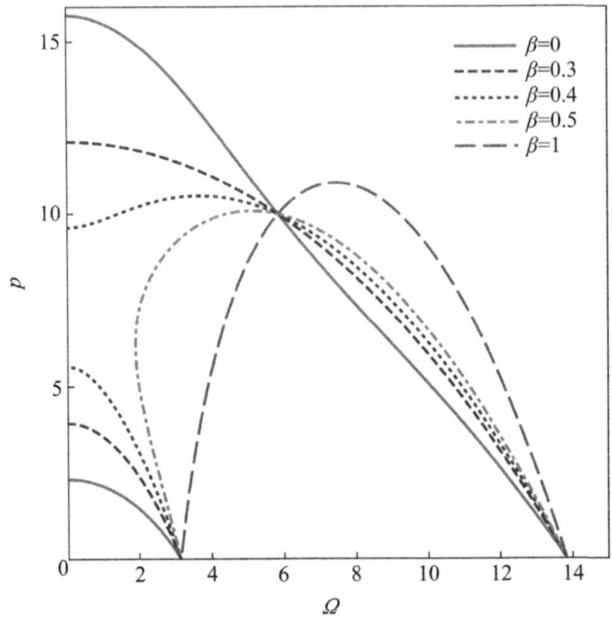

FIGURE 5.12 Load frequency interaction curves between the dimensionless force p and dimensionless frequency Ω for a generalized Beck's column in free space subjected to a sub-tangential follower force, HHB ($s = 10$, $\alpha = 1$).

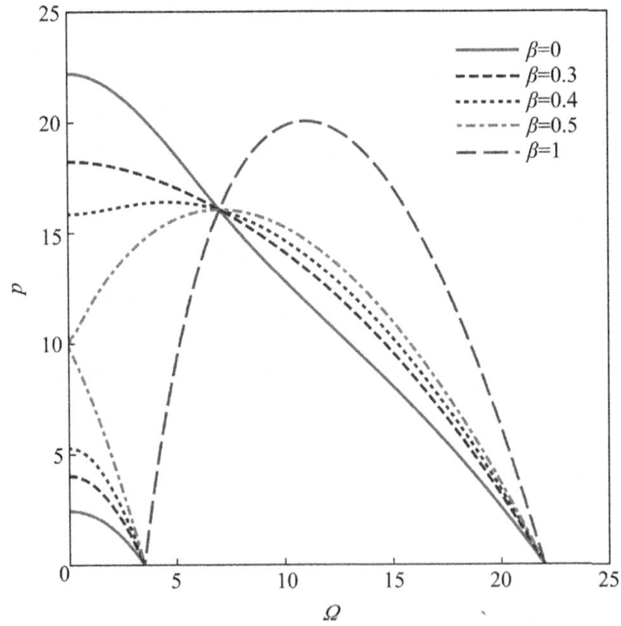

FIGURE 5.13 Load frequency interaction curves between the dimensionless force p and dimensionless frequency Ω for a generalized Beck's column in free space subjected to a sub-tangential follower force, EBB ($s \to \infty$ or $G \to \infty$).

numerical computation. From Figs. 5.11–5.13, it can be seen that irrespective of the HEB or HHB theory, both the natural frequencies (of the first and second order) and the divergence loads have a pronounced reduction compared to those for EBB, which is consistent with the foregoing conclusion. When neglecting shear deformation or considering sufficiently large shear stiffness, this is valid for a sufficiently long cantilever. In this case, the flutter instability occurs when $\beta > 0.5$, and the flutter load takes $p^f_{cr} = 20.05$ with $\Omega = 11$ for a follower force with $\beta = 1$, as shown in Fig. 5.13 for EBB, in agreement with earlier results [49,50]. However, for practical situations, shear stiffness is finite or the slenderness ratio is finite, and the higher-order HEB and HHB theories are preferred. In this case, we find that flutter instability takes place for a tangency coefficient larger than $\beta = 0.41$ if $\alpha = 0$ and $\beta = 0.429$ if $\alpha = 1$, respectively, which are clearly lower than the critical tangency coefficient $\beta = 0.5$ for classical Beck's column based on the EBB theory [39]. Moreover, for a follower force with $\beta = 1$, the flutter loads are $p^f_{cr} = 9.16$ with $\Omega = 6.73$ if $\alpha = 0$ and $p^f_{cr} = 10.88$ with $\Omega = 7.45$ if $\alpha = 1$, respectively, which are both lower than the corresponding value of $p^f_{cr} = 20.05$ with $\Omega = 11$ for classical Beck's columns [39]. It turns out that the classical flutter loads are seriously overestimated.

Fig. 5.14 shows the effect of the tangency coefficient on the divergence and flutter instability boundary map of a generalized Beck's column, where

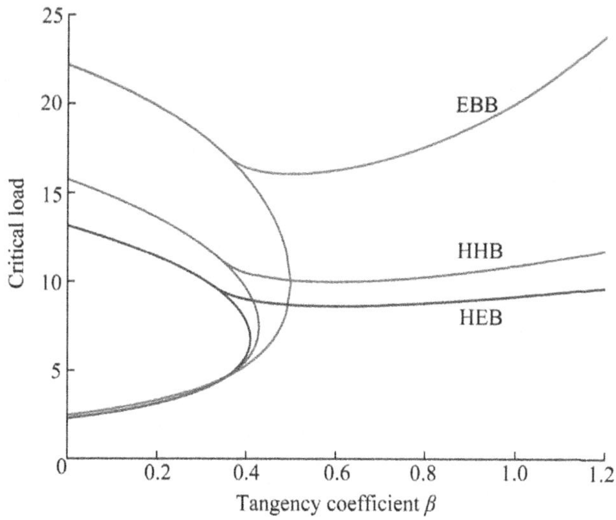

FIGURE 5.14 Comparison of the instability boundary map, the divergence, and flutter loads as a function of the tangency coefficient β for a generalized Beck's column in a free space subjected to a subtangential follower force.

$\beta = 0$ corresponds to a dead force and $\beta = 1$ to a follower force. For EBB with large shear rigidity or long length, the maximum β value reaches 0.5, whereas, for HEB or HHB, the flutter load and divergence load decline significantly.

5.5.2 On a Winkler foundation

Here we consider the case of a one parameter elastic foundation, i.e., a Winkler foundation. When taking the cross section's exponential warping shape, the divergence loads for $\beta = 0$ and flutter loads for $\beta = 1$ are calculated and tabulated in Table 5.7, in which the slenderness ratios $s = 5$, 10, and ∞ are selected. It is clear that when taking $s \to \infty$, our results reduce to the well-known ones [42,47]. It also confirms the validity of our approach. Note that in our numerical calculations, the case of $s \to \infty$ is performed by taking $s = 10^9$. In addition, for those columns with finite slenderness ratios, the divergence and flutter loads are both reduced. In particular, the critical flutter load remains unchanged when the stiffness of Winkler foundations varies, which is called the Herrmann–Smith paradox [51]. However, when adopting the higher-order HEB or HHB theories, the Herrmann–Smith paradox does not occur, as shown in Table 5.7. On the contrary, to our surprise, with an increase in Winkler foundation stiffness, the critical flutter loads have a slight reduction, rather than an increase, while the coalescence flutter frequencies show a modest rise. This remains an intuitively unexpected result, but differs from the Herrmann–Smith paradox. Of course, the critical divergence loads rise with an increase in the Winkler foundation stiffness, consistent with our intuition. This infers that for static stability of cantilever columns resting on a Winkler foundation, an increase in the normal stiffness of elastic foundation can enhance the load-carrying capability of columns even when subjected to a follower force. For dynamic stability, this in turn decreases the critical flutter loads but increases the corresponding coalescence flutter frequency. For clarity, the load frequency interaction curves between the dimensionless force p and dimensionless frequency Ω for Beck's column resting on a Winkler foundation with three different values $k_1 = 0$, 10, and 100 subjected to a follower force ($\beta = 1$) are displayed in Figs. 5.15 and 5.16 for HEB ($\alpha = 0$, $s = 10$) and EBB ($s \to \infty$ or $G \to \infty$), respectively. Also, $G \to \infty$ is treated as $G = 10^9 E$ in numerical implementation. Note that when $k_1 = 0$, the Winkler foundation is absent. As a result, it can be seen from Figs. 5.15 and 5.16 that when $k_1 = 0$, the critical divergence load does not appear, in this case, only flutter instability occurs when a follower force reaches its critical value, whereas when $k_1 > 0$, not only the critical divergence loads but also the critical flutter loads take place. Moreover, the flutter loads remain unchanged for EBB subjected to a follower force. The critical divergence and flutter loads as functions of the elastic foundation stiffness are presented in Figs. 5.17 and 5.18, respectively. By comparing these curves between EBB and HEB, the

TABLE 5.7 Divergence and flutter loads (p_{cr}^{d} and p_{cr}^{f}) of cantilevered circular columns on a Winkler foundation ($k_2 = 0$).

			HEB ($\alpha = 0$)			HEB ($\alpha = 1$)			EBB [R88]
	β	k_1	5	10	∞	5	10	∞	
p_{cr}^{d}	0	0	1.8903	2.2924	2.4674	2.0109	2.3072	2.4674	2.467
		5	2.7500	3.1804	3.3544	2.9286	4.3417	3.3544	3.354
		10	3.2452	3.9283	4.1783	3.5773	3.9640	4.1783	4.178
		50	4.7416	7.1406	8.8614	6.0933	7.4851	8.8614	8.861
		100	5.6085	9.0447	11.9964	7.8720	9.6938	11.9964	11.996
p_{cr}^{d}	1	0	–	–	–	–	–	–	–
		5	3.8535	4.3175	4.4721	4.1092	4.3417	4.4721	–
		10	5.0872	6.0152	6.3236	5.6987	6.0792	6.3236	–
		50	7.5927	12.5954	14.1421	12.1338	13.1531	14.1421	–
		100	7.6264	16.9066	20.0000	16.8133	18.2650	20.0000	–
									[A15]
p_{cr}^{f}	1	0	3.73 (3.48)	9.16 (6.76)	20.05 (11.03)	5.68 (4.99)	10.88 (7.45)	20.05 (11.03)	20.05095 (11.01)
		5	3.70 (4.47)	9.12 (7.10)	20.05 (11.28)	5.02 (5.44)	10.83 (7.77)	20.05 (11.28)	20.05095 (11.24)

Continued

TABLE 5.7 Divergence and flutter loads (p_{cr}^d and p_{cr}^f) of cantilevered circular columns on a Winkler foundation ($k_2 = 0$).—cont'd

			HEB ($\alpha = 0$)			HEB ($\alpha = 1$)			EBB [R88]
	β	k_1	5	10	∞	5	10	∞	
		10	3.60 (5.00)	9.08 (7.46)	20.05 (11.50)	5.44 (5.84)	10.77 (8.08)	20.05 (11.50)	20.05095 (11.46)
		50	2.73 (8.01)	8.75 (9.77)	20.05 (13.08)	3.99 (8.35)	10.34 (10.21)	20.05 (13.08)	
		100	1.87 (10.51)	8.33 (12.00)	20.05 (14.86)	2.80 (10.53)	9.80 (12.36)	20.05 (14.86)	

Remark: $p_{cr}^f(\omega_c^f)$ means the flutter load with the coalescence frequency ω_{cr}^f, R88 = [47], A15 = [42].

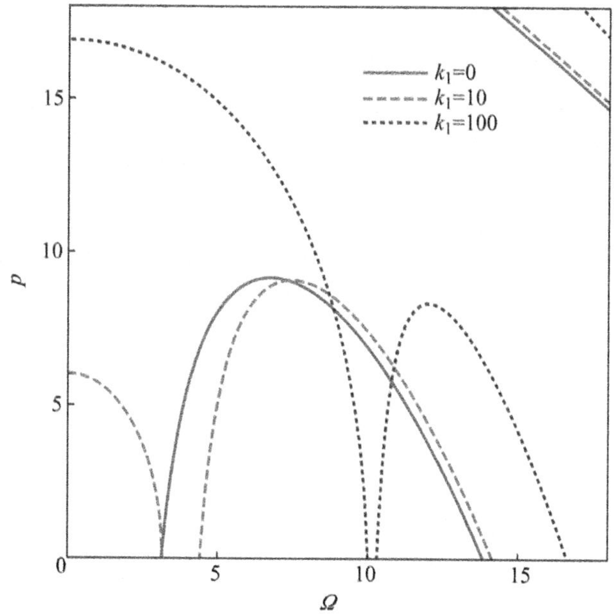

FIGURE 5.15 Load frequency interaction curves between the dimensionless force p and dimensionless frequency Ω for Beck's column resting on a Winkler foundation with $k_2 = 0$ and subjected to a follower force ($\beta = 1$), HEB ($s = 10$, $\alpha = 0$).

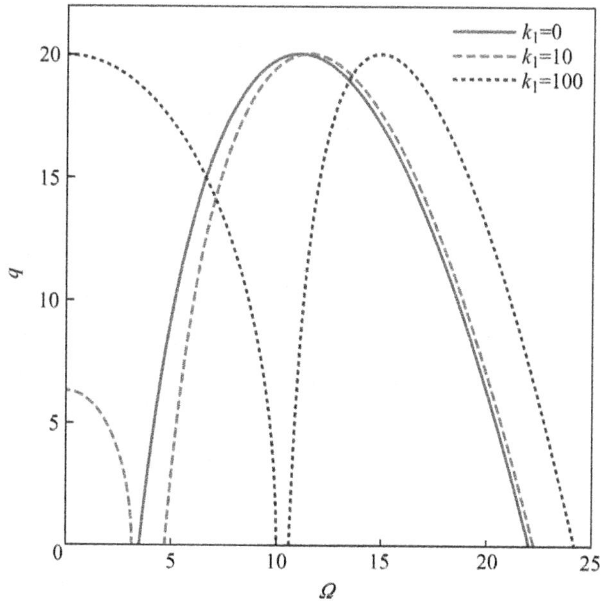

FIGURE 5.16 Load frequency interaction curves between the dimensionless force p and dimensionless frequency Ω for Beck's column resting on a Winkler foundation with $k_2 = 0$ and subjected to a follower force ($\beta = 1$), EBB ($s \rightarrow \infty$ or $G \rightarrow \infty$).

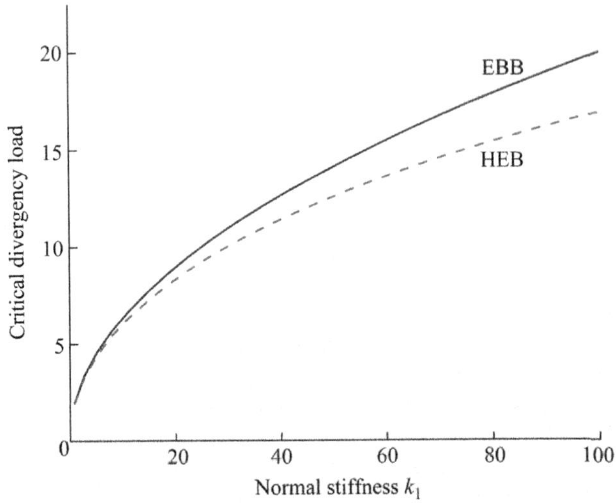

FIGURE 5.17 Critical divergence loads against the Winkler foundation stiffness for a cantilever column with Engesser's hypothesis ($\alpha = 0$) for $\beta = 0$ and $\beta = 1$, respectively.

flutter loads take the same value and are independent of the elastic foundation stiffness for EBB (Fig. 5.16), while the flutter loads decrease and the coalescence flutter frequencies increase with increasing k_1 for HEB (Fig. 5.15).

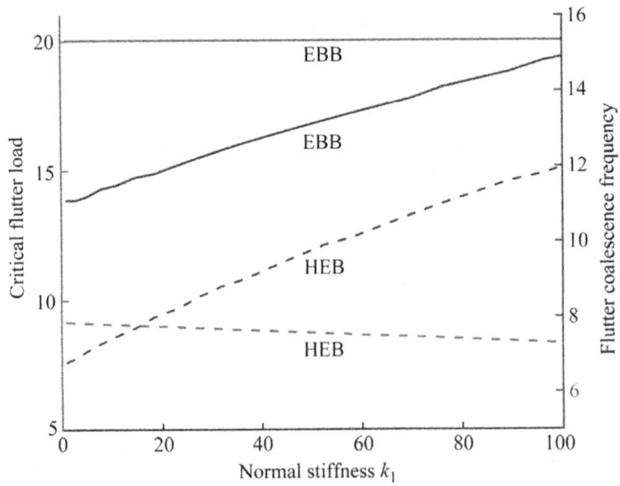

FIGURE 5.18 Flutter loads against the Winkler foundation stiffness for a cantilever column with Engesser's hypothesis ($\alpha = 0$) for $\beta = 0$ and $\beta = 1$, respectively.

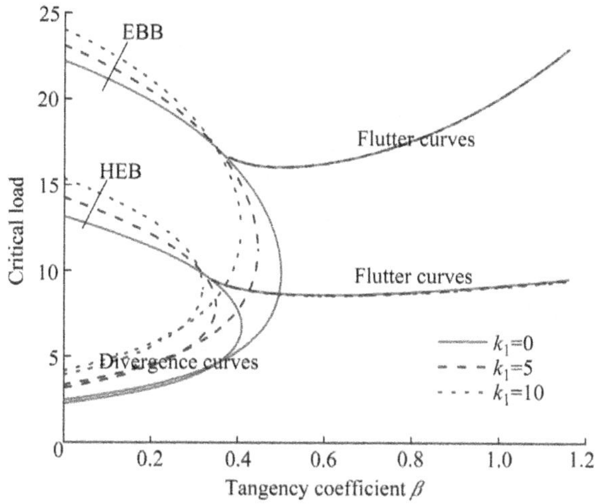

FIGURE 5.19 Effect of the tangency coefficient on the critical loads for a generalized Beck's column on a Winkler foundation with $k_2 = 0$ subjected to a subtangential follower force.

Fig. 5.19 illustrates the effect of the tangency coefficient on the critical loads for a generalized Beck's column subjected to a subtangential follower force. It can be observed that when an elastic foundation is present, both the first and second divergence loads increase with the elastic foundation stiffness increasing, which is in line with the results in Table 5.7. However, the flutter curves for different k_1 values coincide with each other when using the EBB theory, implying that the Herrmann–Smith paradox takes place. However, if using the HEB theory, the flutter loads have a slight discrepancy, which is barely seen in Fig. 5.19. The main reason for this is that the range of k_1 is relatively small. If taking a larger value of k_1, say $k_1 = 10$, in Table 5.7, the flutter load for a follower force is 8.33, which is noticeably lower than the 9.12 for $k_1 = 5$. Furthermore, the tangency coefficient value for flutter behaviors becomes lower, e.g., from $\beta = 0.5$, 0.45, and 0.41 for EBB to $\beta = 0.41$, 0.35, and 0.32 for HEB when taking $k_1 = 0$, 5, and 10, respectively.

5.5.3 On a Winkler–Pasternak foundation

Now we pay attention to a biparameter elastic foundation or Winkler–Pasternak foundation with k_1 and k_2. For simplicity, in the following calculation, the k_1 parameter is fixed as $k_1 = 10$. When a dead force (conservative force) ($\beta = 0$) and a follower force ($\beta = 1$) are applied, the first divergence and flutter loads are calculated for various values of k_2 for HEB ($\alpha = 0$) and HHB ($\alpha = 1$), respectively. Moreover, the numerical results for

HHB are slightly higher than those for HEB, which is consistent with the previous conclusion. In particular, for the case of an applied conservative force ($\beta = 0$), the divergence load reaches the local maximum at $k_2 = 1$. Nonetheless, in the case of a nonconservative force such as a follower force ($\beta = 1$), the shear stiffness of the elastic foundation reduces the first divergence load as well as the flutter load. The coalescence frequency remains almost unchanged, which is different from the effect of k_1 on the coalescence frequency.

A comparison of the load frequency interaction curves between the dimensionless force p and dimensionless frequency Ω for Beck's column resting on a Winkler–Pasternak foundation and subjected to a follower force ($\beta = 1$) is made in Figs. 5.20 and 5.21 for HEB and EBB, respectively. From Figs. 5.20 and 5.21, the shear stiffness of the elastic foundation k_2 lowers not only the divergence loads for static stability but also the flutter loads for dynamic stability, which is different from the effect of the elastic stiffness k_1. Nonetheless, it barely affects the coalescence frequency at the flutter instability, where the coalescence frequencies are given in parentheses in Table 5.8. Here, $s \rightarrow \infty$ in our numerical calculations and is still taken as $s = 10^9$. The stiffness k_1 increases the flutter coalescence frequency, which can be seen in Figs. 5.15 and 5.16 and Table 5.7.

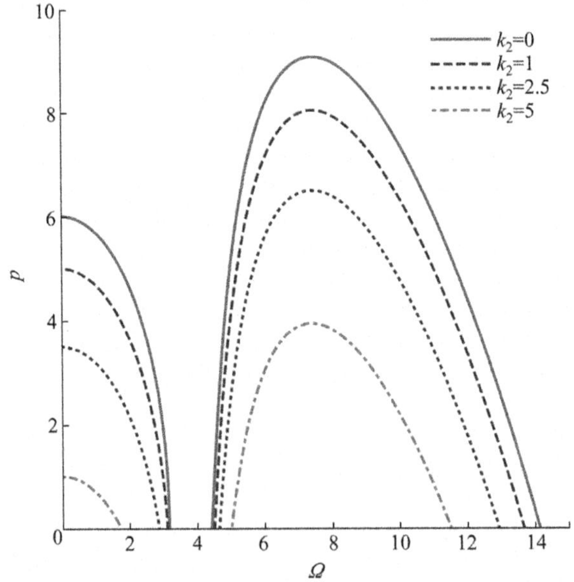

FIGURE 5.20 Load frequency interaction curves between the dimensionless force p and dimensionless frequency Ω for a Beck's column resting on a Winkler–Pasternak foundation with $k_1 = 10$ and subjected to a follower force ($\beta = 1$), HEB ($s = 10$, $\alpha = 0$).

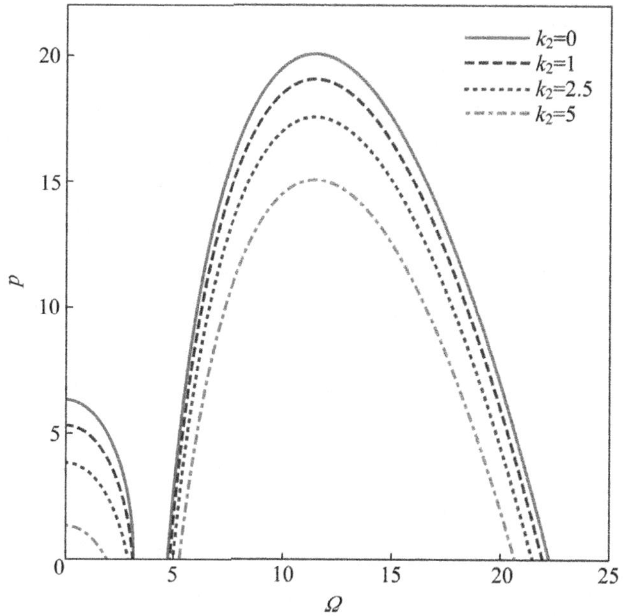

FIGURE 5.21 Load frequency interaction curves between the dimensionless force p and dimensionless frequency Ω for a Beck's column resting on a Winkler-Pasternak foundation with $k_1 = 10$ and subjected to a follower force ($\beta = 1$), EBB ($s \rightarrow \infty$ or $G \rightarrow \infty$).

For a subtangential follower force, say, $\beta = 0.5$, the load frequency interaction curves of HEB and EBB are displayed in Figs. 5.22 and 5.23, respectively. From Figs. 5.15, 5.16, and 5.20–5.23, it can be seen that the shear stiffness reduces the flutter/divergence load, but increases the flutter coalescence frequency, whereas the stiffness k_1 does not affect the flutter load for EBB, although it decreases the flutter loads and increases the coalescence flutter frequencies for HEB. It can be seen again that the Herrmann–Smith paradox takes place for EBB, whereas it does not occur for HEB. Figs. 5.24 and 5.25 give a comparison of the load frequency interaction curves of HEB and EBB for various tangency coefficients when taking $\alpha = 0$, $k_1 = 10$, and $k_2 = 1$.

To examine the influence of the shear stiffness k_2 on the critical load, for several different k_2 values, the critical load curves as a function of the tangency coefficient for a generalized Beck's column connected to a Winkler–Pasternak foundation subjected to a subtangential follower force are illustrated for HEB and EBB in Figs. 5.26 and 5.27, respectively. Figs. 5.28 and 5.29 show the critical divergence and flutter loads against the shear stiffness of the Winkler–Pasternak foundation for a cantilever column under Engesser's hypothesis ($\alpha = 0$). In particular, for static stability, the critical divergence load exists not only for a conservative load $\beta = 0$ but also for a nonconservative follower

TABLE 5.8 Divergence and flutter loads (p_{cr}^d and p_{cr}^f) of cantilevered circular columns on a foundation with $k_1 = 10$.

			HEB ($\alpha = 0$)			HHB ($\alpha = 1$)		
	β	k_2	5	10	∞	5	10	∞
p_{cr}^d	0	0	3.2452	3.9283	4.1783	3.5773	3.9640	4.1783
		1	3.8841	4.3746	4.5322	4.1714	4.3990	4.5322
		2.5	2.5872	3.5152	3.8246	3.0083	3.5561	3.8246
		5	2.6764	1.0152	1.3246	0.1101	1.0283	1.3246
p_{cr}^d	1	0	5.0872	6.0152	6.3246	5.6987	6.0792	6.3246
		1	4.0872	5.0152	5.3246	4.6351	5.0704	5.3246
		2.5	2.5872	3.5152	3.8246	3.0083	3.5561	3.8246
		5	0.0872	1.0152	1.3246	0.1126	1.0283	1.3246
p_{cr}^f	1	0	3.60 (5.00)	9.08 (7.43)	20.05 (11.46)	5.33 (5.84)	10.77 (8.08)	20.05 (11.46)
		1	2.45 (5.00)	8.05 (7.42)	19.05 (11.46)	3.73 (5.62)	9.60 (8.01)	19.05 (11.46)
		2.5	0.72 (4.99)	6.51 (7.41)	17.55 (11.46)	1.17 (5.20)	7.82 (7.91)	17.55 (11.46)
		5	–	3.95 (7.40)	15.05 (11.46)	–	4.80 (7.72)	15.05 (11.46)

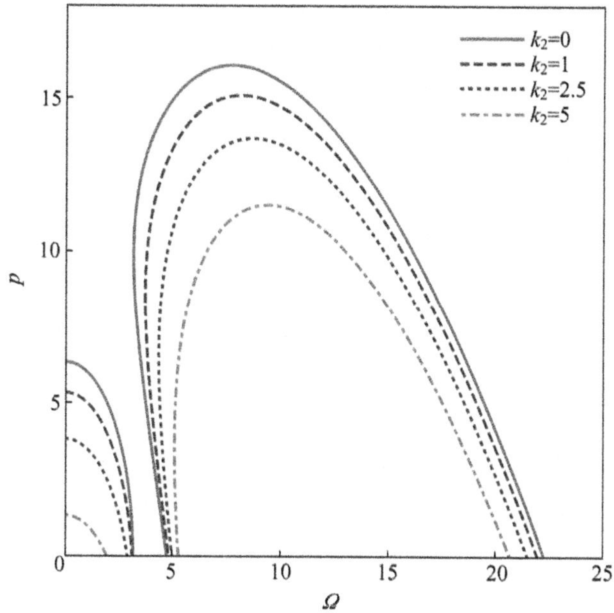

FIGURE 5.22 Load frequency interaction curves between the dimensionless force p and dimensionless frequency Ω for a Beck's column resting on a Winkler–Pasternak foundation with $k_1 = 10$ and subjected to a subtangential follower force ($\beta = 0.5$), HEB ($s = 10$, $\alpha = 0$).

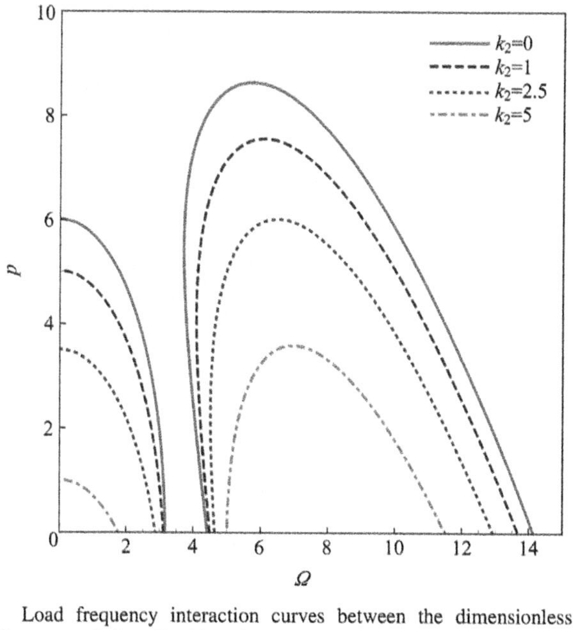

FIGURE 5.23 Load frequency interaction curves between the dimensionless force p and dimensionless frequency Ω for a Beck's column resting on a Winkler–Pasternak foundation with $k_1 = 10$ and subjected to a subtangential follower force ($\beta = 0.5$), EBB ($s \rightarrow \infty$ or $G \rightarrow \infty$).

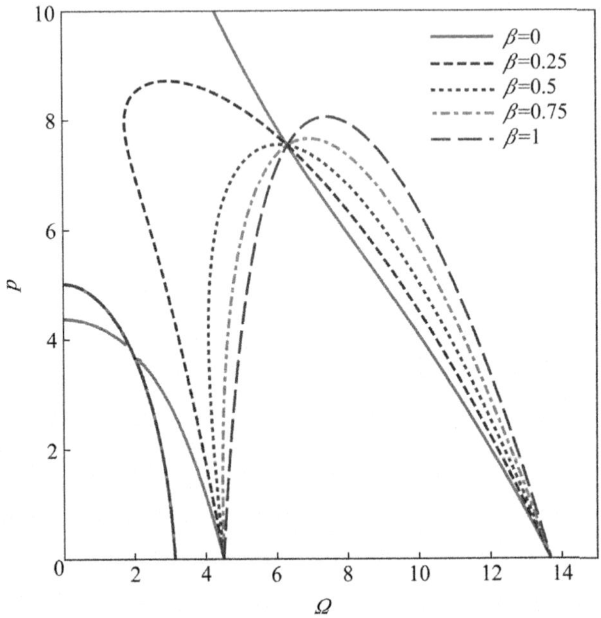

FIGURE 5.24 Load frequency interaction curves between the dimensionless force p and dimensionless frequency Ω for a Beck's column resting on a Winkler–Pasternak foundation with $k_1 = 10$ and $k_2 = 1$ under the Engesser hypothesis ($\alpha = 0$) for various tangency coefficients, HEB ($s = 10$).

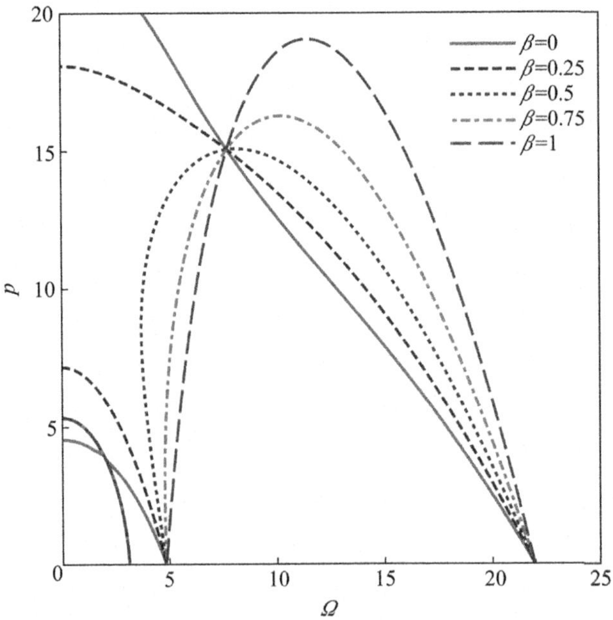

FIGURE 5.25 Load frequency interaction curves between the dimensionless force p and dimensionless frequency Ω for a Beck's column resting on a Winkler–Pasternak foundation with $k_1 = 10$ and $k_2 = 1$ under the Engesser hypothesis ($\alpha = 0$) for various tangency coefficients, EBB ($s \rightarrow \infty$).

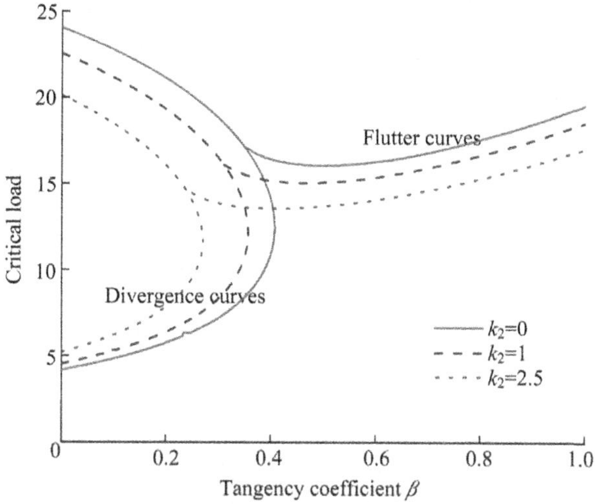

FIGURE 5.26 Effect of the tangency coefficient on the critical loads for a generalized Beck's column on a Winkler–Pasternak foundation with $k_1 = 10$ and different k_2 values, HEB ($\alpha = 1$, $s = 10$).

force $\beta = 1$. In other words, when a follower force is applied, both the divergence instability for static behavior and the flutter instability for dynamic behavior take place when the follower force reaches its respective critical value (Table 5.9).

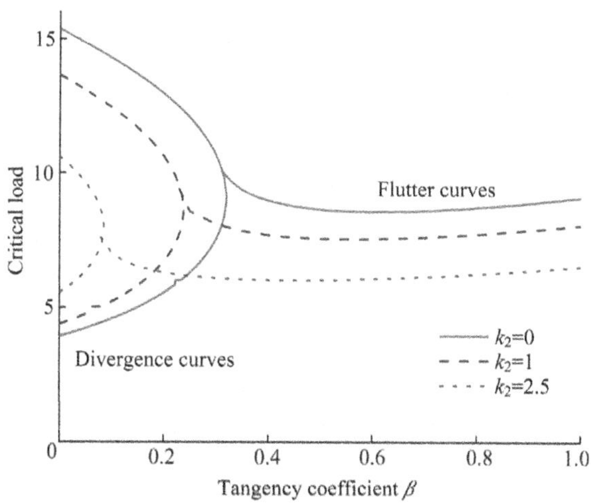

FIGURE 5.27 Effect of the tangency coefficient on the critical loads for a generalized Beck's column on a Winkler–Pasternak foundation with $k_1 = 10$ and different k_2 values, EBB ($s \rightarrow \infty$ or $G \rightarrow \infty$).

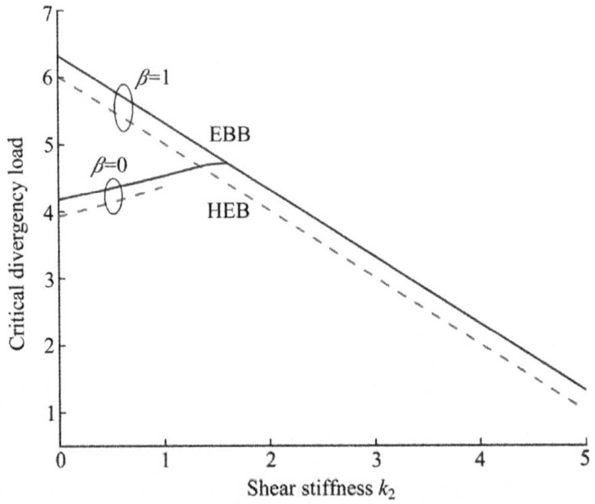

FIGURE 5.28 Critical loads against the shear stiffness of a Winkler–Pasternak foundation with $k_1 = 10$ for a Beck's column with Engesser's hypothesis ($\alpha = 0$), the divergence load for $\beta = 0, 1$.

Finally, we examine the effect of the elastic foundation on the stability region when using the HEB and EBB theories, shown in Figs. 5.30 and 5.31, respectively. Obviously, when using the HEB theory, the divergence region

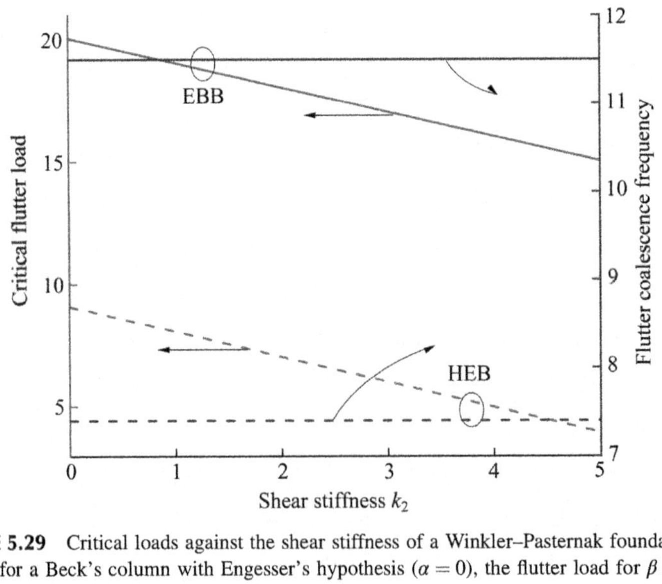

FIGURE 5.29 Critical loads against the shear stiffness of a Winkler–Pasternak foundation with $k_1 = 10$ for a Beck's column with Engesser's hypothesis ($\alpha = 0$), the flutter load for $\beta = 1$.

TABLE 5.9 Divergence and flutter loads (p_{cr}^d and p_{cr}^f) of cantilevered circular columns on a foundation with $k_1 = 10$.

			HEB ($\alpha = 0$)			HHB ($\alpha = 1$)		
	β	k_2	5	10	∞	5	10	∞
p_{cr}^d	0	0	3.2452	3.9283	4.1783	3.5773	3.9640	4.1783
		1	3.8841	4.3746	4.5322	4.1714	4.3990	4.5322
		2.5	2.5872	3.5152	3.8246	3.0083	3.5561	3.8246
		5	2.6764	1.0152	1.3246	0.1101	1.0283	1.3246
p_{cr}^d	1	0	5.0872	6.0152	6.3246	5.6987	6.0792	6.3246
		1	4.0872	5.0152	5.3246	4.6351	5.0704	5.3246
		2.5	2.5872	3.5152	3.8246	3.0083	3.5561	3.8246
		5	0.0872	1.0152	1.3246	0.1126	1.0283	1.3246
p_{cr}^f	1	0	3.60 (5.00)	9.08 (7.43)	20.05 (11.46)	5.33 (5.84)	10.77 (8.08)	20.05 (11.46)
		1	2.45 (5.00)	8.05 (7.42)	19.05 (11.46)	3.73 (5.62)	9.60 (8.01)	19.05 (11.46)
		2.5	0.72 (4.99)	6.51 (7.41)	17.55 (11.46)	1.17 (5.20)	7.82 (7.91)	17.55 (11.46)
		5	–	3.95 (7.40)	15.05 (11.46)	–	4.80 (7.72)	15.05 (11.46)

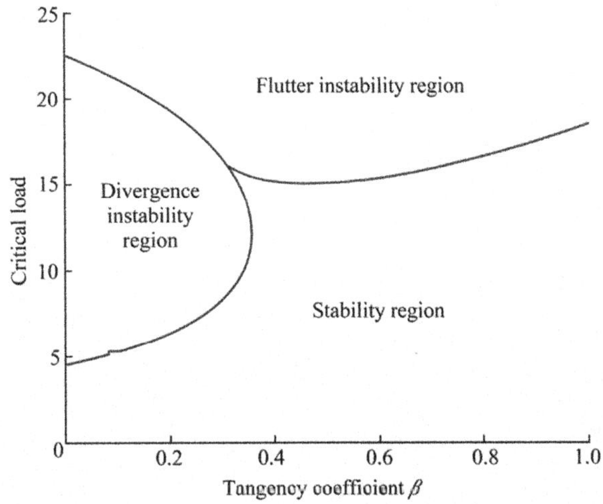

FIGURE 5.30 The comparison of the stability region for a generalized Beck's column on a Winkler–Pasternak foundation with $k_1 = 10$ and $k_2 = 1$, HEB ($\alpha = 0$, $s = 10$).

becomes larger compared with that obtained using the EBB theory. The reason for this is that the second divergence load and the tangency coefficient decrease significantly when the EBB theory is applied for the same elastic foundation.

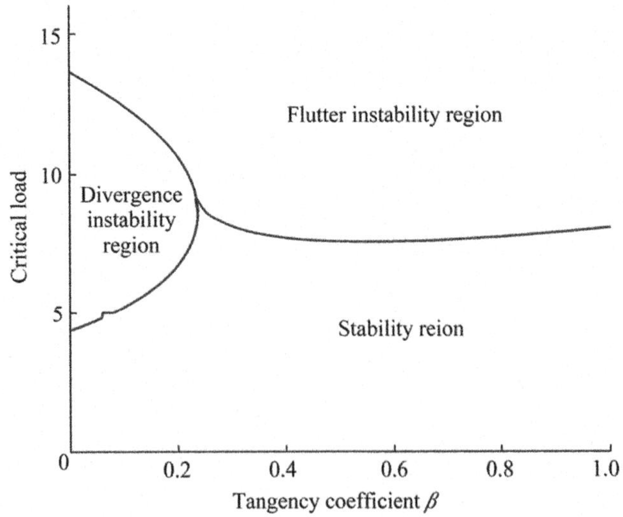

FIGURE 5.31 The comparison of the stability region for a generalized Beck's column on a Winkler–Pasternak foundation with $k_1 = 10$ and $k_2 = 1$, EBB ($s \to \infty$ or $G \to \infty$).

References

[1] Cedolin L. Stability of structures: elastic, inelastic, fracture and damage theories. World Scientific; 2010.

[2] Zhang H, Kang YA, Li XF. Stability and vibration analysis of axially-loaded shear beam-columns carrying elastically restrained mass. Appl Math Model 2013;37(16–17):8237–50.

[3] Ma WL, Jiang ZC, Li XF. Effect of warping shape on buckling of circular and rectangular columns under axial compression. Appl Math Model 2021;89:1475–90.

[4] Levinson M. A new rectangular beam theory. J Sound Vib 1981;74(1):81–7.

[5] Karama M, Afaq KS, Mistou S. Mechanical behaviour of laminated composite beam by the new multi-layered laminated composite structures model with transverse shear stress continuity. Int J Solid Struct 2003;40(6):1525–46.

[6] Touratier M. An efficient standard plate theory. Int J Eng Sci 1991;29(8):901–16.

[7] Huang Y, Li XF. Bending and vibration of circular cylindrical beams with arbitrary radial nonhomogeneity. Int J Mech Sci 2010;52(4):595–601.

[8] Huang Y, Li XF. Buckling of functionally graded circular columns including shear deformation. Mater Des 2010;31(7):3159–66.

[9] Zhaohua F, Cook RD. Beam elements on two-parameter elastic foundations. J Eng Mech 1983;109(6):1390–402.

[10] Kargarnovin MH, Younesian D. Dynamics of Timoshenko beams on Pasternak foundation under moving load. Mech Res Commun 2004;31(6):713–23.

[11] Herisanu N, Marinca V. Free oscillations of Euler-Bernoulli beams on nonlinear Winkler-Pasternak foundation[C]//Acoustics and vibration of mechanical structures—AVMS-2017. In: Proceedings of the 14th AVMS conference, timisoara, Romania, may 25–26, 2017. Cham: Springer International Publishing; 2017. p. 41–8.

[12] Liang X, Cao Z, Sun H, et al. Analytical and semi-analytical methods for the evaluation of dynamic thermo-elastic behavior of structures resting on a Pasternak foundation. J Pressure Vessel Technol 2019;141(1):010908.

[13] Omurtag MH, Kadıoğlu F. Free vibration analysis of orthotropic plates resting on Pasternak foundation by mixed finite element formulation. Comput Struct 1998;67(4):253–65.

[14] Wang TM, Stephens JE. Natural frequencies of Timoshenko beams on Pasternak foundations. J Sound Vib 1977;51(2):149–55.

[15] Shen HS. Postbuckling of orthotropic plates on two-parameter elastic foundation. J Eng Mech 1995;121(1):50–6.

[16] Selvadurai APS. Elastic analysis of soil-foundation interaction. Elsevier; 2013.

[17] Mullapudi R, Ayoub A. Nonlinear finite element modeling of beams on two-parameter foundations. Comput Geotech 2010;37(3):334–42.

[18] Ozgan K, Daloglu AT. Effect of transverse shear strains on plates resting on elastic foundation using modified Vlasov model. Thin-Walled Struct 2008;46(11):1236–50.

[19] Avramidis IE, Morfidis K. Bending of beams on three-parameter elastic foundation. Int J Solid Struct 2006;43(2):357–75.

[20] Hetenyi M. A general solution for the bending of beams on an elastic foundation of arbitrary continuity. J Appl Phys 1950;21(1):55–8.

[21] Morfidis K. Exact matrices for beams on three-parameter elastic foundation. Comput Struct 2007;85(15–16):1243–56.

[22] Morfidis K. Vibration of Timoshenko beams on three-parameter elastic foundation. Comput Struct 2010;88(5–6):294–308.

[23] Limkatanyu S, Prachasaree W, Damrongwiriyanupap N, et al. Exact stiffness for beams on Kerr-type foundation: The virtual force approach. J Appl Math 2013;2013(1):626287.
[24] Wang J, Zhang C. A three-parameter elastic foundation model for interface stresses in curved beams externally strengthened by a thin FRP plate. Int J Solid Struct 2010;47 (7–8):998–1006.
[25] Wang J, Zhang C. Three-parameter, elastic foundation model for analysis of adhesively bonded joints. Int J Adhesion Adhes 2009;29(5):495–502.
[26] Wang L, Ma J, Peng J, et al. Large amplitude vibration and parametric instability of inextensional beams on the elastic foundation. Int J Mech Sci 2013;67:1–9.
[27] Findley WN, Lai JS, Onaran K, et al. Creep and relaxation of nonlinear viscoelastic materials with an introduction to linear viscoelasticity. J Appl Mech 1977;44(2):364.
[28] Andersen L, Nielsen SRK, Kirkegaard PH. Finite element modelling of infinite Euler beams on Kelvin foundations exposed to moving loads in convected co-ordinates. J Sound Vib 2001;241(4):587–604.
[29] Ansari M, Esmailzadeh E, Younesian D. Frequency analysis of finite beams on nonlinear Kelvin-Voight foundation under moving loads. J Sound Vib 2011;330(7):1455–71.
[30] Hardy MSA. The generation of waves in infinite structures by moving harmonic loads. J Sound Vib 1995;180(4):637–44.
[31] Hatada T, Kobori T, Ishida M, et al. Dynamic analysis of structures with Maxwell model. Earthq Eng Struct Dynam 2000;29(2):159–76.
[32] Soukup J. A thin rectangular viscoelastic orthotropic plate under transverse impuls loading. Appl Comput Mech 2007;1:657–66.
[33] Younesian D, Hosseinkhani A, Askari H, et al. Elastic and viscoelastic foundations: a review on linear and nonlinear vibration modeling and applications. Nonlinear Dyn 2019;97 (1):853–95.
[34] Oni ST, Omolofe B. Dynamic response of prestressed Rayleigh beam resting on elastic foundation and subjected to masses traveling at varying velocity. J Vib Acoust 2011;133 (4):041005.
[35] Li XF, Tang AY, Xi LY. Vibration of a Rayleigh cantilever beam with axial force and tip mass. J Construct Steel Res 2013;80:15–22.
[36] Xi LY, Li XF, Tang GJ. Free vibration of standing and hanging gravity-loaded Rayleigh cantilevers. Int J Mech Sci 2013;66:233–8.
[37] Simitses G, Hodges DH. Fundamentals of structural stability. Butterworth-Heinemann; 2006.
[38] Timoshenko SP, Gere JM, Prager W. Theory of elastic stability. J Appl Mech 1962;29 (1):220–1.
[39] Sugiyama Y, Langthjem MA, Katayama K. Dynamic stability of columns under nonconservative forces. Springer International Publishing; 2019.
[40] Chen YZ. Interaction between compressive force and vibration frequency for a varying cross-section cantilever under action of generalized follower force. J Sound Vib 2003;259 (4):991–9.
[41] Bokaian A. Natural frequencies of beams under compressive axial loads. J Sound Vib 1988;126(1):49–65.
[42] Atanackovic TM, Bouras Y, Zorica D. Nano-and viscoelastic Beck's column on elastic foundation. Acta Mech 2015;226(7):2335–45.
[43] Wang CY, Wang CM. Structural vibration: exact solutions for strings, membranes, beams, and plates. CRC Press; 2013.

[44] Huang Y, Yang LE, Luo QZ. Free vibration of axially functionally graded Timoshenko beams with non-uniform cross-section. Compos B Eng 2013;45(1):1493–8.

[45] Esmailzadeh E, Ohadi AR. Vibration and stability analysis of non-uniform Timoshenko beams under axial and distributed tangential loads. J Sound Vib 2000;236(3):443–56.

[46] Attard MM, Lee JS, Kim MY. Dynamic stability of shear-flexible Beck's columns based on Engesser's and Haringx's buckling theories. Comput Struct 2008;86(21–22):2042–55.

[47] Nageswara Rao B, Venkateswara Rao G. Stability of a cantilever column resting on an elastic foundation subjected to a subtangential follower force at its free end. J Sound Vib 1988;125(3):570–7.

[48] Marzani A, Tornabene F, Viola E. Nonconservative stability problems via generalized differential quadrature method. J Sound Vib 2008;315(1–2):176–96.

[49] Li XF, Lee KY. Effects of Engesser's and Haringx's hypotheses on buckling of Timoshenko and higher-order shear-deformable columns. J Eng Mech 2018;144(1):04017150.

[50] Goyal VK, Kapania RK. Dynamic stability of laminated beams subjected to nonconservative loading. Thin-Walled Struct 2008;46(12):1359–69.

[51] Smith TE, Herrmann G. Stability of a beam on an elastic foundation subjected to a follower force. J Appl Mech 1972;39(2):628–9.

Chapter 6

Nonlocal effects in nanoscale

6.1 Introduction of nonlocal elastic theories

According to Eringen [1–3], the stress field at a point x in an elastic continuum not only depends on the strain field at the point (hyperelastic case) but also on strains at all other points of the body. Eringen attributed this fact to the atomic theory of lattice dynamics and to experimental observations on phonon dispersion. Thus, the nonlocal stress tensor σ at point x is expressed as

$$\sigma = \int_V K\left(\left|\boldsymbol{x}' - \boldsymbol{x}\right|, \tau\right) \boldsymbol{t}\left(\boldsymbol{x}'\right) \mathrm{d}\boldsymbol{x}' \tag{6.1}$$

where $\boldsymbol{t}(\boldsymbol{x}')$ is the classical, macroscopic stress tensor at point x and the kernel function $K(|\boldsymbol{x}' - \boldsymbol{x}|, \tau)$ represents the nonlocal modulus, $|\boldsymbol{x}' - \boldsymbol{x}|$ being the distance (in Euclidean norm) and s is a material constant that depends on internal and external characteristic lengths (such as the lattice spacing and wavelength, respectively). The macroscopic stress t at a point x in a Hookean solid is related to the strain ε at the point by the generalized Hooke's law

$$\boldsymbol{t}(\boldsymbol{x}) = \boldsymbol{C}(\boldsymbol{x}) : \varepsilon(\boldsymbol{x}) \tag{6.2}$$

where $\boldsymbol{C}$ is the fourth-order elasticity tensor and ":" denotes the double-dot product.

The constitutive equations [Eqs. (6.1) and (6.2)] together define the nonlocal constitutive behavior of a Hookean solid. Eq. (6.1) represents the weighted average of the contributions of the strain field at all points in the body to the stress field at a point. However, the integral constitutive relation in Eq. (6.1) makes the elasticity problems difficult to solve. Nevertheless, it is possible (see Ref. [2]) to represent the integral constitutive relations in an equivalent differential form as

$$\left(1 - \tau^2 l^2 \nabla^2\right) \sigma = \boldsymbol{t}, \tau = \frac{e_0 a}{l} \tag{6.3}$$

where e_0 is a material constant, and a and l are the internal and external characteristic lengths, respectively.

Higher-Order Theories of Functionally Graded Circular Cylindrical Beams and Columns
https://doi.org/10.1016/B978-0-443-44718-1.00005-0

6.2 Assumption and governing equations

6.2.1 Circular cylindrical beams

For a beam with a circular cross section, similar to the Timoshenko beam theory, there are two independent variables, the lateral deflection and rotation of the cross section about the centroidal axis, which are denoted by w and ψ, respectively. For such a beam, the longitudinal displacement u in the axial direction, denoted as the x-direction, can be induced by the lateral deflection and rotation of the cross section and then may be expressed as follows [4,5]:

$$u = z\psi - \frac{z^3 + zy^2}{3R^2}\left(\frac{\partial w}{\partial x} + \psi\right) \tag{6.4}$$

Consequently, using the kinematic relations, the normal strain ε_{xx} and shear strains γ_{xy}, γ_{xz} can be expressed as follows:

$$\varepsilon_{xx} = z\frac{\partial \psi}{\partial x} - \frac{z^3 + zy^2}{3R^2}\left(\frac{\partial^2 w}{\partial x^2} + \frac{\partial \psi}{\partial x}\right) \tag{6.5}$$

$$\gamma_{xy} = -\frac{2zy}{3R^2}\left(\frac{\partial w}{\partial x} + \psi\right) \tag{6.6}$$

$$\gamma_{xz} = \frac{3R^2 - 3z^2 - y^2}{3R^2}\left(\frac{\partial w}{\partial x} + \psi\right) \tag{6.7}$$

In addition, from the above relations, it is easy to check the shear strain γ_{xr} [4,5]

$$\gamma_{xr} = \frac{z}{R^2\sqrt{z^2 + y^2}}\left(R^2 - z^2 - y^2\right)\left(\psi + \frac{\partial w}{\partial x}\right) \tag{6.8}$$

As a result, at the surface of beams with circular cross section, the shear strain or shear stress vanishes. This means that the selection of the displacement component u given by Eq. (6.4) is appropriate for treating bending problems of beams with circular cross section.

According to the nonlocal theory of elasticity [3], the one dimensional version of the constitutive equations can be written in the following form

$$\sigma_{xx} - l^2\frac{\partial^2 \sigma_{xx}}{\partial x^2} = E\varepsilon_{xx} \tag{6.9}$$

$$\tau_{xy} - l^2\frac{\partial^2 \tau_{xy}}{\partial x^2} = G\gamma_{xy} \tag{6.10}$$

$$\tau_{xz} - l^2\frac{\partial^2 \tau_{xz}}{\partial x^2} = G\gamma_{xz} \tag{6.11}$$

where σ_{xx} is the normal stress, τ_{xy}, τ_{xz} are the shear stresses; E and G are Young's modulus and shear modulus of the beam, respectively; and $l = e_0 a$ is a small-scale parameter, with e_0 being a nondimensional material constant that can be determined by experiment or molecular dynamic simulation and a being the material internal characteristic length that is taken as the C–C bond length. In this chapter, we take $e_0 = 0.8$ and the C–C bond length $a = 0.142$ nm for computation. The bending moment M and shearing force Q at any position x of the beam can be evaluated through the following integrals

$$M = \int_A \sigma_{xx} z \mathrm{d}A \tag{6.12}$$

$$Q = \int_A \tau_{xz} \mathrm{d}A \tag{6.13}$$

where $A = \pi R^2$ is the cross sectional area. With Eqs. (6.5)–(6.7) in mind, substitution of Eqs. (6.9)–(6.11) into Eqs. (6.12) and (6.13) results in

$$M - l^2 \frac{\partial^2 M}{\partial x^2} = \frac{EI}{9}\left(7\frac{\partial \psi}{\partial x} - 2\frac{\partial^2 w}{\partial x^2}\right) \tag{6.14}$$

$$Q - l^2 \frac{\partial^2 Q}{\partial x^2} = \frac{2GA}{3}\left(\frac{\partial w}{\partial x} + \psi\right) \tag{6.15}$$

where $I = \pi R^4/4$ is the second moment of the cross sectional area.

On the other hand, with the aid of the equations of motion, we have:

$$\frac{\partial \sigma_{xx}}{\partial x} + \frac{\partial \tau_{xy}}{\partial y} + \frac{\partial \tau_{xz}}{\partial z} = \rho \frac{\partial^2 u}{\partial t^2} \tag{6.16}$$

$$\frac{\partial \tau_{xz}}{\partial x} + \frac{\partial \tau_{yz}}{\partial y} + \frac{\partial \sigma_{zz}}{\partial z} = \rho \frac{\partial^2 w}{\partial t^2} \tag{6.17}$$

where ρ is the mass density of the beam. Here the equation of motion in the y-direction is neglected since our analysis is limited to the planar bending in the xoz-plane. Now we multiply both sides of Eq. (6.16) by z and then integrate over the cross sectional area A, resulting in

$$\frac{\partial M}{\partial x} - Q = \rho I\left(\frac{7}{9}\frac{\partial^2 \psi}{\partial t^2} - \frac{2}{9}\frac{\partial^3 w}{\partial x \partial t^2}\right) \tag{6.18}$$

The derivation of the above equation makes use of Eqs. (6.12), (6.13), and (6.4). A straightforward integration of both sides of Eq. (6.17) over the cross sectional area A leads to

$$\frac{\partial Q}{\partial x} + q = \rho A \frac{\partial^2 w}{\partial t^2} \tag{6.19}$$

where q is the prescribed distributed loading on a lateral surface of the beam, or equivalently along the axial direction.

Next, eliminating M and Q from Eqs. (6.14), (6.15), (6.18), and (6.19), we can obtain a set of coupled partial differential equations governing the dynamic behavior of beams with a circular cross section:

$$\frac{EI}{6}\left(7\frac{\partial^2\psi}{\partial x^2}-2\frac{\partial^3 w}{\partial x^3}\right)-GA\left(\frac{\partial w}{\partial x}+\psi\right)=\frac{\rho I}{6}\left(1-l^2\frac{\partial^2}{\partial x^2}\right)\left(7\frac{\partial^2\psi}{\partial t^2}-2\frac{\partial^3 w}{\partial x\partial t^2}\right) \tag{6.20}$$

$$\frac{2}{3}GA\left(\frac{\partial^2 w}{\partial x^2}+\frac{\partial\psi}{\partial x}\right)=\left(1-l^2\frac{\partial^2}{\partial x^2}\right)\left(\rho A\frac{\partial^2 w}{\partial t^2}-q\right) \tag{6.21}$$

Here, we make some observations of the above equations. First, it is found that if setting the scale parameter $l=0$, the above resulting equations reduce to those derived in Refs. [4,5]. Second, if adopting the nonlocal Timoshenko beam theory which contains an undetermined parameter, i.e., shear correction factor κ [6–8], this parameter cannot be directly determined by the model itself. This is due to the fact that for the theories of not only classical Timoshenko beams but also nonlocal Timoshenko beams, the shear stress is assumed to be uniformly distributed at any cross section. Thus a shear correction factor κ must be introduced to relax the traction-free surface condition. In this chapter, the above-mentioned uniformly distributed shear stress assumption is abandoned, and the corresponding shear correction factor κ is found to be 6/7.

In order to further simplify the above-resulting coupled equations, we introduce a new unknown function F to convert the above set to a single governing equation. To this end, we express w and ψ as follows:

$$w=F-\frac{7EI}{6GA}\frac{\partial^2 F}{\partial x^2}+\frac{7\rho I}{6GA}\left(1-l^2\frac{\partial^2}{\partial x^2}\right)\frac{\partial^2 F}{\partial t^2} \tag{6.22}$$

$$\psi=-\frac{\partial F}{\partial x}-\frac{EI}{3GA}\frac{\partial^3 F}{\partial x^3}+\frac{\rho I}{3GA}\left(1-l^2\frac{\partial^2}{\partial x^2}\right)\frac{\partial^3 F}{\partial x\partial t^2} \tag{6.23}$$

It is readily checked that when substituting Eqs. (6.22) and (6.23) into Eq. (6.20), it is automatically satisfied. Substituting Eqs. (6.22) and (6.23) into Eq. (6.21) leads to a single governing equation in F as follows:

$$\frac{\partial^4 F}{\partial x^4}+\left(1-l^2\frac{\partial^2}{\partial x^2}\right)\frac{\partial^2}{\partial t^2}\left[\frac{\rho A}{EI}F-\left(\frac{\rho}{E}+\frac{7\rho}{6G}\right)\frac{\partial^2 F}{\partial x^2}+\frac{7\rho^2}{6GE}\left(1-l^2\frac{\partial^2}{\partial x^2}\right)\frac{\partial^2 F}{\partial t^2}\right]$$
$$=\frac{1}{EI}\left(1-l^2\frac{\partial^2}{\partial x^2}\right)q \tag{6.24}$$

Now let us consider several special cases. First, if setting the shear rigidity to be sufficiently high compared to the bending rigidity, i.e., $GA/EI \to \infty$, from Eqs. (6.22) and (6.23) we find

$$F = w, \psi = -\frac{\partial w}{\partial x} \tag{6.25}$$

Then the final governing Eq. (6.24) reduces to

$$EI\frac{\partial^4 w}{\partial x^4} + \left(1 - l^2\frac{\partial^2}{\partial x^2}\right)\left(\rho A\frac{\partial^2 w}{\partial t^2} - \rho I\frac{\partial^4 w}{\partial t^2 \partial x^2}\right) = q - l^2\frac{\partial^2 q}{\partial x^2} \tag{6.26}$$

which is identical to the governing equation for nonlocal Rayleigh beams [9]. In Eq. (6.26), the contribution of shear deformation has been ignored, and the rotary inertia is still included. If further neglecting the rotary inertia, the well-known governing equation of the nonlocal Euler–Bernoulli theory of beams is recovered [10]:

$$EI\frac{\partial^4 w}{\partial x^4} + \rho A\left(\frac{\partial^2 w}{\partial t^2} - l^2\frac{\partial^4 w}{\partial t^2 \partial x^2}\right) = q - l^2\frac{\partial^2 q}{\partial x^2} \tag{6.27}$$

For a general case, shear deformation and rotary inertia are both present, and then the chapter provides an exact analytic approach to tackle related problems.

6.2.2 Cylindrical tube beams with radial nonhomogeneity

To describe the radial nonhomogeneity of functionally graded (FG) cylindrical nanotubes of length L, as shown in Fig. 6.1, the material properties are dominated by the mixed power-law model [5,11].

$$P(r) = P_i + (P_o - P_i)\left(\frac{r - R_i}{R_o - R_i}\right)^N \tag{6.28}$$

where $P(r)$ represents the effective mechanical properties, the subscripts o and

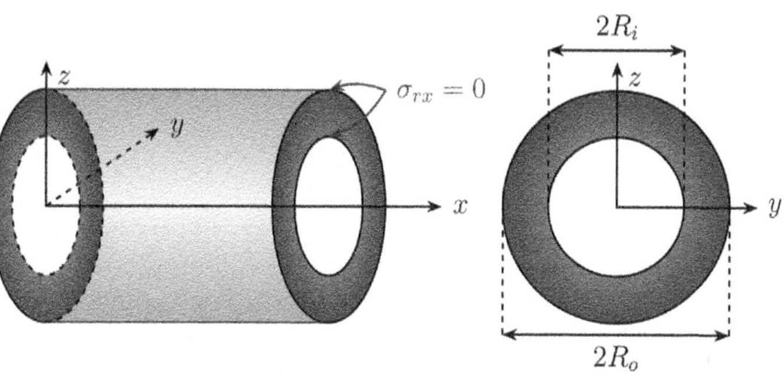

FIGURE 6.1 Schematic of an FG circular cylindrical tube.

i represent the external and internal surfaces of cylindrical tubes, respectively, R represents the radius, and N is a material parameter (the gradient index) that depicts the volume fraction change of the components involved. One can readily see that the material properties of FG tubes rely on the radius r, such as Young's modulus $E(r)$ and shear modulus $G(r)$. According to the geometric characteristics of cylindrical tubes, it is convenient to utilize Cartesian (x, y, z) coordinates and cylindrical coordinates (x, r, θ) simultaneously, meanwhile, the origin point is fixed on the axis of the nanotube. The following relationship exists between the two coordinate systems.

$$y = r\cos\theta, z = r\sin\theta, r = \sqrt{y^2 + z^2} \tag{6.29}$$

With reference to both coordinate systems, we define (u, v, w) and (u, u_r, u_θ) as the displacement components corresponding to (x, y, z) and (x, r, θ), respectively.

Under applied loading, the flexural behavior of an FG nanotube can be characterized by two independent variables: the deflection $w(x, t)$ and rotation angle $\psi(x, t)$. The axial displacement can be written as follows [12]:

$$u(x, y, z, t) = u_0(x, t) - z\frac{\partial w}{\partial x} + f(y, z)\left[\psi(x, t) + \frac{\partial w}{\partial x}\right] \tag{6.30}$$

where $u_0(x, t)$ represents the axial displacement caused by axial loading and disappears if there is no applied axial loading. The known function f is introduced to describe the cross sectional warping deformation. It should be noted that when Cartesian and cylindrical coordinate systems are selected, the warping shape function can be written as $f(y, z)$ and $f(r\cos\theta, r\sin\theta)$ in Cartesian and polar coordinates, respectively. Here, the dynamic behavior only in the xoz-plane of FG nanotubes without axial loading is considered, thus, $u_0(x)$ is negligible. In addition, the warping shape function $f(y, z)$ is assumed to be independent of x and t. Note that if $f(y, z) = 0$, the above relationship [Eq. (6.30)] reduces to the Euler–Bernoulli beam theory, while if $f(y, z) = 1$, the Timoshenko beam theory is recovered. With Eq. (6.30) along with $u_0 = 0$, one obtains the expression for the strain components as follows:

$$s_{xx} = -z\frac{\partial^2 w}{\partial x^2} + f(y, z)\left(\frac{\partial \psi}{\partial x} + \frac{\partial^2 w}{\partial x^2}\right) \tag{6.31}$$

$$\gamma_{xz} = \left[\psi(x, t) + \frac{\partial w}{\partial x}\right]\frac{\partial f}{\partial z} \tag{6.32}$$

$$\gamma_{xr} = \left[\psi(x, t) + \frac{\partial w}{\partial x}\right]\frac{\partial f}{\partial r} \tag{6.33}$$

In the present study, a novel exponential warping is proposed below:

$$f(y,z) = z\exp\left(\frac{R_i^2R_o^2 - (y^2+z^2)^2}{2(R_i^2+R_o^2)(y^2+z^2)}\right) \tag{6.34}$$

or

$$f(r\cos\theta, r\sin\theta) = r\exp\left(\frac{R_i^2R_o^2 - r^4}{2(R_i^2+R_o^2)r^2}\right)\sin\theta \tag{6.35}$$

With Eq. (6.34) or Eq. (6.35), it is readily found that

$$\frac{\partial f}{\partial r} = \frac{z}{r}\exp\left(\frac{R_i^2R_o^2 - r^4}{2(R_i^2+R_o^2)r^2}\right)\left[1 - \frac{r^4 + R_i^2R_o^2}{(R_i^2+R_o^2)r^2}\right] \tag{6.36}$$

from which, it is clear that $\partial f/\partial r = 0$ at $r = R_o$ or $r = R_i$, respectively. Owing to Eq. (6.33), one finds that shear strain γ_{xr} automatically vanishes on the outer and inner surfaces, i.e.,

$$\gamma_{xr} = 0, \text{as } r = R_o, R_i \tag{6.37}$$

By using Eringen's nonlocal elasticity, the form of nonlocal differential constitutive relation can be written as [1]:

$$\sigma_{xx} = E s_{xx} + \varepsilon^2 \frac{\partial^2 \sigma_{xx}}{\partial x^2} \tag{6.38}$$

$$\tau_{xz} = G\gamma_{xz} + \varepsilon^2 \frac{\partial^2 \tau_{xz}}{\partial x^2} \tag{6.39}$$

$$\tau_{xr} = G\gamma_{xr} + \varepsilon^2 \frac{\partial^2 \tau_{xr}}{\partial x^2} \tag{6.40}$$

where $\varepsilon = e_0 a$ is the nonlocal parameter to describe the influence of nonlocal elastic stress field, in which, e_0 is a constant determined experimentally and a is an internal characteristic length such as the carbon–carbon bond length. Considering $f(y,z)$ has no relation with x, from Eqs. (6.37) and (6.40), one readily finds

$$\tau_{xr} = 0, r = R_i \text{ or } R_o \tag{6.41}$$

That is, the shear stress free condition is identically fulfilled on the outer and inner surfaces. Thus, Eq. (6.34) provides us with a suitable warping shape function to adequately describe the flexural behavior of an FG nanotube. This is easily understood since the nonlocality is characterized by the axial variable rather than the cross section. Therefore the warping shape function is independent of the nonlocal parameter. Thus, by substituting Eqs. (6.31)–(6.33) into Eqs. (6.38)–(6.40), the stress components can be rewritten as:

$$\sigma_{xx}=E\left[-z\frac{\partial^2 w}{\partial x^2}+f(y,z)\left(\frac{\partial\psi}{\partial x}+\frac{\partial^2 w}{\partial x^2}\right)\right]+\varepsilon^2\frac{\partial^2\sigma_{xx}}{\partial x^2} \tag{6.42}$$

$$\tau_{xz}=G\frac{\partial f}{\partial z}\left[\psi(x,t)+\frac{\partial w}{\partial x}\right]+\varepsilon^2\frac{\partial^2\tau_{xz}}{\partial x^2} \tag{6.43}$$

$$\begin{aligned}\tau_{xr} &= G\frac{\partial f}{\partial r}\left[\psi(x,t)+\frac{\partial w}{\partial x}\right]+\varepsilon^2\frac{\partial^2\tau_{xr}}{\partial x^2}\\ &= G\frac{\partial f}{\partial r}\left(1+\varepsilon^2\frac{\partial^2}{\partial x^2}\right)\left[\psi(x,t)+\frac{\partial w}{\partial x}\right]+\varepsilon^4\frac{\partial^4\tau_{xr}}{\partial x^4}\end{aligned} \tag{6.44}$$

where ε^4 is a higher-order infinitesimal quantity and can be ignored. By observing Eq. (6.44), it can be seen that the shear stress free boundary condition is still automatically satisfied on the surface of nanotubes when the scale effect is considered. With stress components on hand, one gets

$$Q=\int_A \tau_{xz}\mathrm{d}A=(GA)^*\left(\frac{\partial w}{\partial x}+\psi\right)+\varepsilon^2\frac{\partial^2 Q}{\partial x^2} \tag{6.45}$$

$$M=\int_A z\sigma_{xx}\mathrm{d}A=(EI)^*\left(\frac{\partial^2 w}{\partial x^2}+\frac{\partial\psi}{\partial x}\right)-(EI)_0\frac{\partial^2 w}{\partial x^2}+\varepsilon^2\frac{\partial^2 M}{\partial x^2} \tag{6.46}$$

where

$$(EI)^*=\int_A zf(y,z)E(r)\mathrm{d}A \tag{6.47}$$

$$(EI)_0=\int_A z^2E(r)\mathrm{d}A \tag{6.48}$$

$$(GA)^*=\int_A G(r)\frac{\partial f}{\partial z}\mathrm{d}A \tag{6.49}$$

It should be noted that since both Cartesian and cylindrical coordinate systems are used, the warping shape function f is not only a function of variables y and z, but also a function of r and θ, which are used simultaneously and related by Eq. (6.29).

In the above, $\partial^2 Q/\partial x^2$ on the right-hand side of Eq. (6.45) is still unknown. Similarly, for the bending moment, $\partial^2 M/\partial x^2$ is also unknown. By means of the transversal and bending moment equilibrium of FG nanotubes, one has

$$\frac{\partial Q}{\partial x}=-q+m_0\frac{\partial^2 w}{\partial t^2} \tag{6.50}$$

$$\frac{\partial M}{\partial x}-Q=-m_2\frac{\partial^3 w}{\partial t^2\partial x}+m_{2f}\left(\frac{\partial^3 w}{\partial t^2\partial x}+\frac{\partial^2\psi}{\partial t^2}\right) \tag{6.51}$$

where

$$m_0=\int_A \rho(r)\mathrm{d}A \tag{6.52}$$

$$m_2=\int_A z^2\rho(r)\mathrm{d}A \tag{6.53}$$

$$m_{2f}=\int_A zf(y,z)\rho(r)\mathrm{d}A \tag{6.54}$$

and q represents transversely distributed load. Differentiating Eq. (6.51) with respect to x yields

$$\frac{\partial^2 M}{\partial x^2}-\frac{\partial Q}{\partial x}=-m_2\frac{\partial^4 w}{\partial t^2\partial x^2}+m_{2f}\left(\frac{\partial^4 w}{\partial t^2\partial x^2}+\frac{\partial^3\psi}{\partial t^2\partial x}\right) \tag{6.55}$$

Substituting Eq. (6.50) into Eq. (6.55) leads to

$$\frac{\partial^2 M}{\partial x^2}=-q+m_0\frac{\partial^2 w}{\partial t^2}-m_2\frac{\partial^4 w}{\partial t^2\partial x^2}+m_{2f}\left(\frac{\partial^4 w}{\partial t^2\partial x^2}+\frac{\partial^3\psi}{\partial t^2\partial x}\right) \tag{6.56}$$

Substituting Eq. (6.56) into Eq. (6.46) results in:

$$\begin{aligned}M=&(EI)^*\left(\frac{\partial^2 w}{\partial x^2}+\frac{\partial\psi}{\partial x}\right)-(EI)_0\frac{\partial^2 w}{\partial x^2}\\&+\varepsilon^2\left[m_0\frac{\partial^2 w}{\partial t^2}-m_2\frac{\partial^4 w}{\partial t^2\partial x^2}+m_{2f}\left(\frac{\partial^4 w}{\partial t^2\partial x^2}+\frac{\partial^3\psi}{\partial t^2\partial x}\right)-q\right]\end{aligned} \tag{6.57}$$

Similarly, substituting Eq. (6.50) into Eq. (6.45) gives

$$Q=(GA)^*\left(\frac{\partial w}{\partial x}+\psi\right)+\varepsilon^2\left(m_0\frac{\partial^3 w}{\partial t^2\partial x}-\frac{\partial q}{\partial x}\right) \tag{6.58}$$

Combining Eqs. (6.57), (6.44), and equilibrium equations, we obtain the following coupled differential equations

$$\begin{aligned}&(EI)^*\left(\frac{\partial^3 w}{\partial x^3}+\frac{\partial^2\psi}{\partial x^2}\right)-(EI)_0\frac{\partial^3 w}{\partial x^3}-(GA)^*\left(\frac{\partial w}{\partial x}+\psi\right)\\&=-\left(1-\varepsilon^2\frac{\partial^2}{\partial x^2}\right)\left[m_2\frac{\partial^3 w}{\partial t^2\partial x}-m_{2f}\left(\frac{\partial^3 w}{\partial t^2\partial x}+\frac{\partial^2\psi}{\partial t^2}\right)\right]\end{aligned} \tag{6.59}$$

$$(GA)^*\left(\frac{\partial^2 w}{\partial x^2}+\frac{\partial \psi}{\partial x}\right)+\varepsilon^2\left(-\frac{\partial^2 q}{\partial x^2}+m_0\frac{\partial^4 w}{\partial t^2\partial x^2}\right)=-q+m_0\frac{\partial^2 w}{\partial t^2} \tag{6.60}$$

Once we set $f=0$, which means the neglect of shear deformation, the rotational inertia still exists, leading to

$$(EI)^*=(GA)^*=m_{2f}=0 \tag{6.61}$$

Combining Eqs (6.61), (6.50), (6.51), and (6.57), one readily derives

$$(EI)_0\frac{\partial^4 w}{\partial x^4}+m_0\frac{\partial^2 w}{\partial t^2}-m_2\frac{\partial^4 w}{\partial t^2\partial x^2}-\varepsilon^2\left(-\frac{\partial^2 q}{\partial x^2}+m_0\frac{\partial^2 w}{\partial t^2}-m_2\frac{\partial^6 w}{\partial t^2\partial x^4}\right)=q \tag{6.62}$$

This governing equation actually corresponds to that of the nonlocal Rayleigh beam theory. If one further neglects rotary inertia, i.e., $m_2=0$, the governing equation for nonlocal Euler–Bernoulli beams is obtained

$$(EI)_0\frac{\partial^4 w}{\partial x^4}+m_0\frac{\partial^2 w}{\partial t^2}-\varepsilon^2 m_0\frac{\partial^2 w}{\partial t^2}=q-\varepsilon^2\frac{\partial^2 q}{\partial x^2} \tag{6.63}$$

which is consistent with that in Refs. [13,14]. Although the coupled system of partial differential equations [Eqs. (6.59) and (6.60)] has been derived, its analysis remains inconvenient. Instead, we make an attempt to further simplify them to a single partial differential equation.

Such an attempt is a common mathematical approach that has been frequently used in previous research [11,15]. This is done by selecting the expressions for w and ψ below

$$w=F-\frac{(EI)^*}{(GA)^*}\frac{\partial^2 F}{\partial x^2}+\frac{m_{2f}}{(GA)^*}\frac{\partial^2 F}{\partial t^2}-\frac{\varepsilon^2 m_{2f}}{(GA)^*}\frac{\partial^4 F}{\partial t^2\partial x^2} \tag{6.64}$$

$$\psi=\frac{(EI)^*-(EI)_0}{(GA)^*}\frac{\partial^3 F}{\partial x^3}-\frac{\partial F}{\partial x}+\frac{m_2-m_{2f}}{(GA)^*}\frac{\partial^3 F}{\partial t^2\partial x}+\frac{\varepsilon^2(m_{2f}-m_2)}{(GA)^*}\frac{\partial^5 F}{\partial t^2\partial x^3} \tag{6.65}$$

As with the treatment in Refs. [5,16], after substituting Eqs. (6.64) and (6.65) into coupled Eqs. (6.59) and (6.60), one finds that a final single governing equation can be derived as follows:

$$\begin{aligned}&\frac{\partial^4 F}{\partial x^4}+\left(1-\varepsilon^2\frac{\partial^2}{\partial x^2}\right)\left(a_1\frac{\partial^2 F}{\partial t^2}-a_2\frac{\partial^4 F}{\partial t^2\partial x^2}\right)+a_3\left(1-2\varepsilon^2\frac{\partial^2}{\partial x^2}\right)\frac{\partial^4 F}{\partial t^4}\\&+a_4\frac{\partial^8 F}{\partial t^4\partial x^4}=\left(1-\varepsilon^2\frac{\partial^2}{\partial x^2}\right)\frac{q}{(EI)_0}\end{aligned} \tag{6.66}$$

where

$$\begin{cases} a_1 = \dfrac{m_0}{(EI)_0} \\ a_2 = \dfrac{m_2}{(EI)_0} + \dfrac{m_0}{(EI)_0}\dfrac{(EI)^*}{(GA)^*} \\ a_3 = \dfrac{m_0}{(EI)_0}\dfrac{m_{2f}}{(GA)^*} \\ a_4 = \dfrac{m_{2f}m_0\varepsilon^4}{(EI)_0(GA)^*} \end{cases} \tag{6.67}$$

Eq. (6.66) reduces to that for macro hollow cylindrical nanotubes when setting the nonlocal parameter $\varepsilon = 0$, i.e.,

$$\frac{\partial^4 F}{\partial x^4} - \left[\frac{m_2}{(EI)_0} + \frac{m_0}{(EI)_0}\frac{(EI)^*}{(GA)^*}\right]\frac{\partial^4 F}{\partial t^2 \partial x^2} + \frac{m_0}{(EI)_0}\left[\frac{\partial^2 F}{\partial t^2} + \frac{m_{2f}}{(GA)^*}\frac{\partial^4 F}{\partial t^4}\right] = \frac{q}{(EI)_0} \tag{6.68}$$

which is consistent with the previous results [11]. Therefore, the present model is applicable to both classical FG tubes and micro-/nano-scale FG tubes.

6.3 Bending wave of carbon nanotubes (CNTs)

6.3.1 Single-walled CNTs

Here we analyze the wave propagation of single -walled CNTs (SWCNTs) in a free space or embedded in a surrounding elastic medium. For generality, we first treat the case of the presence of a surrounding elastic medium. For simplicity, only a one parameter elastic foundation is considered—a two-parameter elastic foundation can be similarly analyzed and is therefore omitted for brevity [9]. To do so, a Winkler-type elastic foundation model is employed to simulate the interaction of CNTs with the surrounding elastic medium. According to this model, transverse load q can be stated as:

$$q = -\beta w \tag{6.69}$$

where β is the positive elastic stiffness describing the material property of the surrounding elastic medium, and the negative sign indicates that the pressure q is opposite to the direction of deflection. For transverse waves propagating in CNTs, the introduced auxiliary function F may take the form:

$$F = \widetilde{F}e^{i(kx-\omega t)} \tag{6.70}$$

where $\widetilde{F}$ is the amplitude of the wave motion, and k and ω are the wavenumber and circular frequency, respectively. Substituting Eqs. (6.69) and (6.70) into Eq. (6.24), one gets

$$\left[\frac{\rho A}{EI}\left(1+l^2k^2\right)\omega^2\right]^2-\frac{m_1}{c_1r_g^4}\frac{\rho A}{EI}\left(1+l^2k^2\right)\omega^2+\frac{m_2}{c_1r_g^4}=0 \tag{6.71}$$

with

$$\begin{cases} m_1=1+(c_1+1)r_g^2k^2+c_1\dfrac{\beta r_g^4}{EI}\left(1+l^2k^2\right) \\ m_2=k^4+\dfrac{\beta}{EI}\left(1+l^2k^2\right)\left(1+c_1r_g^2k^2\right) \\ c_1=\dfrac{7(1+\nu)}{3} \\ r_g=\sqrt{\dfrac{I}{A}} \end{cases} \tag{6.72}$$

where r_g is the radius of gyration, and ν is Poisson's ratio. By solving the quadratic equation [Eq. (6.71)], we obtain the dispersion relation as

$$\omega_{1,2}=\sqrt{\frac{EI}{\rho A}\frac{1}{\left(1+l^2k^2\right)}\frac{m_1\mp\sqrt{m_1^2-4c_1r_g^4m_2}}{2c_1r_g^4}} \tag{6.73}$$

From the above, two branches of waves are found to exist. The lower phase velocity

$$v_1=\omega_1/k \tag{6.74}$$

corresponds to the acoustic branch, and the higher phase velocity $v_2=\omega_2/k$ corresponds to the optical branch, respectively. Letting $k\rightarrow 0$, we can easily obtain two critical frequencies, ω_c^1 and ω_c^2, namely

$$\omega_1^c=\sqrt{\frac{\beta}{\rho A}},\ \omega_2^c=\sqrt{\frac{6GA}{7\rho I}} \tag{6.75}$$

The first critical frequency ω_1^c depends only on the surrounding elastic medium, whereas the second critical frequency ω_2^c depends only on the shear deformation. It turns out that two branches of bending waves occur only for the frequencies in excess of ω_1^c and ω_2^c. In the presence of the surrounding elastic medium, transverse waves cannot propagate at frequencies lower than ω_1^c. Moreover, waves can propagate in SWCNTs with two different wave speeds at frequencies higher than ω_2^c, and with one wave speed at frequencies between ω_1^c and ω_2^c.

Next, let us make some observations for a special case, i.e., without the surrounding elastic medium. In this case, we have $\beta=0$. The absence of the surrounding elastic matrix gives rise to $\omega_1^c=0$ and bending waves always occur for any frequencies. Moreover, there is only one acoustic branch at frequencies lower than ω_2^c, and two wave speeds, including acoustic wave speed and optical wave speed, at frequencies higher than ω_2^c. The dispersion relation Eq. (6.73) then simplifies to

$$\omega_{1,2}=\sqrt{\frac{EI}{\rho A}\frac{1}{\left(1+l^2k^2\right)}\frac{1+(c_1+1)r_g^2k^2\mp\sqrt{\left[1+(c_1+1)r_g^2k^2\right]^2-4c_1r_g^4k^4}}{2c_1r_g^4}} \tag{6.76}$$

Furthermore, if the shear rigidity is removed and the rotary inertia remains, one obtains the dispersion relation as

$$\omega_{1,2}=\sqrt{\frac{EI}{\rho A}\frac{1}{\left(1+l^2k^2\right)}\frac{k^4}{1+r_g^2k^2}} \tag{6.77}$$

in agreement with that using the nonlocal Rayleigh beam model. If the shear rigidity and rotary inertia are both neglected in Eq. (6.73), the dispersion relation reduces to

$$\omega=\sqrt{\frac{EI}{\rho A}\frac{k^4}{\left(1+l^2k^2\right)}} \tag{6.78}$$

which corresponds to that of the nonlocal Euler–Bernoulli beam model [10].

To illustrate the effectiveness of the present model in describing the wave propagation of CNTs, numerical results of the phase velocity of flexural waves are calculated and compared with other results. The dependences of the phase velocity $v=\omega/k$ on the wavenumber k are shown in Figs. 6.2 and 6.3 for $(5,5)$ and $(10,10)$ armchair SWCNTs in the absence of a surrounding elastic medium, respectively. In the following computations, we take mass density $\rho=2.237$ g/cm^3, thickness $t=0.342$ nm, C–C bond length $a=0.142$ nm, Young's modulus $E_1=0.39$ TPa, and Poisson's ratio $\nu_1=0.22$ for $(5,5)$ armchair CNTs, and $E_2=0.45$ TPa, $\nu_2=0.20$ for $(10,10)$ armchair CNTs [17]. In general, CNTs are more suitable to be taken as a circular cylindrical shell model due to their hollow thin-walled structures. However, in order to avoid becoming cumbersome and to gain an insight into intrinsic features, it is reasonable to model CNTs as an elastic beam when the slenderness ratio is large enough. For comparison, the results obtained by using molecular dynamics simulation [17] are also displayed in these figures. In Figs. 6.2 and 6.3, two branches of flexural wave adopting the nonlocal higher-order beam theory suggested in the present chapter are labeled with NHBT-1 and -2, respectively. Note that in the nonlocal elasticity theory, the dimensionless parameter e_0 may be determined by other approaches including experiment and molecular dynamics simulation [18]. Here, based on the results of molecular dynamics simulation [17], we find that if choosing $e_0=0.8$ in our computation, our results labeled NHBT-1 are very close to those using molecular dynamics

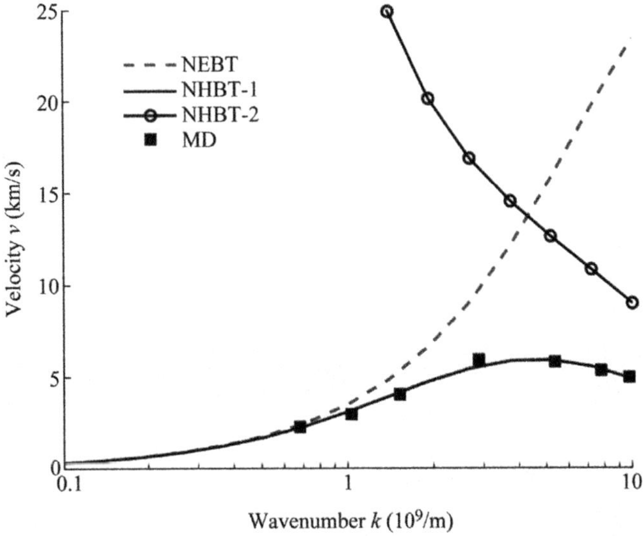

FIGURE 6.2 Comparison of the phase velocity v of bending waves propagating in a (5,5) armchair SWCNT using three different models (NEBT: nonlocal Euler–Bernoulli beam theory; NHBT: nonlocal higher-order beam theory; MD: molecular dynamics simulation).

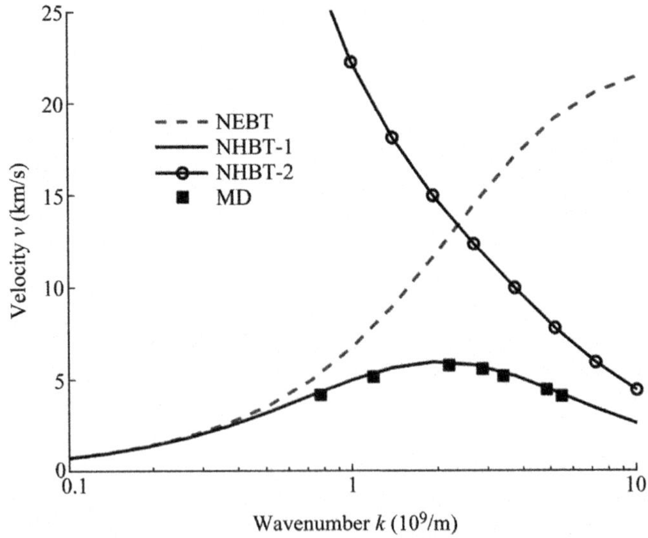

FIGURE 6.3 Comparison of the phase velocity v of bending waves propagating in a (10,10) armchair SWCNT using three different models (NEBT: nonlocal Euler–Bernoulli beam theory; NHBT: nonlocal higher-order beam theory; MD: molecular dynamics simulation).

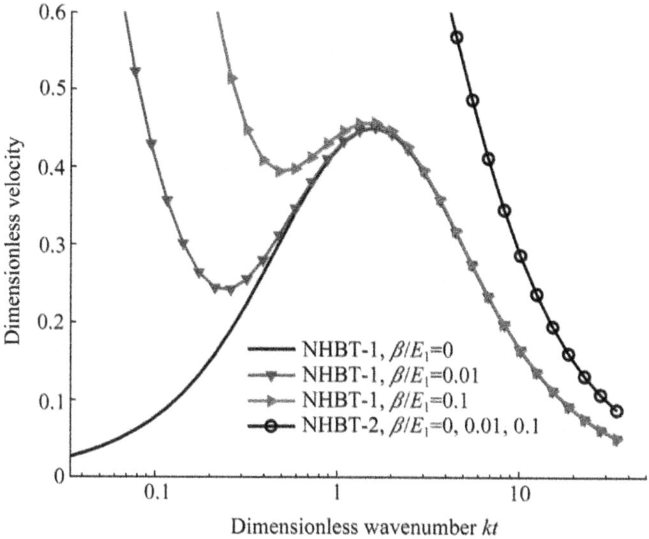

FIGURE 6.4 Effects of a surrounding elastic medium on the acoustic branch of dimensionless phase velocity $\bar{v} = v\big/\sqrt{E/\rho}$ in a (5,5) armchair SWCNT.

simulation [17] for both $(5, 5)$ and $(10, 10)$ armchair SWCNTs. This also implies that the present model is very efficient. For the lower acoustic branch in Figs. 6.2 and 6.3, the wave speed reaches its peak value at some wavenumber, e.g., 5.94 km/s at 4.6/nm for $(5, 5)$ armchair SWCNTs and 7.28 km/s at 3.5/nm for $(10, 10)$ armchair SWCNTs, and then starts to drop with increasing wavenumber. However, if using the nonlocal Euler–Bernoulli beam theory, there is only one branch of flexural wave, labeled with NEBT in Figs. 6.2 and 6.3. Moreover, when the wavenumber is progressively increased, the phase speeds calculated by the nonlocal Euler–Bernoulli beam theory greatly deviate from the results of the molecular dynamics simulation. This is in agreement with our intuition since the results based on the Euler–Bernoulli beam theory are always overestimated.

In order to show the effects of a surrounding elastic medium on the phase velocity of transverse waves, the numerical results of the dimensionless phase velocity $\bar{v} = v\big/\sqrt{E/\rho}$ of a flexural wave for different β values are obtained and displayed in Figs. 6.4 and 6.5 for (5,5) and (10,10) armchair SWCNTs, respectively. As can be seen in Fig. 6.4, the surrounding elastic medium plays a dominant role in determining the acoustic branch of flexural waves at lower wavenumbers. Then we come to a conclusion that the surrounding elastic medium only affects remarkably the phase velocity for small wavenumbers and has little influence on the phase velocity for large wavenumbers. For the

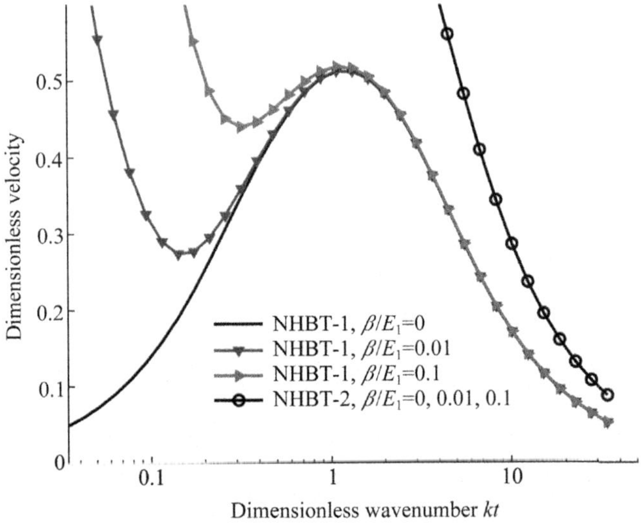

FIGURE 6.5 Effects of a surrounding elastic medium on the acoustic branch of dimensionless phase velocity $\overline{v} = v\big/\sqrt{E/\rho}$ in a (10,10) armchair SWCNT.

optical branch of flexural waves, the surrounding elastic medium has a negligible influence on the phase velocity.

Figs. 6.6 and 6.7 show the dimensionless phase velocity $\overline{v} = v\big/\sqrt{E/\rho}$ against circular frequency ω for an SWCNT in a free space and embedded in an elastic medium. Clearly, from Fig. 6.6, for bending waves in an SWCNT in a free space, an acoustic branch always exists for any frequency. Once the frequency exceeds the critical frequency ω_1^c, the bending wave of the optical branch appears. Fig. 6.7 gives bending waves in an SWCNT embedded in an elastic medium. Compared with Fig. 6.6, we see that in the presence of the surrounding elastic medium, bending waves do not propagate unless the frequency exceeds the first critical frequency ω_1^c. Moreover, between ω_1^c and ω_2^c, there is only one acoustic branch. Once $\omega > \omega_2^c$, not only the acoustic branch but also the optical branch propagates in an SWCNT embedded in an elastic matrix.

A comparison of the present results with those obtained from the Timoshenko beam theory based on stress gradient and strain gradient is made in Figs. 6.8 and 6.9. As pointed out earlier, the shear correction factor κ cannot be determined by the theory itself. Due to this, various values of κ can be found. For example, $\kappa = 0.5$ and $\kappa = 0.8$ are, respectively, chosen in Refs. [6,17]. For these two values, the variation of dimensionless phase velocity of acoustic branch $\overline{v} = v\big/\sqrt{E/\rho}$ with dimensionless wavenumber $\overline{k} = kt$ is plotted in Figs. 6.8 and 6.9 for $\kappa = 0.5$ and $\kappa = 0.8$, respectively. As shown in Figs. 6.8

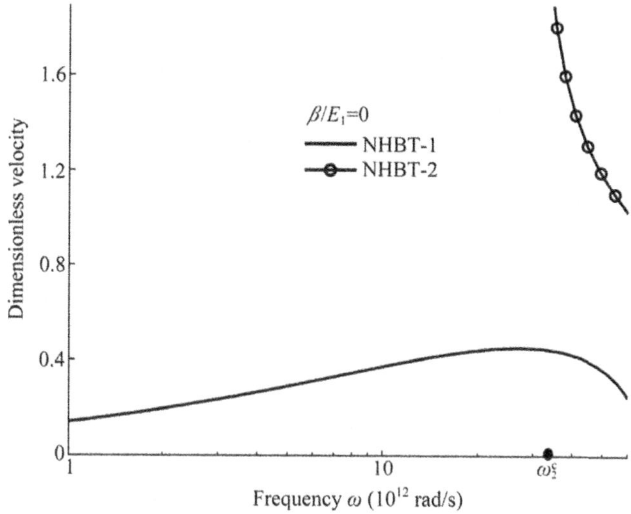

FIGURE 6.6 Variation of the dimensionless phase velocity $\overline{v} = v/\sqrt{E/\rho}$ against circular frequency ω for a (5,5) armchair SWCNT in a free space with β = 0.

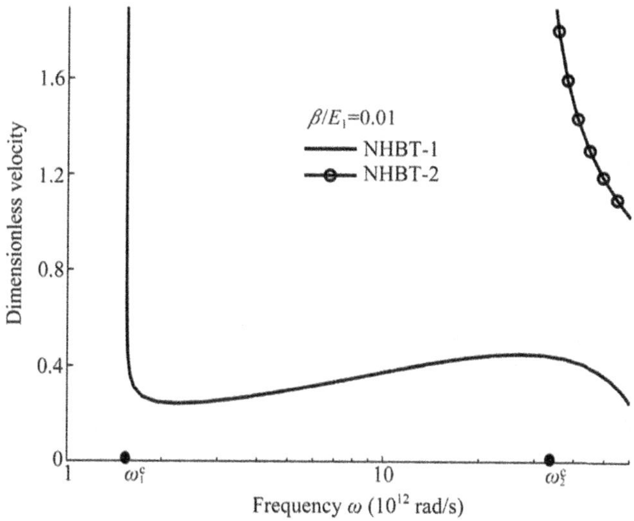

FIGURE 6.7 Variation of the dimensionless phase velocity $\overline{v} = v/\sqrt{E/\rho}$ against circular frequency ω for a (5,5) armchair SWCNT embedded in an elastic matrix with $\beta/E_1 = 0.01$.

and 6.9, the phase velocities have no significant difference at lower wavenumbers, whereas a pronounced discrepancy appears as the wavenumber increases. In particular, it is noted that the phase velocities based on the strain gradient Timoshenko beam theory [17] may become negative when the

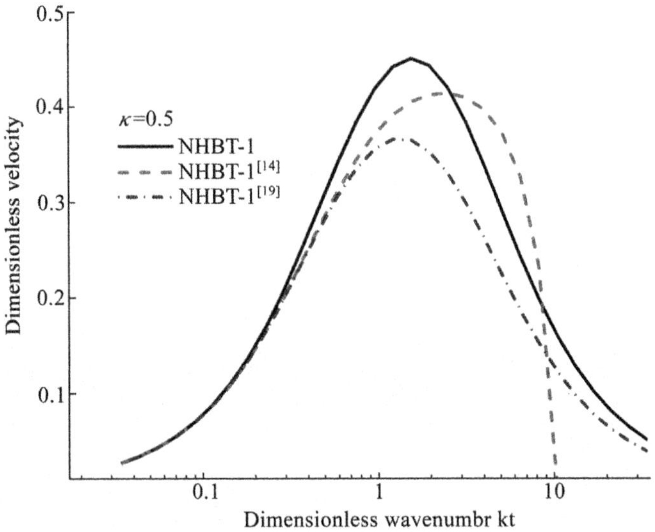

FIGURE 6.8 Comparison of the dimensionless phase velocity $\bar{v} = v/\sqrt{E/\rho}$ of bending waves propagating in a (5,5) armchair SWCNT using three different models; shear correction factor $\kappa = 0.5$.

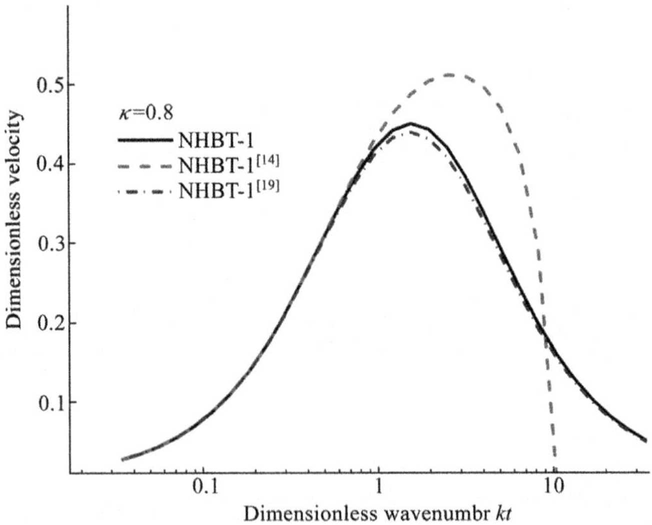

FIGURE 6.9 Comparison of the dimensionless phase velocity $\bar{v} = v/\sqrt{E/\rho}$ of bending waves propagating in a (5,5) armchair SWCNT using three different models; shear correction factor $\kappa = 0.8$.

wavenumber is large enough. The reason for this is that the model becomes unstable [19] since the positive sign before the strain gradient does not result in positive semidefinite energy, which predicts imaginary phase velocities at larger wavenumbers.

6.3.2 Double-walled CNTs

Let us now turn our attention to the wave propagation in multiwalled CNTs. Different from SWCNTs, there is an interaction called the van der Waals force between two adjacent tubes. Here, for simplicity, double-walled CNTs (DWCNTs) are considered. In linear analysis, the van der Waals interaction at any point between the inner and outer tubes is modeled as a linear normal spring. That is, it can be expressed as

$$q = -\alpha [\![w]\!] \tag{6.79}$$

where $[\![w]\!]$ denotes the difference between their lateral deflections, and α is the van der Waals interaction coefficient, which may be estimated by

$$\alpha = \frac{320(2R_1)\mathrm{erg/cm^2}}{0.16d^2}, d = 0.142\ \mathrm{nm} \tag{6.80}$$

where is R_1 the radius of the inner tube. More precisely, if specifying w_j and ψ_j as the lateral deflection and rotation of the inner and outer tubes, corresponding to $j = 1, 2$, respectively, from Eqs. (6.20) and (6.21) we have the following governing equations:

$$\begin{aligned} &\frac{EI_1}{6}\left(7\frac{\partial^2\psi_1}{\partial x^2} - 2\frac{\partial^3 w_1}{\partial x^3}\right) - GA_1\left(\frac{\partial w_1}{\partial x} + \psi_1\right) \\ &= \frac{\rho I_1}{6}\left(1 - l^2\frac{\partial^2}{\partial x^2}\right)\left(7\frac{\partial^2\psi_1}{\partial t^2} - 2\frac{\partial^3 w_1}{\partial x \partial t^2}\right) \end{aligned} \tag{6.81}$$

$$\frac{2}{3}GA_1\left(\frac{\partial^2 w_1}{\partial x^2} + \frac{\partial \psi_1}{\partial x}\right) = \left(1 - l^2\frac{\partial^2}{\partial x^2}\right)\left[\rho A_1\frac{\partial^2 w_1}{\partial t^2} + \alpha(w_1 - w_2)\right] \tag{6.82}$$

for the inner tube, and

$$\begin{aligned} &\frac{EI_2}{6}\left(7\frac{\partial^2\psi_2}{\partial x^2} - 2\frac{\partial^3 w_2}{\partial x^3}\right) - GA_2\left(\frac{\partial w_2}{\partial x} + \psi_2\right) \\ &= \frac{\rho I_2}{6}\left(1 - l^2\frac{\partial^2}{\partial x^2}\right)\left(7\frac{\partial^2\psi_2}{\partial t^2} - 2\frac{\partial^3 w_2}{\partial x \partial t^2}\right) \end{aligned} \tag{6.83}$$

$$\frac{2}{3}GA_2\left(\frac{\partial^2 w_2}{\partial x^2} + \frac{\partial \psi_2}{\partial x}\right) = \left(1 - l^2\frac{\partial^2}{\partial x^2}\right)\left[\rho A_2\frac{\partial^2 w_2}{\partial t^2} + \alpha(w_2 - w_1) + \beta w_2\right] \tag{6.84}$$

for the outer tube. In the above, the inner tube is affected by the outer tube through the van der Waals force, whereas the outer tube is affected by both the inner tube and the surrounding elastic matrix.

Although a derivation of a single governing equation may be achieved by introducing a new function, as shown in the foregoing section, we do not go further along this line. Instead, we directly assume that the deflection and rotation in Eqs. (6.81)–(6.84) can be expressed as follows:

$$\begin{cases} w_j = \widetilde{w}_j e^{i(kx-\omega t)} \\ \psi_j = -i\widetilde{\psi}_j e^{i(kx-\omega t)} \\ j = 1,2 \end{cases} \tag{6.85}$$

where $\widetilde{w}_j$ and $\widetilde{\psi}_j$ are the amplitude of deflection and rotation of the inner and outer walls, respectively. Substituting Eq. (6.85) into Eqs. (6.81)–(6.84), we can obtain a system of linear algebraic equations for $\widetilde{w}_j$ and $\widetilde{\psi}_j$. To obtain a nontrivial solution of the resulting system, the determinant of the coefficient matrix of the system has to vanish. Thus, we obtain the dispersion relation of bending waves propagating in DWCNTs surrounded by an elastic medium as follows:

$$\det\begin{bmatrix} \frac{2\omega^2 k}{7}-\frac{2EI_1k^3-6GA_1k}{7\rho I_1(1+l^2k^2)} & \omega^2-\frac{7EI_1k^2+6GA_1}{7\rho I_1(1+l^2k^2)} & 0 & 0 \\ \omega^2-\frac{2Gk^2}{3\rho(1+l^2k^2)}-\frac{\alpha}{\rho A_1} & \frac{2Gk}{3\rho(1+l^2k^2)} & \frac{\alpha}{\rho A_1} & 0 \\ 0 & 0 & \frac{2\omega^2 k}{7}-\frac{2EI_2k^3-6GA_2k}{7\rho I_2(1+l^2k^2)} & \omega^2-\frac{7EI_2k^2+6GA_2}{7\rho I_2(1+l^2k^2)} \\ \frac{\alpha}{\rho A_2} & 0 & \omega^2-\frac{2GA_2k^2}{3(1+l^2k^2)}-\frac{(\alpha+\beta)}{\rho A_2} & \frac{2}{3}\frac{Gk}{\rho(1+l^2k^2)} \end{bmatrix}=0 \tag{6.86}$$

The effect of the surrounding elastic medium on the phase velocity of transverse waves propagating in a DWCNT is examined. For numerical analysis, we consider a (5,5)@(10,10) armchair DWCNT embedded in an elastic matrix. The variations of dimensionless phase velocity $\overline{v} = v\big/\sqrt{E_2/\rho}$ with dimensionless wavenumber $\overline{k} = kt$ are plotted in Fig. 6.10. From Fig. 6.10, four different phase velocities, labeled NHBT-1, 2, 3, and 4, can be found for a DWCNT. Similar to an SWCNT, the surrounding elastic medium only affects the lowest branch of wave speeds, in particular at lower wavenumbers, and has little influence at large wavenumbers. For the other branches of waves, the phase velocity is almost free from the influence of the surrounding elastic medium.

In order to illustrate the effects of the dimensionless scale coefficient e_0 on the phase velocity of waves propagating in a DWCNT, dimensionless phase velocity $\overline{v} = v\big/\sqrt{E_2/\rho}$ versus the coefficient e_0 is plotted in Figs. 6.11–6.13 at $k = 5 \times 10^8$/m, $k = 10^9$/m, and $k = 2 \times 10^9$/m, respectively. It can be observed that the effect of the scale coefficient on the phase velocity is almost

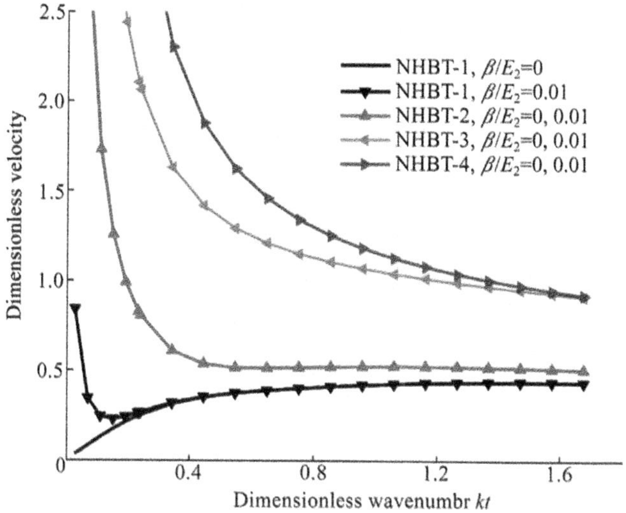

FIGURE 6.10 Dimensionless phase velocity $\overline{v} = v/\sqrt{E_2/\rho}$ of bending waves propagating in a (5,5)@(10,10) armchair DWCNT embedded in an elastic medium.

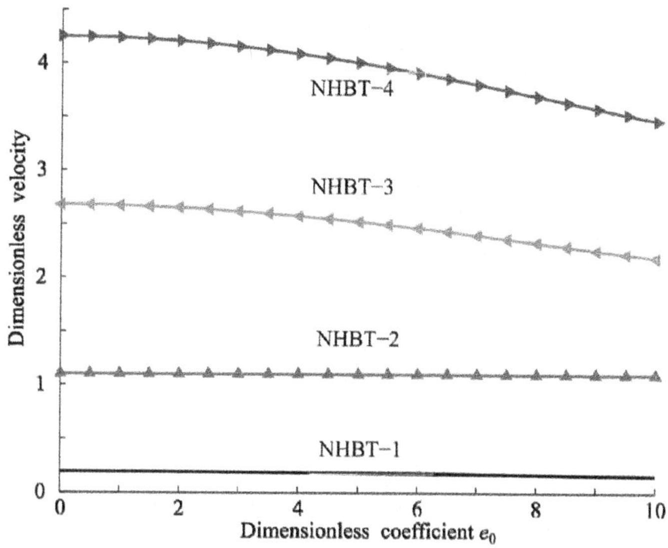

FIGURE 6.11 Dimensionless phase velocity $\overline{v} = v/\sqrt{E_2/\rho}$ versus the dimensionless coefficient $e_0$in a (5,5)@(10,10) armchair DWCNT; $k = 5\times 108/\text{m}$.

negligible at a small wavenumber. As the wavenumber increases, the effect becomes remarkable. In particular, from Fig. 6.13, the phase velocity drops sharply for all four branches, which shows that the phase velocity of bending waves is sensitive to the scale coefficient at a large wavenumber. Furthermore,

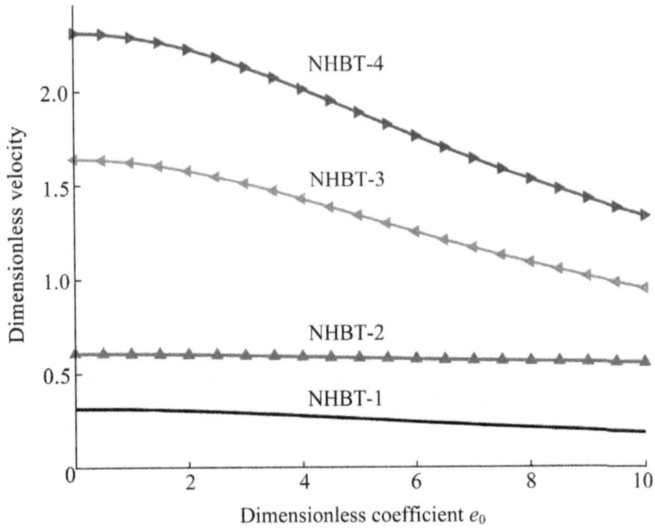

FIGURE 6.12 Dimensionless phase velocity $\overline{v} = v/\sqrt{E_2/\rho}$ versus the dimensionless coefficient e_0 in a (5,5)@(10,10) armchair DWCNT; $k = 109$/m.

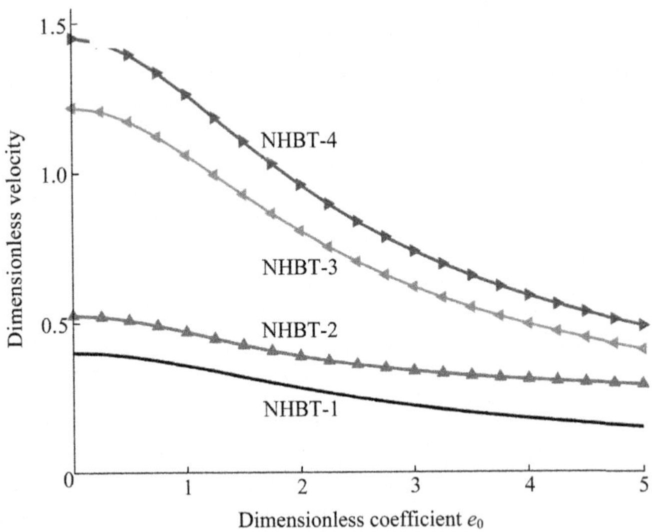

FIGURE 6.13 Dimensionless phase velocity $\overline{v} = v/\sqrt{E_2/\rho}$ versus the dimensionless coefficient e_0 in a (5,5)@(10,10) armchair DWCNT; $k = 2\times109$/m.

we also display the ratio of the phase velocity to the classical value in Figs. 6.14–6.16 for the above-mentioned three wavenumbers. The introduction of the scale coefficient decreases the phase velocity, and this effect is noticeable at large wavenumbers. It is interesting to note that in all cases, the second phase velocity has the smallest influence.

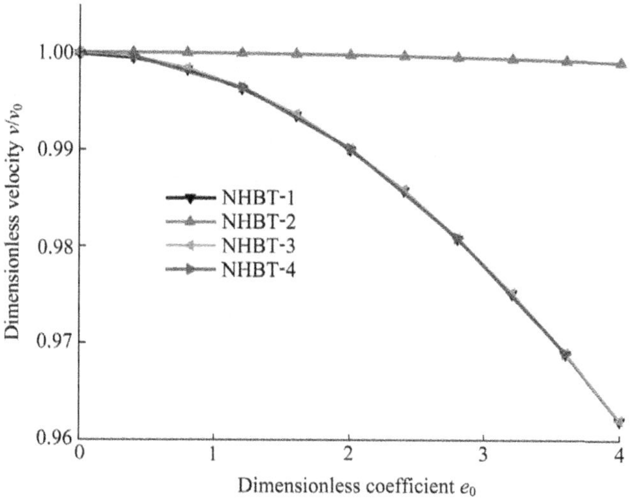

FIGURE 6.14 Ratio of the phase velocity to the classical value versus the dimensionless coefficient e_0 in a (5,5)@(10,10) armchair DWCNT; $k = 5 \times 10^8$/m.

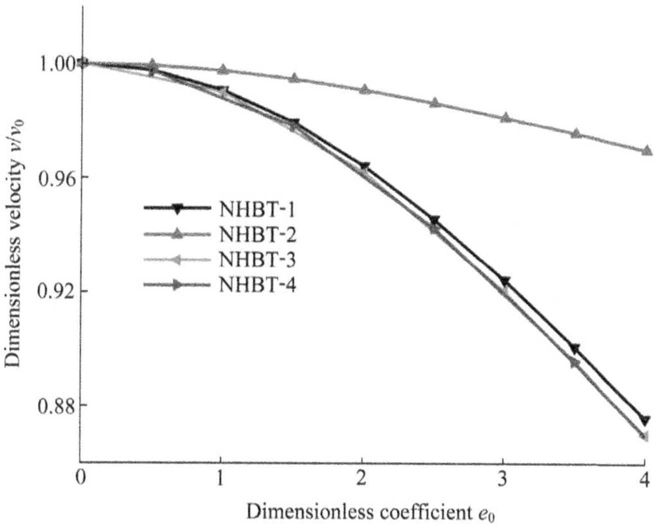

FIGURE 6.15 Ratio of the phase velocity to the classical value versus the dimensionless coefficient e_0 in a (5,5)@(10,10) armchair DWCNT; $k = 10^9$/m.

The influence of the van der Waals force on the dimensionless phase velocity is shown in Figs. 6.17 and 6.18. In Figs. 6.17 and 6.18, the acoustic branch of wave speeds is only displayed for a DWCNT in a free space and embedded in an elastic medium, respectively. From Fig. 6.17, it can be seen

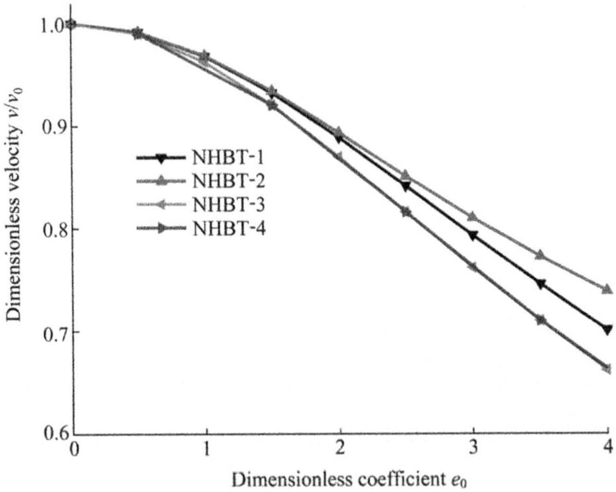

FIGURE 6.16 Ratio of the phase velocity to the classical value versus the dimensionless coefficient e_0 in a (5,5)@(10,10) armchair DWCNT; $k = 2 \times 10^9$/m.

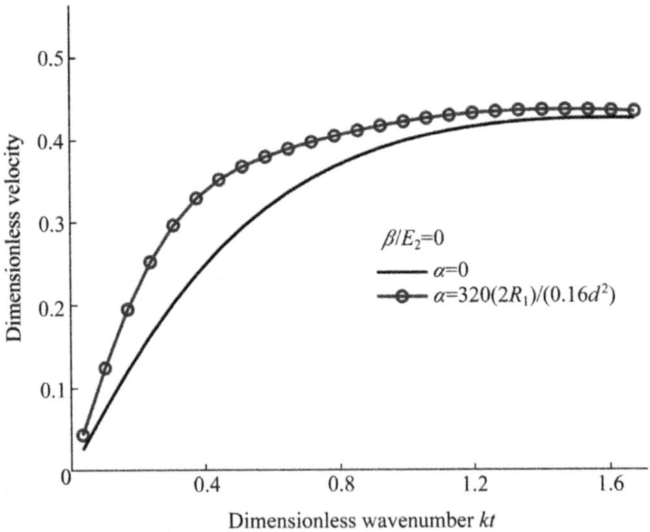

FIGURE 6.17 Influence of van der Waals force on the dimensionless phase velocity $\overline{v} = v/\sqrt{E_2/\rho}$ of a DWCNT in a free space.

that the phase velocity for $\beta = 0$ increases significantly when imposing the van der Waals force compared to the situation without the van der Waals force. However, the dispersion relation for $\beta/E_2 = 0.01$ is different from the waveforms for a DWCNT in a free space, as displayed in Fig. 6.18. This phenomenon shows that the consideration of the van der Waals force is necessary. If neglecting it, the obtained results can contain errors.

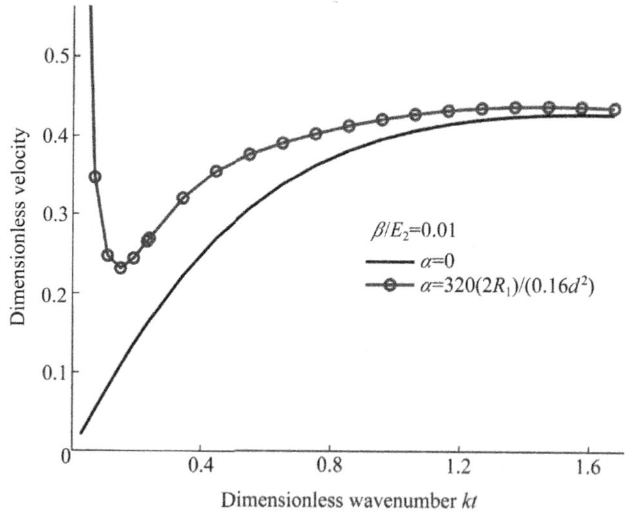

FIGURE 6.18 Influence of van der Waals force on the dimensionless phase velocity $\bar{v} = v/\sqrt{E_2/\rho}$ of a DWCNT embedded in an elastic medium.

6.4 Vibration analysis of tube beams

As for the free vibration problem for nonhomogeneity cylindrical tube beams, the unknown function $F(x, t)$ may take the form of:

$$F(x,t) = f(x)e^{i\omega t} \tag{6.87}$$

where the transverse distributed load disappears for free vibration, i.e., $q = 0$. Substituting Eq. (6.87) into Eq. (6.66) leads to

$$\left(1 - a_2\omega^2\varepsilon^2 + a_4\omega^4\right)\frac{d^4f}{dx^4} + \omega^2\left(a_2 + a_1\varepsilon^2 - \frac{2a_4}{\varepsilon^2}\omega^2\right)\frac{d^2f}{dx^2} + \left(\frac{a_4}{\varepsilon^4}\omega^2 - a_1\right)\omega^2 = 0 \tag{6.88}$$

The vibration behavior discussed here is limited to lower-order frequencies, which are the most common in engineering applications. The general solution to the above equation can be given as

$$f(x) = C_1\cos(\lambda_1 x) + C_2\sin(\lambda_1 x) + C_3\cosh(\lambda_2 x) + C_4\sinh(\lambda_2 x) \tag{6.89}$$

where

$$\lambda_{1,2} = \sqrt{\frac{1}{2A}\left(\sqrt{B^2 - 4AC} \pm B\right)} \tag{6.90}$$

$$A = 1 + \frac{m_{2f} m_0 \varepsilon^4 \omega^4}{(EI)_0 (GA)^*} - \frac{\omega^2 \varepsilon^2}{(EI)_0}\left[m_2 + \frac{m_0 (EI)^*}{(GA)^*}\right] \tag{6.91}$$

$$B = \frac{\omega^2}{(EI)_0}\left[m_2 + \frac{m_0 (EI)^*}{(GA)^*}\right] + \frac{m_0 \varepsilon^2 \omega^2}{(EI)_0}\left[1 - \frac{2 m_{2f} \omega^2}{(GA)^*}\right] \tag{6.92}$$

$$C = -\left[1 - \frac{m_{2f}\omega^2}{(GA)^*}\right]\frac{m_0 \omega^2}{(EI)_0} \tag{6.93}$$

In the above, $C_j (j = 1, \ldots, 4)$ are unknown coefficients to be determined by substituting Eq. (6.89) into appropriate boundary conditions. Substituting Eq. (6.89) into Eqs. (6.64) and (6.65), one readily obtains the expressions of deflection w and rotation angle ψ in terms of $\lambda_{1,2}$:

$$\begin{aligned} w = & (1 + \Pi_1)[C_1 \cos(\lambda_1 x) + C_2 \sin(\lambda_1 x)] \\ & + (1 - \Pi_2)[C_3 \cosh(\lambda_2 x) + C_4 \sinh(\lambda_2 x)] \end{aligned} \tag{6.94}$$

$$\begin{aligned} \psi = & \lambda_1 (1 + \Lambda_1)[C_1 \sin(\lambda_1 x) - C_2 \cos(\lambda_1 x)] \\ & - \lambda_2 (1 - \Lambda_2)[C_3 \sinh(\lambda_2 x) + C_4 \cosh(\lambda_2 x)] \end{aligned} \tag{6.95}$$

where

$$\begin{cases} \Pi_1 = \dfrac{(EI)^* \lambda_1^2 - (1 + \varepsilon^2) m_{2f} \omega^2}{(GA)^*}, \Lambda_1 = \dfrac{[(EI)_0 - (EI)^*]\lambda_1^2 - (1 + \varepsilon^2 \lambda_1^2)(m_2 - m_{2f})\omega^2}{(GA)^*} \\ \Pi_2 = \dfrac{(EI)^* \lambda_2^2 + (1 - \varepsilon^2) m_{2f} \omega^2}{(GA)^*}, \Lambda_2 = \dfrac{[(EI)_0 - (EI)^*]\lambda_2^2 + (1 - \varepsilon^2 \lambda_2^2)(m_2 - m_{2f})\omega^2}{(GA)^*} \end{cases} \tag{6.96}$$

Furthermore, substituting Eqs. (6.94) and (6.95) into Eqs. (6.57) and (6.58), one obtains

$$\begin{aligned} M = & \left\{(EI)^*(\Lambda_1 - \Pi_1) + (EI)_0(1 + \Pi_1) - \varepsilon^2 \right. \\ & \left. \left[(m_0 + m_2\omega^2)(1 + \Pi_1) + \omega^2 m_{2f}(\Lambda_1 - \Pi_1)\right]\right\} \\ & \lambda_1^2 [C_1 \cos(\lambda_1 x) + C_2 \sin(\lambda_1 x)] + \lambda_2^2 [C_3 \cosh(\lambda_2 x) + C_4 \% \sinh(\lambda_2 x)] \\ & \left\{(EI)^*(\Lambda_2 - \Pi_2) + (EI)_0(1 - \Pi_2) + \varepsilon^2\left[(m_0 + m_2\omega^2)(1 - \Pi_2) + \omega^2 m_{2f}(\Lambda_2 - \Pi_2)\right]\right\} \end{aligned} \tag{6.97}$$

$$\begin{aligned} Q = & \lambda_1 [C_1 \sin(\lambda_1 x) - C_2 \cos(\lambda_1 x)]\left[(GA)^*(\Lambda_1 - \Pi_1) + \varepsilon^2 m_0 \omega^2 (1 + \Pi_1)\right] + \\ & \lambda_2 [C_3 \sinh(\lambda_2 x) + C_4 \cosh(\lambda_2 x)]\left[(GA)^*(\Lambda_2 - \Pi_2) - \varepsilon^2 m_0 \omega^2 (1 - \Pi_2)\right] \end{aligned} \tag{6.98}$$

Next, consider two typical boundary conditions: pinned–pinned nanotubes and clamped–clamped nanotubes. For pinned–pinned nanotubes, one has boundary conditions as follows:

$$w(0)=0, M(0)=0, w(L)=0, M(L)=0 \tag{6.99}$$

The application of the relevant boundary conditions at $x=0$ in Eqs. (6.94) and (6.97) allows one to obtain $C_1=C_3=0$. In addition, using the remaining two boundary conditions at $x=L$, one obtains

$$(1+\Pi_1)C_2+(1-\Pi_2)C_4=0 \tag{6.100}$$

$$\begin{aligned}&\left\{(EI)^*(\Lambda_1-\Pi_1)+(EI)_0(1+\Pi_1)-\varepsilon^2\right.\\&\left.\left[\left(m_0+m_2\omega^2\right)(1+\Pi_1)+\omega^2 m_{2f}(\Lambda_1-\Pi_1)\right]\right\}\lambda_1^2C_2\\&+\left\{(EI)^*(\Lambda_2-\Pi_2)+(EI)_0(1-\Pi_2)+\varepsilon^2\right.\\&\left.\left[\left(m_0+m_2\omega^2\right)(1-\Pi_2)+\omega^2 m_{2f}(\Lambda_2-\Pi_2)\right]\right\}\lambda_2^2C_4=0\end{aligned} \tag{6.101}$$

The value of the coefficient matrix determinant must be zero to ensure the above system has a nontrivial solution, which, for low-frequency vibration, results in

$$\sin(\lambda_1 L)=0 \tag{6.102}$$

and with Eqs. (6.90)–(6.93) one obtains

$$b_0\omega^4-b_1\omega^2+\left(\frac{n\pi}{L}\right)^4=0 \tag{6.103}$$

with

$$\begin{cases}b_0=\left[1+\left(\dfrac{n\pi}{L}\right)^2\varepsilon^2\right]^2\dfrac{m_{2f}m_0}{(EI)_0(GA)^*}\\ b_1=\left[1+\left(\dfrac{n\pi}{L}\right)^2\varepsilon^2\right]\left\{1+\left(\dfrac{n\pi}{L}\right)^2\left[\dfrac{m_2}{m_0}+\dfrac{(EI)^*}{(GA)^*}\right]\right\}\dfrac{m_0}{(EI)_0}\end{cases} \tag{6.104}$$

By solving the above quadratic differential equation, an explicit expression for the natural frequency can be provided as follows:

$$\omega=\left(\frac{n\pi}{L}\right)^2\sqrt{\frac{2b_0}{b_1+\sqrt{b_1^2-4b_0\left(\dfrac{n\pi}{L}\right)^4}}} \tag{6.105}$$

As a result of Eq. (6.102), one immediately calculates the value of natural frequencies, and the mode shapes can be derived subsequently by

$$w = \sin\left(\frac{n\pi x}{L}\right) \tag{6.106}$$

By comparing the mode shapes with the classical ones, it can be seen that they are identical [20].

For clamped–clamped nanotubes, the following boundary conditions

$$w(0) = 0, \psi(0) = 0, w(L) = 0, \psi(L) = 0 \tag{6.107}$$

are used to reflect the geometric constraints. Substituting the first three constraints in Eq. (6.107) into Eqs. (6.94) and (6.95), one gets

$$\begin{cases} C_3 = -\dfrac{1+\Pi_1}{1-\Pi_2}C_1 \\ C_4 = -\dfrac{\lambda_1(1+\Lambda_1)}{\lambda_2(1-\Lambda_2)}C_2 \\ C_2 = -\dfrac{\cos(\lambda_1 L) - \cosh(\lambda_2 L)}{\sin(\lambda_1 L) - \Psi\sinh(\lambda_2 L)}C_1 \end{cases} \tag{6.108}$$

where

$$\Psi = \frac{\lambda_1(1+\Lambda_1)(1-\Pi_2)}{\lambda_2(1-\Lambda_2)(1+\Pi_1)} \tag{6.109}$$

Using the last constraint in Eq. (6.107), after some algebra, we finally get the desired frequency equation as follows:

$$2 - 2\cos(\lambda_1 L)\cosh(\lambda_2 L) + \left(\frac{1}{\Psi} - \Psi\right)\sin(\lambda_1 L)\sinh(\lambda_2 L) = 0 \tag{6.110}$$

It should be noted that there is no analytical solution to the above equation, however, one can determine the natural frequency by numerically solving Eq. (6.110). Moreover, with the boundary conditions in Eq. (6.107), one can derive the following mode shapes:

$$w = \cos(\lambda_1 x) - \cosh(\lambda_2 x) - \frac{\cos(\lambda_1 L) - \cosh(\lambda_2 L)}{\sin(\lambda_1 L) - \Psi\sinh(\lambda_2 L)}[\sin(\lambda_1 x) - \Psi\sinh(\lambda_2 x)] \tag{6.111}$$

For other boundary conditions such as clamped–free nanotubes and free–free nanotubes, the frequency equations can be similarly derived but are omitted here to save space. It is worth noting that although the above-mentioned frequency equations and mode shapes are consistent with those in Ref. [11], the coefficients in these equations are completely different; the dependence of the frequencies on the nonlocal parameters is particularly interesting as described in this chapter.

6.4.1 Numerical results and discussion

To examine the effectiveness of the proposed model, we compare the first three dimensionless natural frequency ratios $\sqrt{\omega/\omega_0}$ for homogeneous nonlocal Euler–Bernoulli nanotubes, where

$$\omega_0 = \frac{1}{L^2}\sqrt{\frac{E_o I}{\rho_o A}} \tag{6.112}$$

Two typical boundary conditions are included, i.e., pinned–pinned and clamped–clamped ends. Here we take the materials at the inner and outer surfaces as the same such that the FG nanotube degenerates into homogeneous nanotubes. The size of the selected nanotube is taken as $R_o = 0.372$ nm, $R_i = 0.306$ nm, and $L = 6.78$ nm, and its material properties are $E = 5.5$ TPa and $\rho = 2.3$ g/cm^3 unless otherwise stated.

In Table 6.1, we evaluate the frequency parameter for three cases: the nonlocal Euler–Bernoulli beam theory (NEBT), the nonlocal Rayleigh beam theory (NRBT), and the nonlocal higher-order beam theory (NHBT). If one neglects both shear deformation and rotary inertia (i.e., $G \to \infty, \rho I = 0$), the proposed NHBT degenerates to NEBT, whereas for the NRBT, the shear deformation is neglected but the rotary inertia is taken into account. From Table 6.1, our evaluated results are in good agreement with the previous ones when adopting the NEBT. Additionally, because of the consideration of rotary inertia, the results based on NRBT are slightly lower than those based on NEBT [21,22]. This is attributed to the fact that the rotary inertia includes a slight addition of equivalent mass. Through careful observation, it can be seen that due to the consideration of shear deformation, the results based on NHBT are slightly lower than those based on NRBT. The results in Table 6.1 further show that the proposed theory can effectively degenerate into NRBT and NEBT. However, it should be pointed out that for practical cases, due to the neglect of cross sectional shear deformation and/or rotary inertia, the natural frequencies based on the NRBT and NEBT are accurate enough for tubes with high aspect ratios or slenderness ratios.

Table 6.2 provides a comparison of the first three values of the frequency parameter $\sqrt{\omega/\omega_0}$ for nanotubes based on the proposed NHBT and the NTBT. In order to consider the influence of shear deformation, the shear modulus is no longer taken as infinity. Instead, we take Poisson's ratio as $\nu = 0.19$ and the shear modulus is estimated by $G = E/2(1+\nu)$. The evaluated results are given in Table 6.2. One can readily see that the results from the present model are generally less than those based on the NTBT. This phenomenon can be explained as follows. Since the NTBT considers the influence of shear deformation and rotary inertia, it still assumes the cross section to maintain plane, not warp, which leads to a slight overestimation of the calculated results. Although the NTBT captures the cross sectional shear deformation and rotary inertia, the cross sectional warping deformation is prohibited and the cross section is still planar after bending; therefore, the shear stress on the

TABLE 6.1 Comparison of the first three frequency parameters $\sqrt{\omega/\omega_0}$ using the proposed model and NEBT. *B.C.*, boundary conditions; *C.-C.*, clamped-clampled; *P.-P.*, pinned-pinned.

	$\varepsilon/L = 0$					$\varepsilon/L = 0.1$					$\varepsilon/L = 0.3$				
	Present					Present					Present				
B.C.	NEBT	NRBT	NHBT	[21]	[22]	NEBT	NRBT	NHBT	[21]	[22]	NEBT	NRBT	NHBT	[21]	[22]
P.-P.	3.1416	3.1319	3.0889	3.1416	3.1416	3.0685	3.0591	3.0170	3.0685	3.0685	2.6800	2.6717	2.6350	2.6800	2.6800
	6.2832	6.2073	5.9147	6.2832	6.2832	5.7817	5.7118	5.4426	5.7817	5.7817	4.3013	4.2494	4.0490	4.3013	4.3013
	9.4248	9.1778	8.3812	9.4248	9.4248	8.0400	7.8293	7.1497	8.0400	8.0400	5.4422	5.2996	4.8396	5.4422	5.4422
C.-C.	4.7300	4.7118	4.4274	4.7300	4.7300	4.5945	4.5723	4.3055	4.5935	4.5945	3.9184	3.8855	3.6855	3.9071	3.9184
	7.8525	7.7430	6.8932	7.8525	7.8525	7.1402	7.0157	6.2977	7.1454	7.1402	5.1963	5.0807	4.6162	5.1917	5.1963
	10.996	10.677	9.0599	10.996	10.996	9.2583	8.9302	7.6933	9.2580	9.2583	6.2317	5.9745	5.2120	6.2662	6.2317

TABLE 6.2 First three frequency parameters $\sqrt{\omega/\omega_0}$ for various values of ε/L using the proposed model and NTBT. *B.C.*, boundary conditions; *C.-C.*, clamped-clamped; *P.-P.*, pinned-pinned.

	$\varepsilon/L=0$		$\varepsilon/L=0.1$		$\varepsilon/L=0.3$		$\varepsilon/L=0.5$		$\varepsilon/L=0.7$	
B.C.	**Present**	**[23]**	**Present**	**[23]**	**Present**	**[23]**	**Present**	**[23]**	**Present**	**[23]**
P.-P.	3.0889	3.0929	3.0170	3.0243	2.6350	2.6538	2.2636	2.2867	1.9873	2.0106
	5.9147	5.9399	5.4426	5.5304	4.0490	4.2058	3.2574	3.4037	2.7849	2.9159
	8.3812	8.4444	7.1497	7.4699	4.8396	5.2444	3.8186	4.1644	3.2445	3.5453
C.-C.	4.4274	4.4733	4.3055	4.3471	3.6855	3.7895	3.1216	3.2420	2.7208	2.8383
	6.8932	6.9524	6.2977	6.4952	4.6162	4.9428	3.6967	3.9940	3.1553	3.4192
	9.0599	9.1626	7.6933	8.1969	5.2120	5.8460	4.1277	4.6769	3.5144	3.9961

surface does not vanish. In contrast, our assumption ensures the fulfillment of a shear stress free condition on the inner and outer surfaces of the nanotube. From this standpoint, the determination of exact natural frequencies is quite necessary. From this standpoint, the present NHBT model based on the exponential warping shape has higher accuracy than the existing NEBT and NTBT.

6.4.2 Effect of nonlocal parameters and gradient index

In this subsection, the effects of nonlocal parameters, gradient index, and aspect ratio on the dimensionless natural frequencies of an FG nanotube are discussed. An FG nanotube with $R_i = 0.306$ nm, $R_o = 0.372$ nm, and $L = 6.78$ nm is considered. For convenience, steel and Al_2O_3 are chosen as component materials on the inner and outer surfaces and their material properties are [24,25]:

$$\begin{cases} \text{Steel}: E_i = 210\ \text{GPa}, \nu_i = 0.3\ \text{GPa}, \rho_i = 7800\ \text{kg/m}^3 \\ \text{Al}_2\text{O}_3: E_o = 390\ \text{GPa}, \nu_o = 0.24\ \text{GPa}, \rho_o = 3960\ \text{kg/m}^3 \end{cases} \tag{6.113}$$

In Tables 6.3 and 6.4 we list the first three frequency parameters of pinned–pinned and clamped–clamped FG nanotubes with $L/(R_o + R_i) = 5$,

TABLE 6.3 Frequency parameter of pinned–pinned FG nanotubes with different gradient indexes.

		N				
n	*ε/L*	0	0.2	1	5	10
1	0	2.9518	2.7976	2.5253	2.2678	2.2067
	0.1	2.8832	2.7326	2.4666	2.2151	2.1554
	0.3	2.5181	2.3866	2.1543	1.9346	1.8825
	0.5	2.1632	2.0501	1.8506	1.6619	1.6171
2	0	5.2343	4.9578	4.4689	4.0084	3.9002
	0.1	4.8165	4.5621	4.1122	3.6885	3.5889
	0.3	3.5833	3.3940	3.0593	2.7441	2.6700
	0.5	2.8827	2.7304	2.4612	2.2076	2.1480
3	0	6.9727	6.6017	5.9450	5.3282	5.1841
	0.1	5.9482	5.6317	5.0715	4.5453	4.4224
	0.3	4.0263	3.8121	3.4328	3.0767	2.9935
	0.5	3.1768	3.0078	2.7086	2.4276	2.3619

TABLE 6.4 Frequency parameter of clamped–clamped FG nanotubes with different gradient indexes.

		N				
n	ε/L	0	0.2	1	5	10
1	0	3.8532	3.6465	3.2822	2.9432	2.8649
	0.1	3.7519	3.5509	3.1965	2.8664	2.7901
	0.3	3.2348	3.0626	2.7584	2.4738	2.4076
	0.5	2.7531	2.6070	2.3490	2.1068	2.0502
2	0	5.6148	5.3124	4.7789	4.2825	4.1681
	0.1	5.1499	4.8733	4.3846	3.9293	3.8242
	0.3	3.8160	3.6120	3.2510	2.9137	2.8355
	0.5	3.0663	2.9025	2.6128	2.3418	2.2788
3	0	7.1347	6.7512	6.0731	5.4418	5.2948
	0.1	6.0813	5.7548	5.1773	4.6384	4.5137
	0.3	4.1389	3.9168	3.5240	3.1570	3.0720
	0.5	3.2818	3.1059	2.7946	2.5613	2.4360

respectively, for various gradient indexes ($N = 0$, 0.2, 1, 5, 10) and nonlocal parameters ($\varepsilon/L = 0$, 0.1, 0.3) based on the proposed model. It can be seen in Tables 6.3 and 6.4 that the frequencies decrease as the gradient index N increases. This is because the gradient parameter N changes the material distribution, increasing the stiffness of the structure itself, which in turn leads to higher natural frequencies. As the nonlocal coefficient ε/L increases, the frequency results obtained by the present NHBT gradually decrease. In addition, it can be observed that this phenomenon of frequency reduction is more significant in higher-order frequencies as compared with those in lower-order frequencies, which is consistent with Ref. [23]. If one needs to accurately predict the higher-order frequencies of nanobeams, the results based on the NHBT are more accurate. The influence trends of nonlocal parameters ε/L and gradient index N on natural frequencies are consistent.

Figs. 6.19–6.21 show the variations of the first three frequency parameters of a pinned–pinned FG tube with $L/(R_o + R_i) = 5$ against gradient index N with various nonlocal parameters. By observing the data curves in Figs. 6.19–6.21, we can see that the natural frequencies rely on the values of size parameters ε and gradient index N, and the natural frequencies decrease as the size parameters ε increase, proving that the nonlocal parameter ε reduces the

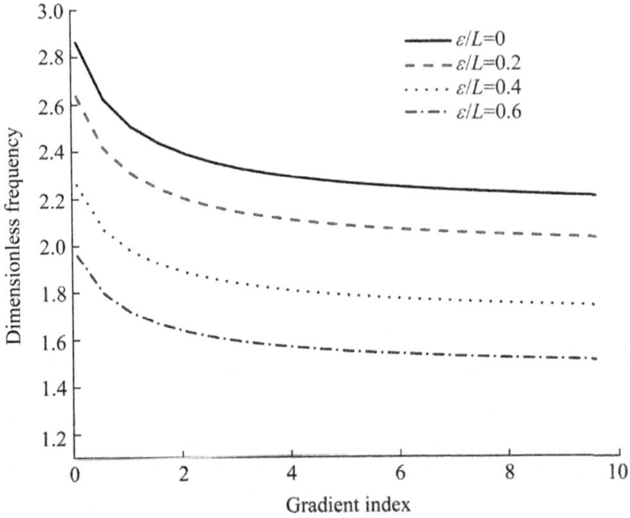

FIGURE 6.19 The variation of the first dimensionless frequency of a pinned–pinned FG nanotube with $L/(R_o + R_i) = 5$ against the gradient index N.

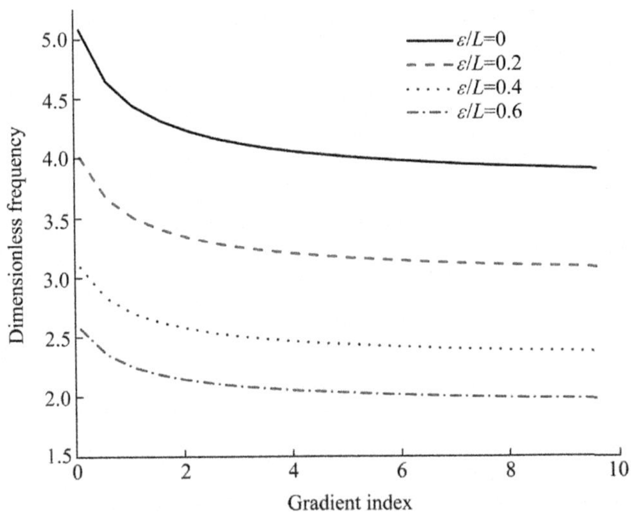

FIGURE 6.20 The variation of the second dimensionless frequency of a pinned–pinned FG nanotube with $L/(R_o + R_i) = 5$ against the gradient index N.

stiffness of the FG nanotubes. From Fig. 6.19 one can see that the fundamental frequency decreases more rapidly when the gradient index N ranges from 0 to 4 than that which ranges from 4 to 10, which illustrates that the change of component materials of FG tubes has little effect on the fundamental

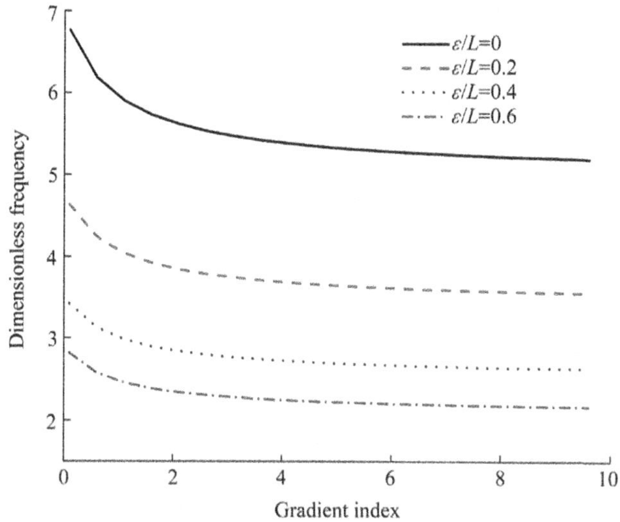

FIGURE 6.21 The variation of the third dimensionless frequency of a pinned-pinned FG nanotube with $L/(R_o + R_i) = 5$ against the gradient index N.

frequency when the gradient index $N > 4$. The second frequency decreases at a high rate for the gradient index N ranges from 0 to 3, as shown in Fig. 6.20. Meanwhile the descent speed of the third frequency is high for N ranges from 0 to 2, which can be confirmed in Fig. 6.21. In addition, the natural frequencies decrease as the nonlocal parameter increases in Figs. 6.19–6.21.

One can also find that the value of fundamental frequency decreases as the nonlocal parameter increases, consistent with the trend in Fig. 6.19. However, for higher-order frequencies, the effect of the nonlocal parameter is more noticeable when ε/L increases from 0 to 0.4 than when it varies from 0.4 to 0.6. The stated characteristic becomes more obvious with the increase of frequency order, as demonstrated in Figs. 6.20 and 6.21.

With the increase of the gradient index N, the natural frequencies gradually stabilize to a specific value, and the stabilization rate enhances with the increase of frequency order. As shown in Fig. 6.19, the gradient index N is effective only in the interval between 0 and 6 for the fundamental frequency, whereas the gradient index N becomes almost ineffective for the third and fourth frequencies when the gradient index $N > 2$.

Fig. 6.22 plots the variation of frequency ratios ω_{NH}/ω_{LH} of a pinned–pinned FG nanotube with $L/(R_o + R_i) = 5$ (i.e., ratio of the frequencies derived from proposed model ω_{NH} to the corresponding data obtained from local higher-order shear deformation beam theory ω_{LH}) versus the nonlocal parameter ε/L. All frequency ratios are less than 1, indicating that the frequency calculated from the presented model ω_{NH} is less than the corresponding ω_{LH}, which means that the neglect of the scale effect overestimates

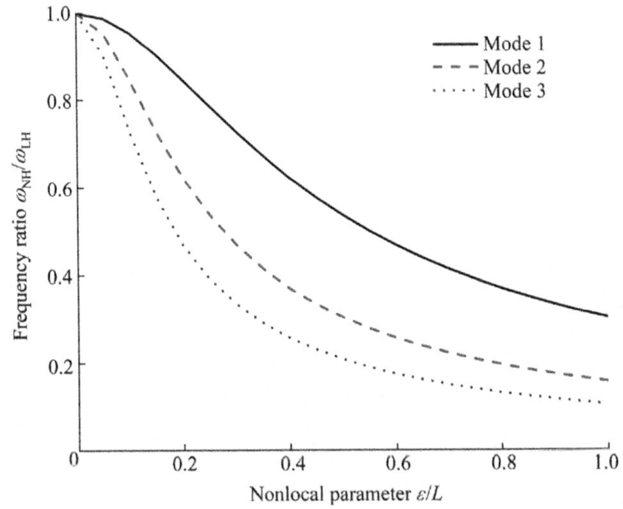

FIGURE 6.22 Effects of the nonlocal parameter ε on the first three frequency ratios $\omega_{NH}/\ \omega_{LH}$ of a pinned–pinned FG nanotube with $L/(R_o + R_i) = 5$.

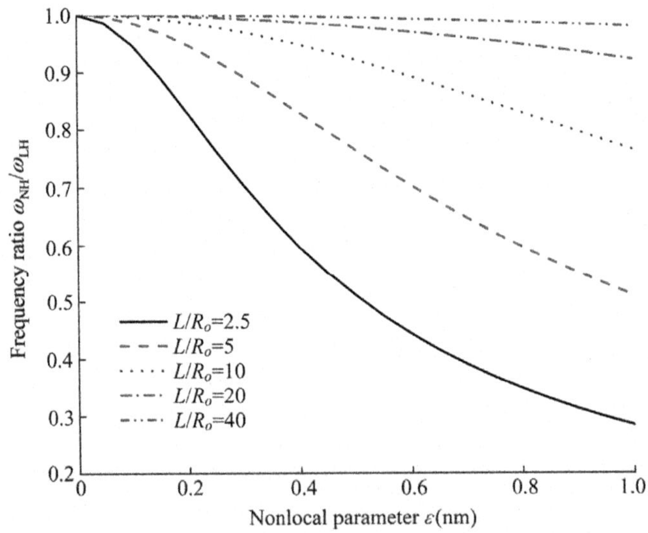

FIGURE 6.23 Effects of the nonlocal parameter ε on the fundamental frequency ratio $\omega_{NH}/\ \omega_{LH}$ of a pinned–pinned FG nanotube with different aspect ratios.

the natural frequencies. Meanwhile, the influence of this overestimation increases when the nonlocal parameter and frequency order increase.

Fig. 6.23 illustrates the fundamental frequency ratio ω_{NH}/ω_{LH} of a pinned–pinned FG nanotube with different aspect ratios against the nonlocal

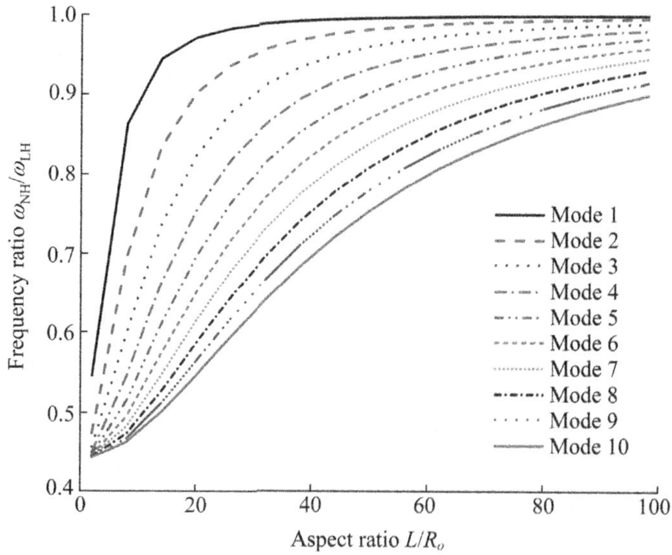

FIGURE 6.24 Effects of shear deformation and rotary inertia on the natural frequencies of a pinned–pinned FG nanotube with $\varepsilon = 0.1$ nm.

parameter ε. It is shown that the influence of the nonlocal parameter on the fundamental frequency of an FG nanotube is significant for small aspect ratios and almost disappears for high aspect ratios. Fig. 6.24 compares the frequency ratio ω_{NH}/ω_{NE} of a pinned–pinned FG nanotube with the aspect ratio L/R_o for an invariant nonlocal parameter $\varepsilon = 0.1$ nm. The frequencies obtained are smaller than those derived from nonlocal Euler–Bernoulli beam theory, demonstrating the above-mentioned conclusion that shear deformation and rotary inertia reduce the natural frequencies. With an increase of the frequency order and an increase or decrease of the aspect ratio, the influence of shear deformation and rotary inertia is significantly amplified. Another interesting characteristic is that the frequency ratio does not vary when the frequency order rises for an extraordinarily short tube. This means that the influence of shear deformation and rotary inertia on the frequency will not be amplified by the increase of frequency order when the aspect ratio remains sufficiently small.

In closing this section, we note that the present approach is based on the analytic treatment of the NHBT. It is still inconvenient for complicated cases, e.g., axially graded beams of arbitrary axial gradient. In the future, this approach may be explored in combination with the finite element method to solve more practical and complicated problems.

References

[1] Eringen AC. Nonlocal polar elastic continua. Int J Eng Sci 1972;10(1):1–16.
[2] Eringen AC. On differential equations of nonlocal elasticity and solutions of screw dislocation and surface waves. J Appl Phys 1983;54:4703–10.
[3] Eringen AC, Wegner JL. Nonlocal continuum field theories[M]. Springer; 2003.
[4] Huang Y, Li XF. Buckling of functionally graded circular columns including shear deformation. Mater Des 2010;31(7):3159–66.
[5] Huang Y, Li XF. Bending and vibration of circular cylindrical beams with arbitrary radial nonhomogeneity. Int J Mech Sci 2010;52(4):595–601.
[6] Lu P, Lee HP, Lu C, et al. Application of nonlocal beam models for carbon nanotubes. Int J Solid Struct 2007;44(16):5289–300.
[7] Reddy JN, Pang SD. Nonlocal continuum theories of beams for the analysis of carbon nanotubes. J Appl Phys 2008;103(2).
[8] Li XF, Wang BL. Vibrational modes of Timoshenko beams at small scales. Appl Phys Lett 2009;94(10).
[9] Wu JX, Li XF, Tang GJ. Bending wave propagation of carbon nanotubes in a bi-parameter elastic matrix. Phys B Condens Matter 2012;407(4):684–8.
[10] Wang Q. Wave propagation in carbon nanotubes via nonlocal continuum mechanics. J Appl Phys 2005;98(12).
[11] Ma WL, Jiang ZC, Lee KY, et al. A refined beam theory for bending and vibration of functionally graded tube-beams. Compos Struct 2020;236:111878.
[12] Ma WL, Li XF, Yong Lee K. Third-order shear deformation beam model for flexural waves and free vibration of pipes. J Acoust Soc Am 2020;147(3):1634–47.
[13] Civalek Ö, Demir Ç. Bending analysis of microtubules using nonlocal Euler-Bernoulli beam theory. Appl Math Model 2011;35(5):2053–67.
[14] Narendar S, Gupta SS, Gopalakrishnan S. Wave propagation in single-walled carbon nanotube under longitudinal magnetic field using nonlocal Euler-Bernoulli beam theory. Appl Math Model 2012;36(9):4529–38.
[15] Ma WL, Cheng C, Chen X, et al. Free vibration of radially graded hollow cylinders subject to axial force via a higher-order shear deformation beam theory. Compos Struct 2021;255:112957.
[16] Li XF. A unified approach for analyzing static and dynamic behaviors of functionally graded Timoshenko and Euler-Bernoulli beams. J Sound Vib 2008;318(4–5):1210–29.
[17] Wang L, Hu H. Flexural wave propagation in single-walled carbon nanotubes. Phys Rev B Condens Matter 2005;71(19):195412.
[18] Duan WH, Wang CM, Zhang YY. Calibration of nonlocal scaling effect parameter for free vibration of carbon nanotubes by molecular dynamics. J Appl Phys 2007;101(2).
[19] Askes H, Aifantis EC. Gradient elasticity and flexural wave dispersion in carbon nanotubes. Phys Rev B Condens Matter 2009;80(19):195412.
[20] Weaver JW, Timoshenko SP, Young DH. Vibration problems in engineering. John Wiley and Sons; 1991.
[21] Zhang DP, Lei YJ, Wang CY, et al. Vibration analysis of viscoelastic single-walled carbon nanotubes resting on a viscoelastic foundation. J Mech Sci Technol 2017;31(1):87–98.
[22] Lu P, Lee HP, Lu C, et al. Dynamic properties of flexural beams using a nonlocal elasticity model. J Appl Phys 2006;99(7).
[23] Wang CM, Zhang YY, He XQ. Vibration of nonlocal Timoshenko beams. Nanotechnology 2007;18(10):105401.

[24] Rahmani O, Pedram O. Analysis and modeling the size effect on vibration of functionally graded nanobeams based on nonlocal Timoshenko beam theory. Int J Eng Sci 2014;77:55–70.

[25] Li X, Li L, Hu Y, et al. Bending, buckling and vibration of axially functionally graded beams based on nonlocal strain gradient theory. Compos Struct 2017;165:250–65.

Chapter 7

Some advanced topics and applications

7.1 Introduction

Beam theories have evolved beyond the basic forms to address complex phenomena in various engineering and scientific applications. In recent years, there has been a growing interest in incorporating multiple-physical effects into these theories. In this chapter, we investigate the thermoelastic coupling effects, surface effects, the coupled stress effects, as well as the piezoelectric and flexoelectric effects briefly. Thermoelastic coupling effects are of great significance, especially in environments where temperature variations are common. Structures in aerospace, nuclear, and other industries are often subjected to varying temperatures, leading to the development of mechanical stresses and strains. By integrating thermoelastic coupling into beam theories, we can more accurately predict the combined thermal-mechanical response of these structures. Surface effects become prominent at the micro and nanoscales. As beam dimensions shrink, the influence of surface energy and stress on the overall mechanical behavior cannot be ignored. Understanding these effects is crucial for applications in MEMS and nanocomposite design, where surface-related phenomena can dominate the devices' performance. Piezoelectric and flexoelectric effects offer promising opportunities for energy conversion, sensing, and actuation. Piezoelectric materials can convert mechanical energy into electrical energy and vice versa, while flexoelectricity couples strain gradients to electric polarization. The coupled stress effects, which account for higher-order stress gradients, are essential for accurately modeling the behavior of micro and nano scale structures as well as composite materials with micro inhomogeneities. These effects provide a more comprehensive understanding of the mechanical response. In summary, exploring these advanced topics in the context of the classical and higher-order beam theories not only enriches the theoretical foundation but also paves the way for innovative applications across multiple disciplines.

Higher-Order Theories of Functionally Graded Circular Cylindrical Beams and Columns
https://doi.org/10.1016/B978-0-443-44718-1.00001-3

7.2 Thermoelastic coupling effects

Owing to the widespread application of micro/nano resonators, sensitivity, reliability, and stability are the main factors in evaluating their performances. Thermoelastic damping (TED), which accounts for the primary energy dissipation in micro/nanostructures, is of great significance.

Previously, the thermoelastic damping of the Euler-Bernoulli beam was investigated and derived [1,2]. After that, the thermoelastic damping for different structures has been analyzed, including beam [3,4], plate [5,6], ring [7,8], and shells [9,10]. Li et al. [11] analyzed the TED of the higher-order beams with rectangular cross sections.

To accurately describe the heat process, several theoretical models have been developed. At previous time, the Fourier heat conduction theory is proposed. The Fourier heat conduction law, due to its assumption of infinite thermal wave velocity, struggles to accurately describe heat transfer at the micro/nanoscale or during rapid transient processes, showing significant limitations in TED analysis. For instance, it fails to account for thermal relaxation time, nonlocal effects, and size dependent behaviors in microstructures. As a result, the non-Fourier heat condition has emerged, including single-phase-lag model [12], double-phase-lag model [13], three-phase-lag model [5], Cattaneo-Veinotte model [14,15], Lord-Shulman model [16], Guyer and Krumhansl (G-K) model [17,18] and fractional heat model [19]. It is noteworthy that the G-K model addresses the infinite-speed defect of Fourier's law by introducing thermal relaxation time and nonlocal length parameters, thereby revealing the influence of thermal viscous effects on heat flux distribution. Given the notable advantages of the G-K model, numerous researchers have conducted studies based on it. For example, Deng et al. [20] further extended its application by employing the G-K model to analyze the impact of phase-lagging on thermoelastic damping in graphene-based nano-beam resonators. The G-K model was used to investigate nonlocal phenomena in radial heat transport within thin films and graphene sheet [21], as well as cylindrical nanowires [22]. Li et al. employed the G-K model to analyze TED in cylindrical nano-shells [9], considering size dependent effects. In this section, we will study TED of the nanobeam with a circular cross section using G-K heat conduction model. The constitutive equations with thermoelastic coupling effects will be derived based on the higher-order beam theory.

7.2.1 Higher-order beam theory with circular cross section

Consider a circular cross section beam with radius r_0 and length L, we adopt a Cartesian coordinate system (x, y, z), associated with elastic displacement components (u, v, w), in combination with a cylindrical coordinate system (x, r, θ), associated with displacements (u, u_r, u_θ) (Fig. 7.1).and the relationships between them are written as:

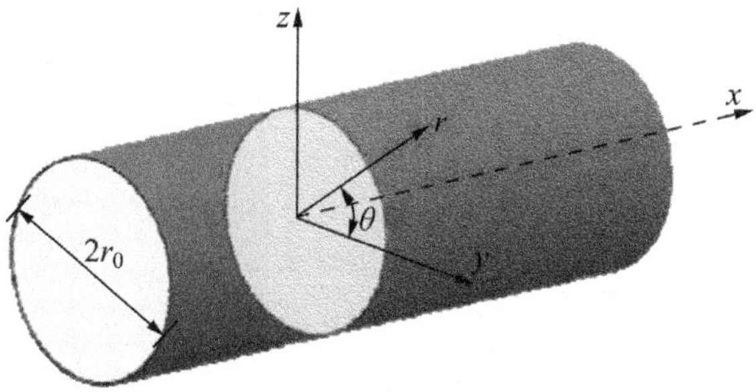

FIGURE 7.1 The schematic of a circular beam.

$$\begin{cases} y = r\cos\theta \\ z = r\sin\theta \end{cases} \tag{7.1}$$

$$u_r = v\cos\theta + w\sin\theta \tag{7.2}$$

The beam is assumed to be homogeneous, continuous and isotropic, restricted to move transversely within the xoz-plane. It should be noted that the displacement component v in y-direction can be approximately assumed to be independent of x, while the deflection w in the z-direction depends only on x, this is to say $v = v(y, z, t)$ and $w = w(x, t)$. The vibration theory of the higher-order shear deformation beam theory is employed. The axial displacement is expressed as:

$$u = -z\frac{\partial w}{\partial x} + f(y,z)\left[\varphi + \frac{\partial w}{\partial x}\right] \tag{7.3}$$

where φ is the rotation of the cross section about the centroidal axis, and $f(y, z)$ is the warping shape function. There are three warping shapes of the cross-section during bending deformation named polynomial warping shape, trigonometric, and exponential warping shape, as listed in Table 7.1 [23].

The shear strain γ_{xz} can be expressed as

$$\gamma_{xz} = \frac{\partial f}{\partial z}\left[\varphi(x) + \frac{\partial w(x)}{\partial x}\right] \tag{7.4}$$

To analyze the behaviors of circular cylindrical beams, it is expedient to introduce the radial displacement and shear strain γ_{xr} over an arbitrary circular cross section, which is defined as:

$$\gamma_{xr} = \frac{\partial U_r}{\partial x} + \frac{\partial u}{\partial r} = \frac{\partial f}{\partial r}\left[\varphi(x) + \frac{\partial w(x)}{\partial x}\right] \tag{7.5}$$

One uses the Hooke's law to obtain the normal stress σ_{xx} and shear stress τ_{xz} as follows:

TABLE 7.1 The warping function $f(y,z)$and related quantities for circular cylindrical beams[23].

Wrapping function	$g(r)$	$f(y, z)$	$\frac{\partial f(y,z)}{\partial z}$	$\frac{A_f}{A}$	$\frac{I_f}{I}$	$\frac{1}{k}=\frac{AI_f}{IA_f}$
Polynomial	$1-\frac{r^2}{3r_0^2}$	$zg(r)$	$1-\frac{y^2+3z^2}{3r_0^2}$	$\frac{2}{3}$	$\frac{7}{9}$	1.1667
Trigonometric	$\frac{2r_0}{\pi r}\sin\frac{\pi r}{2r_0}$	$zg(r)$	$\frac{2r_0y^2}{\pi r^3}\sin\frac{\pi r}{2r_0}+\frac{z^2}{r^2}\cos\frac{\pi r}{2r_0}$	$\frac{2}{\pi}$	$\frac{64(\pi-2)}{\pi^4}$	1.1782
Exponential	$e^{-r^2/2r_0^2}$	$zg(r)$	$\left(1-\frac{z^2}{r_0^2}\right)e^{-r^2/2r_0^2}$	$\frac{1}{\sqrt{e}}$	$8-\frac{12}{\sqrt{e}}$	1.1898

$$\sigma_{xx} = E\left\{ -z\frac{\partial^2 w(x)}{\partial x^2} + f(y,z)\left[\varphi(x) + \frac{\partial w(x)}{\partial x}\right]\right\} \tag{7.6}$$

$$\tau_{xz} = G\left[\frac{\partial f}{\partial z}\left(\varphi + \frac{\partial w}{\partial x}\right)\right] \tag{7.7}$$

$$\tau_{xr} = G\left\{\frac{\partial f}{\partial r}\left[\varphi(x) + \frac{\partial w(x)}{\partial x}\right]\right\} \tag{7.8}$$

where E and G are the Young's modulus and shear modulus, respectively. In addition, the bending moment M and shear force Q at any position x of the higher-order shear beam can be derived:

$$M = \int_A z\sigma_{xx}\mathrm{d}A = -EI\frac{\partial^2 w}{\partial x^2} + EI_f\left(\frac{\partial \varphi}{\partial x} + \frac{\partial^2 w}{\partial x^2}\right) \tag{7.9}$$

$$Q = \int_A \tau_{xz}\mathrm{d}A = GA_f\left(\varphi + \frac{\partial w(x)}{\partial x}\right) \tag{7.10}$$

$$\mathrm{dx}\ A_f = \int_A \frac{\partial f(y,z)}{\partial z}\mathrm{d}A, I_f = \int_A zf(y,z)\mathrm{d}A \tag{7.11}$$

where A stands for the area of the cross section. Incidentally, in the case of $(z) = z$, the Timoshenko hypothesis is retrieved and uniform shear strain can be found and $A_f = A, I_f = I$. The uniform shear stress on the surface of circular beams violates traction-free conditions on the surface; thereby the shear correction factor κ in the Timoshenko beam theory is introduced to remedy this shortage.

$$M = \int_A z\sigma_{xx}\mathrm{d}A = EI\frac{\partial \varphi}{\partial x} \tag{7.12}$$

$$Q = \kappa\int_A \tau_{xz}\mathrm{d}A = -\kappa GA\left(\varphi + \frac{\partial w}{\partial x}\right) \tag{7.13}$$

Because the cylindrical beam reflects in the z-direction rather than the y-direction under applied transverse loading or bending moment, the motion along the y-direction is neglected. Thus, with reference to flexural deformation in the xoz-plane, only two equations of motion are involved:

$$\begin{cases} \dfrac{\partial Q}{\partial x} = \rho A\dfrac{\partial^2 w}{\partial t^2} \\ \dfrac{\partial M}{\partial x} - Q = \rho\displaystyle\int_A z\frac{\partial^2 u}{\partial x^2}\mathrm{d}A \end{cases} \tag{7.14}$$

where ρ is the mass density. Substituting Eq. (7.9) and Eq. (7.10) into Eq. (7.14), one can obtain

$$GA_f\frac{\partial}{\partial x}\left(\varphi+\frac{\partial w}{\partial x}\right)=\rho A\frac{\partial^2 w}{\partial t^2} \tag{7.15}$$

$$\begin{aligned}\frac{\partial}{\partial x}\left[-EI\frac{\partial^2 w}{\partial x^2}+EI_f\left(\frac{\partial\varphi}{\partial x}+\frac{\partial^2 w}{\partial x^2}\right)\right]-GA_f\left(\varphi+\frac{\partial w}{\partial x}\right)\\ =\rho I\frac{\partial^2}{\partial t^2}\frac{\partial w}{\partial x}+\rho I_f\frac{\partial^2}{\partial t^2}\left[\varphi+\frac{\partial w(x)}{\partial x}\right]\end{aligned} \tag{7.16}$$

By setting

$$\begin{cases} w=\left(1-\dfrac{EI_f}{GA_f}\dfrac{\partial^2}{\partial x^2}+\dfrac{\rho I_f}{GA_f}\dfrac{\partial^2}{\partial t^2}\right)F=\left(1-\dfrac{EI}{kGA}\dfrac{\partial^2}{\partial x^2}+\dfrac{\rho I}{kGA}\dfrac{\partial^2}{\partial t^2}\right)F \\ \varphi=-\left(1+\dfrac{EI-EI_f}{GA_f}\dfrac{\partial^2}{\partial x^2}-\dfrac{\rho I-\rho I_f}{GA_f}\dfrac{\partial^2}{\partial t^2}\right)\dfrac{\partial}{\partial x}F \end{cases} \tag{7.17}$$

Eq. (7.16) is satisfied naturally, then, substituting (7.17) into Eq. (7.15), we can obtain

$$\frac{\partial^4}{\partial x^4}F-\frac{\rho}{E}\left(1+\frac{E}{kG}\right)\frac{\partial^4}{\partial x^2 t^2}F+\frac{\rho A}{EI}\left(1+\frac{\rho I}{kGA}\frac{\partial^2}{\partial t^2}\right)\frac{\partial^2}{\partial t^2}F=0 \tag{7.18}$$

with

$$\frac{1}{k}=\begin{cases}\dfrac{AI_f}{IA_f} & \text{(HSBT)}\\ \dfrac{1}{\kappa} & \text{(TBT)}\end{cases} \tag{7.19}$$

7.2.2 Thermoelastic fields

To capture the two-dimensional nonlocal thermoelastic damping behaviors in micro/nano systems, Guyer-Krumhansl (G-K) heat conduction model is employed and the heat flux is given as:

$$\left(1+\tau\frac{\partial}{\partial t}\right)\boldsymbol{q}=-k\nabla T+l^2\left[\nabla^2\boldsymbol{q}+2\nabla(\nabla\cdot\boldsymbol{q})\right] \tag{7.20}$$

where $\boldsymbol{q}$ represents heat flux, T denotes the temperature, τ represents the thermal relaxation time, and l represents the thermal nonlocal length. Considering the following conservation equation of energy

$$-\nabla\cdot\boldsymbol{q}=\rho C_E\frac{\partial T}{\partial t}+\frac{E\alpha T_0}{1-2\nu}\frac{\partial\varepsilon_V}{\partial t} \tag{7.21}$$

the heat condition equation can be expressed as:

$$\left(1+\tau\frac{\partial}{\partial t}-3l^2\nabla^2\right)\left(\frac{\partial T}{\partial t}+\frac{\Delta_E}{\alpha(1-2\nu)}\frac{\partial \varepsilon_V}{\partial t}\right)=\chi\nabla^2 T \tag{7.22}$$

with

$$\Delta_E=\frac{E\alpha^2 T_0}{\rho C_E},\chi=\frac{k}{\rho C_E}$$

where T_0 is the reference temperature, α indicates the thermal expansion coefficient, and C_E is the specific heat capacity, ∇^2 denotes the Laplace operator. In addition, ε_V denotes volumetric or cubical strain, and ν refers to Poisson's ratio.

The constitutive equations of the coupled thermoelastic fields for a beam can be expressed as:

$$\varepsilon_{xx}=\frac{\sigma_{xx}}{E}+\alpha\Delta T=-z\frac{\partial^2 w}{\partial x^2}+f(y,z)\left[\frac{\partial\varphi(x)}{\partial x}+\frac{\partial^2 w}{\partial x^2}\right] \tag{7.23}$$

$$\varepsilon_{yy}=\nu(\alpha\Delta T-\varepsilon_{xx})+\alpha\Delta T \tag{7.24}$$

$$\varepsilon_{zz}=-\nu\frac{\sigma_{xx}}{E}+\alpha\Delta T=\nu(\alpha\Delta T-\varepsilon_{xx})+\alpha\Delta T \tag{7.25}$$

The cubical strain can be expressed as:

$$\varepsilon_V=(1-2\nu)\left\{-z\frac{\partial^2 w}{\partial x^2}+f(y,z)\left[\frac{\partial\varphi(x)}{\partial x}+\frac{\partial^2 w}{\partial x^2}\right]-\alpha\Delta T\right\}+2(1+\nu)\Delta T \tag{7.26}$$

Substituting Eq. (7.26) into Eq. (7.22), recalling that $2\Delta_E\frac{1+\nu}{1-2\nu}\ll 1$, we have:

$$\chi\nabla^2 T=\left(1+\tau\frac{\partial}{\partial t}-3l^2\nabla^2\right)\frac{\partial}{\partial t}\left\{T+\frac{\Delta_E}{\alpha}\left[-z\frac{\partial^2 w}{\partial x^2}+f(y,z)\left(\frac{\partial\varphi(x)}{\partial x}+\frac{\partial^2 w}{\partial x^2}\right)\right]\right\} \tag{7.27}$$

with

$$\nabla^2=\frac{\partial^2}{\partial r^2}+\frac{1}{r}\frac{\partial}{\partial r}+\frac{\partial^2}{r^2\partial\theta^2}$$

Assuming that free vibration of the beam harmonic and kinematic parameters together with the temperature share the same complex-valued frequency ω, we write:

$$\{F(x,t),T(r,\theta,t)\}=\{\widetilde{F}(x),\widetilde{T}(r,\theta)\}e^{i\omega t} \tag{7.28}$$

Substituting Eq. (7.28) into Eqs. (7.18) and (7.27), we obtain

$$\frac{d^4}{dx^4}\widetilde{F}+\frac{\rho}{E}\left(1+\frac{E}{kG}\right)\omega^2\frac{d^2}{dx^2}\widetilde{F}-\frac{\rho A}{EI}\omega^2\left(1-\frac{\rho I}{kGA}\omega^2\right)\widetilde{F}=0 \tag{7.29}$$

$$\nabla^2\widetilde{T}-\frac{i\omega-\tau\omega^2}{\chi+3i\omega l^2}\widetilde{T}=\frac{\Delta_E}{\alpha}\frac{i\omega-\tau\omega^2}{\chi+3i\omega l^2}r\sin\theta\left[-\frac{\partial^2\widetilde{w}}{\partial x^2}+g(r)\left(\frac{\partial\widetilde{\varphi}}{\partial x}+\frac{\partial^2\widetilde{w}}{\partial x^2}\right)\right] \tag{7.30}$$

and, the corresponding deflection, rotation, bending moment, and transverse shear force can be expressed as follows

$$\widetilde{w}=\left(1-\frac{EI}{kGA}\frac{\partial^2}{\partial x^2}-\frac{\rho I}{kGA}\omega^2\right)\widetilde{F} \tag{7.31}$$

$$\widetilde{\varphi}=-\left(1+\frac{EI-EI_f}{GA_f}\frac{\partial^2}{\partial x^2}+\frac{\rho I-\rho I_f}{GA_f}\omega^2\right)\frac{\partial}{\partial x}\widetilde{F} \tag{7.32}$$

$$\widetilde{M}=-EI\frac{\partial^2}{\partial x^2}\widetilde{F} \tag{7.33}$$

$$\widetilde{Q}=-\left(EI\frac{\partial^2}{\partial x^2}+\rho I\omega^2\right)\frac{\partial}{\partial x}\widetilde{F} \tag{7.34}$$

Based on the standard theory of ordinary differential equations, a general solution to the above Eq. (7.29) can be immediately obtained as:

$$\widetilde{F}(x)=C_1\cos(\lambda_1 x)+C_2\sin(\lambda_1 x)+C_3\cosh(\lambda_2 x)+C_4\sinh(\lambda_2 x) \tag{7.35}$$

for frequency $\omega<\omega_c$ or $\omega R\sqrt{\rho/G}<2\sqrt{k}$, and

$$\widetilde{F}(x)=C_1\cos(\lambda_3 x)+C_2\sin(\lambda_3 x)+C_3\cos(\lambda_4 x)+C_4\sin(\lambda_4 x) \tag{7.36}$$

for frequency $\omega\geq\omega_c$ or $\omega R\sqrt{\rho/G}\geq 2\sqrt{k}$, where $\omega_c=\sqrt{kGA/(\rho I)}$ is the critical frequency, C_1, C_2, C_3 and C_4 are unknown constants,

$$\lambda_{1,2}=\omega\sqrt{\frac{1}{2}\left[\sqrt{\left(\frac{\rho}{E}+\frac{\rho}{kG}\right)^2+\frac{4\rho A}{EI}\left(\frac{1}{\omega^2}-\frac{1}{\omega_c{}^2}\right)}\pm\left(\frac{\rho}{E}+\frac{\rho}{kG}\right)\right]} \tag{7.37}$$

and

$$\lambda_{3,4}=\omega\sqrt{\frac{1}{2}\left[\pm\sqrt{\left(\frac{\rho}{E}+\frac{\rho}{kG}\right)^2+\frac{4\rho A}{EI}\left(\frac{1}{\omega^2}-\frac{1}{\omega_c{}^2}\right)}+\left(\frac{\rho}{E}+\frac{\rho}{kG}\right)\right]} \tag{7.38}$$

Thus, the expressions of $\widetilde{w}(x)$ and $\widetilde{\varphi}(x)$ can be derived as

$$\begin{bmatrix} \widetilde{w}(x) \\ \widetilde{\varphi}(x) \end{bmatrix} = \begin{bmatrix} \alpha_1 \cos(\lambda_1 x) & \alpha_1 \sin(\lambda_1 x) & \alpha_2 \cosh(\lambda_1 x) & \alpha_2 \sinh(\lambda_1 x) \\ \gamma_1 \sin(\lambda_1 x) & -\gamma_1 \cos(\lambda_1 x) & \gamma_2 \sinh(\lambda_2 x) & \gamma_2 \cosh(\lambda_2 x) \end{bmatrix} \begin{bmatrix} C_1 \\ C_2 \\ C_3 \\ C_4 \end{bmatrix} \tag{7.39}$$

for frequency $\omega < \omega_c$,

$$\begin{bmatrix} \widetilde{w}(x) \\ \widetilde{\varphi}(x) \end{bmatrix} = \begin{bmatrix} \alpha_3 \cos(\lambda_3 x) & \alpha_3 \sin(\lambda_3 x) & \alpha_4 \cos(\lambda_4 x) & \alpha_4 \sin(\lambda_4 x) \\ \gamma_3 \sin(\lambda_3 x) & -\gamma_3 \cos(\lambda_3 x) & \gamma_4 \sin(\lambda_4 x) & -\gamma_4 \cos(\lambda_4 x) \end{bmatrix} \begin{bmatrix} C_1 \\ C_2 \\ C_3 \\ C_4 \end{bmatrix} \tag{7.40}$$

for frequency $\omega \geq \omega_c$. where α_j, β_j and γ_j ($j = 1, \ldots, 4$) are defined as below:

$$\alpha_j = 1 + \frac{\eta_j}{kGA}, (j = 1, 3, 4), \alpha_2 = 1 - \frac{\eta_2}{kGA}. \tag{7.41}$$

$$\beta_j = -\lambda_j \eta_j, (j = 1, 2, 3, 4) \tag{7.42}$$

$$\gamma_j = \lambda_j - \frac{\lambda_j \eta_j}{GA}\left(\frac{A}{A_f} - \frac{1}{k}\right), (j = 1, 3, 4), \gamma_2 = -\lambda_2 - \frac{\lambda_2 \eta_2}{GA}\left(\frac{A}{A_f} - \frac{1}{k}\right) \tag{7.43}$$

with

$$\eta_j = EI\lambda_j^2 - \rho I\omega^2 (j = 1, 3, 4), \eta_2 = EI\lambda_j^2 + \rho I\omega^2 \tag{7.44}$$

For the Timoshenko beam, the above expressions are still valid, and the γ can be simplified and expressed as follows:

$$\gamma_j = \lambda_j, (j = 1, 3, 4), \gamma_2 = -\lambda_2 \tag{7.45}$$

The thermoelastic damping behaviors can be obtained by solving Eq. (7.30), we omit the subsequent derivation process for brevity.

7.3 Surface effects

This section presents a theoretical analysis of free vibration and forced vibration of nanowires with surface effects. Using the Timoshenko beam theory incorporated with the surface effects, exact frequency equations are derived for various end conditions. Natural frequencies and mode shapes are determined for simply supported, clamped-clamped, and clamped-free ends. Forced vibration of nanowires is also treated by superposing vibration modes of free vibration. An inverse problem is further investigated to determine the size-dependent effective Young's modulus of nanowires. Obtained results for a nanowire with residual stress and surface elasticity are confirmed by comparing them with the results obtained from the finite element method.

Theoretical natural frequencies and finite element method simulation agree well with the results from experimental data and molecular dynamics simulation. The effect of surface stress on forced vibration is analyzed. Residual surface tension has a significant influence on the frequencies and a somewhat effect on the mode shapes.

Many experimental evidences and atomistic/molecular simulations show that mechanical properties of materials at nanoscale, such as measured Young's modulus and bending stiffness of nanotubes (NTs) or nanowires (NWs), etc., have a significant size dependent effect [24–26]. To characterize this size effect, some continuum models have been proposed. Considering a very large specific surface area of NWs or NTs [27], surface effects have been recognized as an important factor responsible for experimentally measured size dependent mechanical properties. From this viewpoint, much theoretical and experimental attention has been directed toward the investigation of surface effects.

Accurate determination of the mechanical properties of NWs, particular Young's modulus, is essential for their practical applications in nanotechnology. However, constrained by the current techniques, measurement of Young's modulus relies primarily on some indirect means by using some powerful tools such as atomic force microscope (AFM) [28,29], scanning electron microscope (SEM) [26,30] and transmission electron microscope (TEM) [31–33], and molecular dynamics (MD) simulations [31–34]. The electric field-induced resonance method proposed by Poncharal et al. [25] has been widely applied to measure the Young's modulus of NWs. For example, Ding et al. [35] studied the mechanical response of crystalline boron NWs with the mechanical resonance method and tensile testing. Babaei et al. [36] determined the effective Young's modulus of silicon nitride cantilevers for thicknesses in the range of 20–684 nm by measuring resonance frequencies from thermal noise spectra. Zeng et al. [37] measured the fundamental resonant frequencies of a Si NW cantilever stimulated in an electrical field with an extra support localized at different positions along the NW to determine the Young's modulus of Si NWs. Due to size effects, the mechanical behaviors of nanoscale materials are quite different from those of their bulk counterparts [38–41]. For the determination of Young's modulus of an NW by measuring resonance frequencies, the frequency expressions of the classical cantilever beam are always introduced and corrected properly. However, we note that the predicted or effective Young's modulus is considered to be modified by surface elasticity, while the residual stress is absent.

In theory, due to large aspect ratio, NTs or NWs can be understood as a continuum beam [42], and many studies including bending and stability analysis of NTs and NWs with the surface effects have been conducted [43–49]. In particular, vibration analysis of NWs is more interesting for some practical applications. In this field, Wang and Feng [50,51] firstly presented a theoretical model to consider the effects of both surface elasticity and residual surface tension on the natural frequencies. Farshi et al. [52] also used the Timoshenko beam theory with the surface effect to analyze the vibration of NTs and calculated their natural frequencies. Lee and Chang [53] studied the natural frequency of NTs with consideration of surface effect using the nonlocal Timoshenko beam theory. He and Lilley [54] studied the influence of both surface and shear effect on the resonant frequency of NWs using the Timoshenko beam theory. Lei et al. [55] investigated free vibration of double-walled carbon NTs using the Timoshenko beam model incorporating the nonlocal elasticity theory and surface effect. Askari et al. [56] gave an analysis of nonlinear vibration of nonlocal NWs based on the Timoshenko beam theory. In the above-mentioned researches, on the one hand, simply-supported NWs or NTs are considered in almost all of the studies because they are the simplest type of beams since the mode shapes are always invariant, that is,sin $n\pi x$. On the other hand, the curvature of Timoshenko beams in all reported investigations of transverse vibration of NWs is described by $\partial^2 w/\partial x^2$, w being the deflection of NWs, which is only suitable when shear deformation is neglected. Thus, for Timoshenko beams, such a chosen curvature cannot provide sufficiently accurate analysis.

Due to a large specific surface area of NWs, the contribution of surface effects should be considered in studying the mechanical behavior and structural responses. To capture these features, we invoke the Gibbsian formulation of the thermodynamics of surfaces. The surface stress σ_{ij}^s can be expressed as [57,58]

$$\sigma_{ij}^s = \gamma\delta_{ij} + \frac{\partial\gamma}{\partial\varepsilon_{ij}^s}, i,j = 1,2 \tag{7.46}$$

where γ represents the excess free energy per unit area owing to the existence of a surface, ε_{ij}^s represents the surface elastic strain tensor, and δ_{ij} is the Kronecker delta. For a one dimensional NW, the above relation, (7.46) may be rewritten as a convenient form

$$\sigma^s = \tau_0 + E\varepsilon^s \tag{7.47}$$

where σ^s, ε^s denote the axial surface stress and axial surface strain, respectively, τ_0 represents the residual surface tension, E^s represents the surface elasticity modulus.

For an NW, it is more suitable to employ the Timoshenko beam theory (TBT) than the usual Euler-Bernoulli beam theory (EBT) since the shear deformation and rotary moment of inertia of cross sections are both taken into account in the former theory. Based on the Timoshenko beam theory, we have the following equations of motion

$$\frac{\partial M}{\partial x} - Q = \rho I \frac{\partial^2 \phi}{\partial t^2} \tag{7.48}$$

$$\frac{\partial Q}{\partial x}+p=\rho A\frac{\partial^2 w}{\partial t^2} \tag{7.49}$$

where w denotes the transverse deflection, ϕ denotes the rotation of the cross-section about the neutral axis (the positive direction of ϕ is opposite to that of $\partial w/\partial x$), ρ denotes the mass density of the NW, and A and I are the cross-sectional area and its moment of inertia, respectively. For the present situation, bending moment M, shear Q, and distributed load P, respectively, take [49]

$$M=(EI)^*\frac{\partial \phi}{\partial x} \tag{7.50}$$

$$Q=\kappa GA\left(\phi+\frac{\partial w}{\partial x}\right) \tag{7.51}$$

$$p=-H\frac{\partial \phi}{\partial x} \tag{7.52}$$

with

$$(EI)^*=EI+E^sJ \tag{7.53}$$

In the above Eqs. (7.50) to (7.52), for the classical Timoshenko beams, it is well-known that the shear strain is $\phi+\partial w/\partial x$ and the curvature is $-\partial\phi/\partial x$. It is worth noting that the surface stress defined in Eq. (7.52) is different from that used in other models such as He and Lilley [54] and Lei et al. [55], who use the relation $p=H\partial^2 w/\partial x^2$. The reason for our selection is that from the one dimensional version of the generalized Young–Laplace equation

$$\Delta\sigma_{ij}n_i n_j=\tau_{\alpha\beta}\kappa_{\alpha\beta}\ (i,j=1,2,3;\ \alpha,\beta=1,2) \tag{7.54}$$

we find that distributed loading $P(x)$ along the whole NW can be expressed by [46,51].

$$p(x)=H\kappa_0 \tag{7.55}$$

where $\Delta\sigma_{ij}$ is the stress jump across an interface/surface, n_i is the unit vector normal to the surface, $\tau_{\alpha\beta}$ is the surface stress tensor, and $\kappa_{\alpha\beta}(\kappa_0)$ is the curvature tensor.

$$I=\begin{cases}\pi D^4/64 & \text{(circular)}\\ WB^3/12 & \text{(rectangular)}\end{cases},\quad J=\begin{cases}\pi D^3/8 & \text{(circular)}\\ WB^2/2+B^3/6 & \text{(rectangular)}\end{cases} \tag{7.56}$$

$$H=\begin{cases}2\tau_0 D & \text{(circular)}\\ 2\tau_0 W & \text{(rectangular)}\end{cases},\quad \kappa=\begin{cases}6(1+v)^2/(7+12v+4v^2) & \text{(circular)}\\ 5(1+v)/(6+5v) & \text{(rectangular)}\end{cases} \tag{7.57}$$

where D represents the diameter of a circular NW, and W and B represent the width and height of a rectangular cross section, respectively (see Fig. 7.2(a)), and ν represents Poisson's ratio.

Obviously, two choices are identical if the Euler-Bernoulli beam theory is used, whereas $p = -H\partial\phi/\partial x$ in Eq. (7.52) [49] is somewhat different from $p = H\partial^2 w/\partial x^2$ [54,55] if the Timoshenko beam theory is used.

Substitution of Eqs. (7.50) to (7.52) into Eqs. (7.48) and (7.49) leads to the following system of governing equations

$$(EI)^*\frac{\partial^2\phi}{\partial x^2} - \kappa GA\left(\phi + \frac{\partial w}{\partial x}\right) = \rho I\frac{\partial^2\phi}{\partial t^2} \tag{7.58}$$

$$\kappa GA\left(\frac{\partial\phi}{\partial x} + \frac{\partial^2 w}{\partial x^2}\right) - H\frac{\partial\phi}{\partial x} - p_0 = \rho A\frac{\partial^2 w}{\partial t^2} \tag{7.59}$$

where P_0 is the applied distributed force.

To further simplify the above coupled governing equations, we introduce a new auxiliary function F that satisfies the following equation

$$\frac{\partial^4 F}{\partial x^4} - \frac{H}{(EI)^*}\frac{\partial^2 F}{\partial x^2} + \frac{\rho A}{(EI)^*}\frac{\partial^2 F}{\partial t^2} - \left[\frac{\rho I}{(EI)^*} + \frac{\rho A}{\kappa GA}\right]\frac{\partial^4 F}{\partial x^2\partial t^2} + \frac{\rho A}{(EI)^*}\frac{\rho I}{\kappa GA}\frac{\partial^4 F}{\partial t^4} = p_0 \tag{7.60}$$

If the transverse deflection w and rotation ϕ take the following form in terms of F

$$w(x) = F - \frac{(EI)^*}{\kappa GA}\frac{\partial^2 F}{\partial x^2} + \frac{\rho I}{\kappa GA}\frac{\partial^2 F}{\partial t^2} \tag{7.61}$$

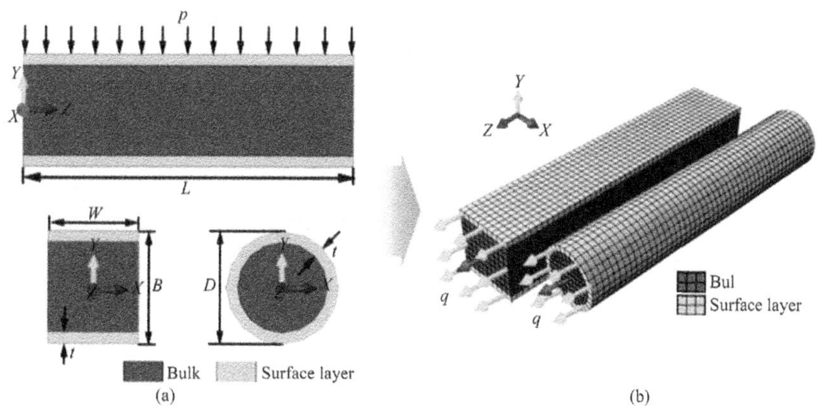

FIGURE 7.2 The core-shell composite model of an individual NW and corresponding FEM model. (a) Core-shell composite model of an individual NW and cross sectional view of a rectangular and circular NW. (b) FEM model of a rectangular and a circular NW.

$$\phi(x) = -\frac{\partial F}{\partial x} \tag{7.62}$$

It is readily found that Eqs. (7.58) and (7.59) are automatically satisfied. In other words, the system of governing equations has been converted to a single governing equation to describe the dynamic behavior of Timoshenko beams with the surface effects.

Using a single governing equation is more convenient for the analysis of free vibration problems of Timoshenko beams with the surface effects than using the coupled system of governing Eqs. (7.58) and (7.59). Once the solution of Eq. (7.60) is gained through appropriate conditions, the deflection and rotation as well as the moment and shear force will be determined by Eqs. (7.61) and (7.62) as well as Eqs. (7.50) and (7.51), respectively.

In closing this section, let us observe several special cases. From the above governing equation and the corresponding results, the governing equation for Rayleigh beams and corresponding results can be recovered when $\kappa GA \to \infty$. Secondly, the shear beam theory and corresponding results can be recovered if $\rho I = 0$. Thirdly, the simplest Euler-Bernoulli beam theory with the size effects [50] is recovered when setting $\kappa GA \to \infty$ and $\rho I = 0$ simultaneously. Finally, it is noted that by disregarding the surface effects, that is, $H = 0$ and $(EI)^* = EI$, the governing equations for classical Timoshenko beams can be recovered from the present.

7.3.1 Free vibration

In order to analyze the bending vibration of Timoshenko beams with the surface effects, in Eq. (7.60), we take $F = \overline{F}(\xi)e^{i\omega t}$, $i = \sqrt{-1}$, $\xi = x/L$, where $\overline{F}$ is a function, ω is the circular frequency. Thus, the governing Eq. (7.60) with $p_0 = 0$ becomes

$$\overline{F}^{IV}(\xi) - \overline{F}''(\xi)\left[h - \Omega^2(s^2 + r^2)\right] + \Omega^2(\Omega^2 s^2 r^2 - 1)\overline{F}(\xi) = 0 \tag{7.63}$$

where the prime denotes differentiation with respect to the argument. For the convenience of later analysis, dimensionless variables are introduced in Eq. (7.63) as follows

$$h = \frac{HL^2}{(EI)^*}, \Omega^2 = \frac{\rho AL^4\omega^2}{(EI)^*}, r^2 = \frac{I}{AL^2}, s^2 = \frac{(EI)^*}{\kappa GAL^2} \tag{7.64}$$

To solve Eq. (7.63) under various boundary conditions, it is pertinent to give expressions for the bending moment, shear force, rotation, and deflection in terms of $\overline{F}$.

$$M(\xi) = -\frac{(EI)^*}{L^2}\overline{F}''(\xi) \tag{7.65}$$

$$Q(\xi) = -\frac{\kappa GA}{L}\left(\Omega^2 r^2 r^4 \overline{F}(\xi) + \overline{F}'''(\xi)\right) \tag{7.66}$$

$$\phi(\xi) = -\frac{1}{L}\overline{F}'(\xi) \tag{7.67}$$

$$w(\xi) = \left[\left(1 - \Omega^2 r^2 r^4\right)\overline{F}(\xi) - \overline{F}''(\xi)\right] \tag{7.68}$$

where the time factor $e^{i\omega t}$ has been suppressed.

By expressing a solution of Eq. (7.63) in the following form $\overline{F}(\xi) = e^{\gamma\xi}$, one easily finds that the constant γ has to satisfy the following algebraic equation

$$\gamma^4 - \left[h - \Omega^2\left(r^2 + r^4\right)\right]\gamma^2 + \Omega^2\left(\Omega^2 r^2 r^4 - 1\right) = 0 \tag{7.69}$$

Solving the above algebraic equation, one gets two pairs of distinct roots $\gamma_{1,2} = \pm\delta_1, \gamma_{3,4} = \pm\delta_2$ where

$$\delta_{1,2} = \sqrt{\left|\frac{\left[h - \Omega^2(r^2 + r^4)\right]^2 - 4\Omega^2\left(\Omega^2 r^2 r^4 - 1\right) \pm \left[\Omega^2(r^2 + r^4) - h\right]}{2}\right|} \tag{7.70}$$

From the above, $\Omega_0 = 1/(sr)$ is called the cut-off frequency. Here we restrict our attention to low-frequency vibration, that is, $\Omega < 1/(sr)$. For free vibration at high frequencies exceeding the cut-off frequency Ω_0, a similar procedure can be applied and is therefore omitted here. In this case, a general solution to Eq. (7.63) for low-frequency free vibration is given by

$$\overline{F}(\xi) = C_1 \sin \delta_1\xi + C_2 \cos \delta_1\xi + C_3 \sinh \delta_2\xi + C_4 \cosh \delta_2\xi \tag{7.71}$$

where $C_i (i = 1, 2, 3, 4)$ are unknown constants. With this $\overline{F}(\xi)$, the bending moment, shear force, rotation, and deflection can be expressed as

$$M(\xi) = -\frac{(EI)^*}{L^2}\left[-C_1\delta_1^2 \sin \delta_1\xi - C_2\delta_1^2 \cos \delta_1\xi + C_3\delta_2^2 \sinh \delta_2\xi + C_4\delta_2^2 \cosh \delta_2\xi\right] \tag{7.72}$$

$$\begin{aligned} Q(\xi) = -\frac{\kappa GA}{L}\Big[&\left(\Omega^2 r^2 - \delta_1^2\right)s^2\delta_1 C_1 \cos \delta_1\xi - \left(\Omega^2 r^2 - \delta_1^2\right)s^2\delta_1 C_2 \sin \delta_1\xi \\ &+ \left(\Omega^2 r^2 + \delta_2^2\right)s^2\delta_2 C_3 \cosh \delta_2\xi + \left(\Omega^2 r^2 + \delta_2^2\right)s^2\delta_2 C_4 \sinh \delta_2\xi\Big] \end{aligned} \tag{7.73}$$

$$\phi(\xi) = -\frac{1}{L}[C_1\delta_1 \cos \delta_1\xi - C_2\delta_1 \sin \delta_1\xi + C_3\delta_2 \cosh \delta_2\xi + C_4\delta_2 \sinh \delta_2\xi] \tag{7.74}$$

$$w(\xi) = \left(1 - \Omega^2 r^2 s^2 + s^2\delta_1^2\right)C_1 \sin \delta_1\xi + \left(1 - \Omega^2 r^2 s^2 + s^2\delta_1^2\right)C_2 \cos \delta_1\xi + \left(1 - \Omega^2 r^2 s^2 - s^2\delta_2^2\right)C_3 \sinh \delta_2\xi + \left(1 - \Omega^2 r^2 s^2 - s^2\delta_2^2\right)C_4 \cosh \delta_2\xi \tag{7.75}$$

7.3.2 Pinned-pinned NWs

For pinned-pinned NWs, associated boundary conditions at two ends read

$$w(\xi) = 0, M(\xi) = 0, \xi = 0, 1 \tag{7.76}$$

Applying the above boundary conditions Eq. (7.76) to Eqs. (7.72) and (7.75), after some algebra one readily obtains the frequency equation

$$\sin \delta_1 \sinh \delta_2 = 0 \tag{7.77}$$

which gives $\delta_1 = n\pi$ ($n = 1, 2, 3, \ldots$). Substituting the result into Eq. (7.70) yields an algebraic equation, the solution of which gives the circular frequencies $\omega_n = \eta n^2\pi^2\sqrt{EI/(\rho A)}\Big/L^2$, with η denoting a correction factor relative to the classical counterpart

$$\eta = \frac{1}{n\pi}\sqrt{\frac{(EI)^*}{EI}\frac{2(n^2\pi^2 + h)}{1 + n^2\pi^2(s^2 + r^2) + \sqrt{[1 + n^2\pi^2(s^2 + r^2)]^2 - 4s^2r^2n^4\pi^4(n^2\pi^2 + h)}}} \tag{7.78}$$

Alternatively, the expression for the circular frequencies can be written as $\omega_n = \Omega_n\sqrt{(EI)^*/(\rho A)}\Big/L^2$. Furthermore, we can obtain the corresponding mode shapes of free vibration as follows

$$w_n(\xi) = \sin n\pi\xi, \phi_n(\xi) = \cos n\pi\xi \tag{7.79}$$

From the above, it is easily found that the mode shapes of free vibration are identical to those obtained from the classical theory. Nevertheless, the natural frequencies change due to the consideration of the surface effects. This influence can be described by the factor η. Obviously, from Eq. (7.78), in addition to shear deformation and rotary moment of inertia, η depends on the surface Young's modulus and surface stress as well.

7.3.3 Clamped-clamped NWs

For clamped-clamped beams, associated boundary conditions read

$$w(\xi)=0, \phi(\xi)=0, \xi=0,1 \tag{7.80}$$

Applying the above boundary conditions to Eqs. (7.74) and (7.75), and after some algebraic manipulation, one finally obtains the desired frequency equation

$$2-2\cosh\delta_2\cos\delta_1-\left[\frac{\delta_1\left(1-\Omega^2r^2s^2-\delta_2^2s^2\right)}{\delta_2\left(1-\Omega^2r^2s^2+\delta_1^2s^2\right)}+\frac{\delta_2\left(1-\Omega^2r^2s^2+\delta_1^2s^2\right)}{\delta_1\left(1-\Omega^2r^2s^2-\delta_2^2s^2\right)}\right]\sin\delta_1\sinh\delta_2=0 \tag{7.81}$$

This frequency equation is more complex than the one corresponding to pinned-pinned beams as an exact analytic solution to Eq. (7.81) is unlikely because it is a transcendental equation. Consequently, with the aid of commercial software, one may determine the numerical natural frequencies up to a desired accuracy. Once the natural frequencies are obtained through solving the above transcendental equation, we can substitute these numerical results into Eqs. (7.74) and (7.75) to give the mode shapes as follows

$$w(\xi)=\frac{\delta_2\sinh\delta_2+\delta_1\chi\sin\delta_1}{\delta_1(\cosh\delta_2-\cos\delta_1)}\left(\sin\delta_1\xi-\frac{\delta_1\chi}{\delta_2}\sinh\delta_2\xi\right)+\chi(\cosh\delta_2\xi-\cos\delta_1\xi) \tag{7.82}$$

$$\phi(\xi)=\frac{\delta_2\sinh\delta_2+\delta_1\chi\sin\delta_1}{\delta_1(\cosh\delta_2-\cos\delta_1)}(\cos\delta_1\xi-\cosh\delta_2\xi)+\chi\sin\delta_1\xi+\frac{\delta_2}{\delta_1}\sinh\delta_2\xi \tag{7.83}$$

where

$$\chi=\frac{1-\Omega^2r^2s^2-\delta_2^2s^2}{1-\Omega^2r^2s^2+\delta_1^2s^2} \tag{7.84}$$

As a check, by setting $E^s=0$ and $h=0$, the frequency Eq. (7.81) derived here becomes

$$2(1-\cos\delta_1\cosh\delta_2)+\frac{\Omega}{\sqrt{1-\Omega^2r^2s^2}}\times\left[\Omega^2r^2\left(r^2-s^2\right)^2+\left(3s^2-r^2\right)\right]\sin\delta_1\sinh\delta_2=0 \tag{7.85}$$

identical to that of classical clamped-clamped Timoshenko beams [59].

7.3.4 Clamped-free NWs

For clamped-free beam, associated boundary conditions read

$$w(0)=0, \phi(0)=0 \tag{7.86}$$

$$M(1)=0, Q(1)=0 \tag{7.87}$$

Applying the above boundary conditions to Eqs. (7.74) and (7.75), and after some algebraic manipulations, one finally obtains the desired frequency equation to be

$$\begin{aligned} &\chi\delta_1^2-\delta_2^2\frac{\Omega^2r^2+\delta_2^2}{\Omega^2r^2-\delta_1^2}+\left(\chi\frac{\Omega^2r^2+\delta_2^2}{\Omega^2r^2-\delta_1^2}\right)\delta_2\delta_1\sin\delta_1\sinh\delta_2\\ &-\left(\chi\delta_1^2\frac{\Omega^2r^2+\delta_2^2}{\Omega^2r^2-\delta_1^2}-\delta_2^2\right)\cos\delta_1\cosh\delta_2=0\end{aligned} \tag{7.88}$$

where χ is given by Eq. (7.84).

In this case, the mode shape expressions can be similarly obtained, respectively, as

$$\begin{aligned} w(\xi)=&\frac{\delta_2^2\cosh\delta_2+\delta_1^2\chi\cos\delta_1}{\delta_1(\delta_2\sinh\delta_2+\delta_1\sin\delta_1)}\left(\sin\delta_1\xi-\frac{\delta_1\chi}{\delta_2}\sinh\delta_2\xi\right)\\ &+\chi(\cosh\delta_2\xi-\cos\delta_1\xi)\end{aligned} \tag{7.89}$$

$$\phi(\xi)=\frac{\delta_2^2\cosh\delta_2+\delta_1^2\chi\cos\delta_1}{\delta_1(\delta_2\sinh\delta_2+\delta_1\sin\delta_1)}(\cos\delta_1\xi-\cosh\delta_2\xi)+\chi\sin\delta_1\xi+\frac{\delta_2}{\delta_1}\sinh\delta_2\xi \tag{7.90}$$

As a check, by setting $E^s=0$ and $h=0$, one obtains

$$2+\left[2+\left(r^2-s^2\right)\Omega^2\right]\cos\delta_1\cosh\delta_2-\frac{\Omega^2\left(r^2+s^2\right)}{\sqrt{1-r^2s^2\Omega^2}}\sin\delta_1\sinh\delta_2=0 \tag{7.91}$$

identical to that of clamped-free classical Timoshenko beams [59].

7.3.5 Forced vibration

For the forced vibration, the nonhomogeneous term $P_0(x, t)$ on the right-hand side of Eq. (7.60) is no longer zero. For the forced vibration of NWs, based on superposition, it is convenient to take the desired function F in the form

$$F(x,t)=\sum_{n=1}^{\infty}y_n(\xi)u_n(t), x=\xi L \tag{7.92}$$

where $y_n(\xi)$ is the eigenfunction or mode shape corresponding to the n-th order vibration mode, which depends on associated boundary conditions. Substitution of Eq. (7.92) into Eq. (7.60) yields

$$\sum_{n=1}^{\infty}\frac{\Omega_n^2}{\omega_n^2}\left\{\frac{\Omega_n^2 s^2 r^2 y_n(\xi)}{\omega_n^2}u_n^{IV}(t)+\left[y_n(\xi)-\left(s^2+r^2\right)y_n''(\xi)\right]u_n''(t)-\right.$$
$$\left.\omega_n^2\left[\left(s^2+r^2\right)y_n''(\xi)+\left(\Omega_n^2 s^2 r^2-1\right)y_n(\xi)\right]u_n(t)\right\}=\frac{L^4}{(EI)^*}p_0(\xi,t) \tag{7.93}$$

Next, for simplicity, we only consider a simply-supported NW subjected to an excitation loading $p_0(\xi,t)=p_0(\xi)\sin\omega t$. Under such circumstances, we have

$$y_n(\xi)=\sin n\pi\xi \tag{7.94}$$

In view of the orthogonality property of eigenfunctions, multiplying Eq. (7.93) by $y_l(\xi)$ and integrating with respect to ξ from 0 to one results in

$$\frac{\Omega_n^2}{\omega_n^2}\left\{\frac{\Omega_n^2 r^2 s^2}{\omega_n^2}u_n^{IV}(y)+\left[1+(n\pi)^2\left(s^2+r^2\right)\right]u_n''(t)-\omega_n^2\left[\Omega_n^2 r^2 s^2\right.\right.$$
$$\left.\left.-1-\left(s^2+r^2\right)(n\pi)^2\right]u_n(t)\right\}=z_n\sin\omega t \tag{7.95}$$

where

$$z_n=\frac{2L^4}{(EI)^*}\int_0^1 p_0(\xi)\sin n\pi\xi d\xi \tag{7.96}$$

Obviously, Eq. (7.95) has a particular solution of

$$u_n^*=\frac{z_n\sin\omega t}{\Omega_n^2\left\{1+(s^2+r^2)(n\pi)^2-\Omega_n^2 r^2 s^2-\left[1+(n\pi)^2(s^2+r^2)\right](\omega/\omega_n)^2+\Omega_n^2 r^2 s^2(\omega/\omega_n)^4\right\}} \tag{7.97}$$

where

$$\Omega_n=n\pi\sqrt{\frac{2(n^2\pi^2+h)}{1+n^2\pi^2(s^2+r^2)+\sqrt{\left[1+n^2\pi^2(s^2+r^2)\right]^2-4s^2r^2n^2\pi^2(n^2\pi^2+h)}}} \tag{7.98}$$

as derived previously for a pinned-pinned NW. Therefore, a general solution of Eq. (7.95) can be obtained through adding the particular solution to the general solution of the corresponding homogeneous Eq. (7.95), which contains four unknown constants. The determination of these four constants can be done by applying initial conditions, such as a state of rest at initial time, i.e.,

$$w=0,\phi=0,\frac{\partial w}{\partial t}=0,\frac{\partial\phi}{\partial t}=0,\text{ at }t=0 \tag{7.99}$$

On the other hand, if we only consider the steady response of harmonic vibration excitation, the corresponding initial conditions are neglected, and attention is restricted to the response of the forcing term.

Once a definite solution Eq. (7.92) is determined, the deflection, rotation, bending moment and shear force at any time are obtainable by substituting F given in Eq. (7.92) into Eqs. (7.65) to (7.68), where the time factor is also omitted.

7.3.6 Size dependent effective Young's modulus

Effective Young's modulus of NWs can be determined by measuring resonance frequencies using experimental techniques. From Eq. (7.64), one rewrites the natural frequencies f_n as follows

$$f_n = \frac{\beta_n^2}{2\pi L^2}\sqrt{\frac{EI}{\rho A}\left(1 + \frac{E_s J}{EI}\right)},\ h_n = \frac{H\beta_n^4}{4\rho A \pi^2 L^2 f_n^2} \tag{7.100}$$

where $\beta_n = \sqrt{\Omega_n}$, with Ω_n being the nth root of Eq. (7.77) for pinned-pinned NWs, Eq. (7.81) for clamped-clamped NWs and Eq. (7.88) for clamped-free NWs. It is of great importance to point out that β_n is a constant when the residual stress is neglected (e.g., $\beta_1 = 1.875$, $\beta_2 = 4.694$). Different from previous studies, β_n here also depends on the residual stress τ_0. Thus, from Eq. (7.100) it is seen that the effective Young's modulus is relevant to both the residual stress and the surface elasticity, rather than on the surface elasticity alone. This method plays a significant role not only in experimental investigations but also in FEM simulations of NWs that consider both residual stress and surface elasticity, as will be demonstrated in the following section.

From Eq. (7.100), the surface elasticity and effective Young's modulus are directly obtained, respectively

$$E_s = \frac{EI}{J}\left(\frac{4\pi^2 \rho A L^4 f_n^2}{EI\beta_n^4} - 1\right),\ E_{\text{eff}} = \frac{4\pi^2 \rho A L^4 f_n^2}{I\beta_n^4}\ \text{or}\ E_{\text{eff}} = \frac{HL^2}{Ih_n} \tag{7.101}$$

7.4 Piezoelectric and flexoelectric effects

This section studies the bending behavior of a nanoscale elastic beam with flexoelectric and piezoelectric effects. Based on a two dimensional theory of piezoelectricity and flexoelectricity, an exact elasticity solution is derived when applied transverse mechanical loading and/or electric potential difference between the top and bottom surfaces are prescribed. The obtained results may apply not only to pure piezoelectric beams (by setting all flexoelectric coefficients to zero) but also to isotropic beams exhibiting flexoelectricity (by setting all piezoelectric coefficients to zero), etc. This solution can be taken as a benchmark solution for related beam models used in actuators and generators, based on which the influences of piezoelectricity and flexoelectricity on the transverse deflection, stress distribution, and electric polarization are analyzed. Flexoelectricity plays a crucial role in altering bending deflection, electric polarization for a single homogeneous simply supported beam. The deflection and electric polarization of a beam with both piezoelectric and flexoelectric effects are not simply the sum of those of a purely piezoelectric and purely flexoelectric beam.

The electromechanical coupling of materials has been widely used in nanomechanical systems such as resonators, sensors, actuators, and energy harvesters. The traditional electromechanical coupling includes piezoelectricity and electrostriction. Piezoelectricity is a well-known electromechanical coupling phenomenon occurring in noncentrosymmetric dielectric materials, which centrosymmetric dielectrics do not polarize under uniform mechanical strain. Piezoelectricity generally assumes a linear relationship between electric field and strain. However, the strain gradient or nonuniform strain distributions could break the local symmetry, which induces polarization even in centrosymmetric dielectric materials. This phenomenon is called flexoelectricity, which exists in liquid crystals [60], solid dielectrics [61] and living membranes [62]. This effect can be viewed as a higher-order electromechanical coupling phenomenon compared to the piezoelectric effect. The latter exhibits dielectric polarization in response to a mechanical strain and vice versa.

Analogous to the definition of the direct and converse piezoelectric effects, the electric polarization due to the strain gradient is called the direct flexoelectric effect, while the phenomenon of induced mechanical stress or strain due to the electric field gradient is termed as the inverse or converse flexoelectric effect. Harris [63] observed electric polarization in centrosymmetric solids due to the strain gradient caused by shock waves. Mindlin [64] introduced a gradient model of polarization into a conventional linear electromechanical coupling theory based on the long wavelength limit of the shell-model of lattice dynamics.

Tagantsev [65] developed a microscopic theory of ionic flexoelectricity based on a rigid-ion model, theoretically confirmed that the flexoelectric effect in crystalline solids is different from the piezoelectric effect, and derived a simple model for computing the flexoelectric coefficients. A series of experiments have been carried out to measure the flexoelectric coefficients of some dielectrics, and giant flexoelectric coefficients for ferroelectrics with high dielectric permittivity have been reported [61,66–68]. Theoretically, Maranganti et al. [69] established a mathematical framework for dielectrics involving the coupling between strain gradients and polarization, and conversely between strain and polarization gradients. Within the framework of the linear piezoelectricity theory, Majdoub et al. [70] analyzed the bending behavior of a nanoscale cantilever beam and discussed the role of flexoelectricity in both piezoelectric and nonpiezoelectric nanostructures. They showed that effective piezoelectric coefficient and elastic modulus have significant size dependence due to the flexoelectric effect. Using a variational principle, Shen and Hu [71] established a theoretical framework of elasticity, flexoelectricity, and surface effects for nanosized dielectrics. Yan and Jiang [72,73] investigated the influence of the flexoelectric effect on the bending of piezoelectric nanobeams. Liang et al. [74,75] further gave an analysis of the effects of surface and flexoelectricity on the bending of a piezoelectric cantilever nanobeam. Later, Zhou et al. [76] also gave an electromechanical coupling analysis of flexoelectric–piezomagnetic nanobeams. The above analyses are based on the Euler-Bernoulli beam theory. If taking shear deformation into account, according to the Timoshenko beam theory, the bending of a Timoshenko beam with the flexoelectric effect has been addressed [77,78]. Using a two dimensional theory, Ray [79] coped with the bending of an isotropic flexoelectric beam. On the other hand, flexoelectricity/piezoelectricity nanogenerators are a new technique for energy harvcsting [80,81], based on the piezoelectric and flexoelectricity, a theoretical analysis of nanoscale energy harvesting has been formulated by Wang and Wang [82], Liang et al. [83], and Pang et al. [84]. From the above, although researchers have paid considerable attention to studying the flexoelectric effect in beams within the framework of Euler-Bernoulli and Timoshenko theories of beams, adequate analysis within a two dimensional theory of piezoelectricity and flexoelectricity remains limited.

7.4.1 Basic equations

Consider a nanoscale beam with the piezoelectric and flexoelectric effects, where in addition to electric fields E_i, electric polarization components P_i may be induced by strains and strain gradients, and conversely, elastic strains depend not only on elastic stresses but also on electric polarization and their gradients. Usually, strain gradient-induced polarization or electric gradient-induced strain in crystalline solids are referred to as the flexoelectric effect. When the piezoelectric and flexoelectric effects are present simultaneously in an elastic dielectric, the linear constitutive equations read [85]:

$$\sigma_{ij} = c_{ijkl}\varepsilon_{kl} - e_{kij}E_k + \mu_{ijkl}\varepsilon_{k,l} \tag{7.102}$$

$$P_i = \chi_{ij}E_j + e_{ikl}\varepsilon_{kl} + \mu_{ijk}\varepsilon_{kl,j} \tag{7.103}$$

where P_i and E_i are the polarization vector and the electric field vector, respectively; σ_{kl} and ε_{kl} are the stress tensor and the strain tensor; c_{ijkl}, μ_{ijkl}, e_{kij}, and χ_{ij} are the fourth rank elastic stiffness tensor, the fourth rank flexoelectric coefficient tensor, the third rank piezoelectric coefficient tensor, and the third rank dielectric susceptibility tensor, respectively. A comma in a subscript represents partial differentiation with respect to the spatial variable following the comma. The subscripts i, j, k, l run from 1 to 3. In the absence of the flexoelectric effect, that is,$\mu_{ijkl} = 0$, the above constitutive equations reduce to those for conventional piezoelectric materials. In particular, the piezoelectric effect occurs only for a class of noncentrosymmetric dielectric materials, while in the absence of the piezoelectric effect, that is, $e_{kij} = 0$, the above constitutive equations reduce to those for elastic beams with the flexoelectric effect alone. For completeness, we consider a general case in which both piezoelectric and flexoelectric effects are present. In fact, the latter is more universal than the former because flexoelectricity takes place in all classes of dielectric materials including centrosymmetric crystals.

Similar to the number of elastic stiffness, the number of independent, nonvanishing piezoelectric and flexoelectric coefficients is dependent on the point group of the crystals [86,87]. For simplicity, a transversely isotropic piezoelectric material of point group 6 mm exhibiting the flexoelectric effect is analyzed here. For this class of piezoelectric materials, by choosing a Cartesian coordinate system such that the z-axis is along the polarization direction, and the x-y plane represents the isotropic plane, the material properties are given as follows

$$[\boldsymbol{c}_{ijkl}] = \begin{bmatrix} c_{11} & c_{12} & c_{13} & 0 & 0 & 0 \\ c_{12} & c_{11} & c_{13} & 0 & 0 & 0 \\ c_{13} & c_{13} & c_{33} & 0 & 0 & 0 \\ 0 & 0 & 0 & c_{44} & 0 & 0 \\ 0 & 0 & 0 & 0 & c_{44} & 0 \\ 0 & 0 & 0 & 0 & 0 & c_{66} \end{bmatrix} \tag{7.104}$$

$$[\boldsymbol{\chi}_{ij}] = \begin{bmatrix} \chi_{11} & 0 & 0 \\ 0 & \chi_{11} & 0 \\ 0 & 0 & \chi_{33} \end{bmatrix} \tag{7.105}$$

$$[\boldsymbol{e}_{kij}] = \begin{bmatrix} 0 & 0 & 0 & 0 & e_{15} & 0 \\ 0 & 0 & 0 & e_{15} & 0 & 0 \\ e_{31} & e_{31} & e_{33} & 0 & 0 & 0 \end{bmatrix} \tag{7.106}$$

$$[\boldsymbol{\mu}_{ijkl}] = \begin{bmatrix} \mu_{11} & 0 & 0 & 0 & \mu_{12} & 0 & 0 & 0 & \mu_{13} \\ \mu_{12} & 0 & 0 & 0 & \mu_{11} & 0 & 0 & 0 & \mu_{13} \\ \mu_{31} & 0 & 0 & 0 & \mu_{31} & 0 & 0 & 0 & \mu_{33} \\ 0 & 0 & 0 & 0 & 0 & \mu_{41} & 0 & \mu_{42} & 0 \\ 0 & 0 & \mu_{41} & 0 & 0 & 0 & \mu_{42} & 0 & 0 \\ 0 & \mu_{43} & 0 & \mu_{43} & 0 & 0 & 0 & 0 & 0 \end{bmatrix} \tag{7.107}$$

for convenience, the 3- and 4-suffix subscripts *kij*, *ijkl* have been replaced by a 2-suffix subscript according to the Voigt notation, $\mu_{1111} = \mu_{2222} =$

$\mu_{11}, \mu_{3333} = \mu_{33}, \mu_{1122} = \mu_{2211} = \mu_{12}, \mu_{1133} = \mu_{2233} = \mu_{13}, \quad \mu_{3311} = \mu_{3322} = \mu_{31}, \mu_{2323} = \mu_{1313} = \mu_{41}, \mu_{2332} = \mu_{1331} = \mu_{42}, \mu_{1212} = \mu_{1221} = \mu_{43}$, and $c_{66} = (c_{11} - c_{12})/2$, $\mu_{43} = \mu_{11} - \mu_{12}$. In particular, by setting all piezoelectric coefficients to zero, and $c_{11} = c_{33} = \lambda + 2G$, $c_{12} = c_{13} = \lambda$, $c_{44} = c_{66} = G$, $\mu_{12} = \mu_{13} = \mu_{31}$, $\mu_{33} = \mu_{11}$, $\mu_{41} = \mu_{42} = \mu_{43}$, the above constitutive relations reduce to those for isotropic elastic material with flexoelectricity [88–90]. In this case, five elastic stiffness constants reduce to two Lamé constants (λ, G), seven flexoelectric constants reduce to two flexoelectric constants (μ_{11}, μ_{12}), and no piezoelectric constants are included.

In the Cartesian coordinate system, we consider a plane problem. Thus, we simply discard the variable y and y-dependent variables in the above constitutive equations. This actually corresponds to a sufficiently wide plate along the y-direction or a so-called plane strain problem with respect to y. Also, an adequate description of a quite slender beam or narrow plate may be achieved by treating it as a plane stress problem, in which the coefficients involved are replaced with the corresponding ones for that case. Accordingly, owing to the xoz-plane, the constitutive relations of a transversely isotropic piezoelectric material of point group 6 mm with the flexoelectric effect read

$$\sigma_{xx} = c_{11}\varepsilon_{xx} + c_{13}\varepsilon_{zz} - e_{31}E_z + \mu_{11}E_{x,x} + \mu_{13}E_{z,z} \tag{7.108}$$

$$\sigma_{zz} = c_{13}\varepsilon_{xx} + c_{33}\varepsilon_{zz} - e_{33}E_z + \mu_{31}E_{x,x} + \mu_{33}E_{z,z} \tag{7.109}$$

$$\sigma_{xz} = 2c_{44}\varepsilon_{xz} - e_{15}E_x + \mu_{41}E_{x,z} + \mu_{42}E_{z,x} \tag{7.110}$$

and

$$P_x = \chi_{11}E_x + 2e_{15}\varepsilon_{xz} + \mu_{11}\varepsilon_{xx,x} + \mu_{31}\varepsilon_{zz,x} + 2\mu_{41}\varepsilon_{xz,z} \tag{7.111}$$

$$P_z = \chi_{33}E_z + e_{31}\varepsilon_{xx} + e_{33}\varepsilon_{zz} + \mu_{13}\varepsilon_{xx,z} + \mu_{33}\varepsilon_{zz,z} + 2\mu_{42}\varepsilon_{xz,x} \tag{7.112}$$

where $\sigma_{xx}, \varepsilon_{xx}$ and E_x are the physical normal stress, normal strain and electric field along the x-direction, respectively; $\sigma_{zz}, \varepsilon_{zz}$ and E_z are those along the z-direction, respectively; σ_{xz} and ε_{xz} are the shear stress and shear strain, respectively.

In addition, the elastic strains and strain gradients are expressed in terms of elastic displacement vector $\boldsymbol{u}_j$ by the following relations:

$$\varepsilon_{kl} = \frac{1}{2}\left(u_{k,l} + u_{l,k}\right), \varepsilon_{kl,j} = \frac{1}{2}\left(u_{k,lj} + u_{l,kj}\right) \tag{7.113}$$

and the electric fields along with the electric-field gradients are given by electric potential ϕ below:

$$E_j = -\phi_{,j}, E_{k,l} = -\phi_{,kl} \tag{7.114}$$

Furthermore, using the following equations for equilibrium of force and conservation law of electric charge, we have

$$\frac{\partial \sigma_{xx}}{\partial x}+\frac{\partial \sigma_{xz}}{\partial z}=0 \tag{7.115}$$

$$\frac{\partial \sigma_{xz}}{\partial x}+\frac{\partial \sigma_{zz}}{\partial z}=0 \tag{7.116}$$

$$\frac{\partial D_x}{\partial x}+\frac{\partial D_z}{\partial z}=0 \tag{7.117}$$

with

$$D_j=\kappa_0 E_j+P_j, j=x,z \tag{7.118}$$

where κ_0 is the permittivity of vacuum or air. In the above, we have introduced electric displacement in place of electric polarization. Since κ_0 is sufficiently small as compared to dielectric susceptibility, electric polarization can be approximated by electric displacement. That is, the influence of the vacuum part is neglected.

Next, substituting Eqs. (7.113) and (7.114) along with Eq. (7.118) into Eqs. (7.108)-(7.112), and then into the equilibrium equations for forces and Gauss equation for electric charge conservation Eqs. (7.115)-(7.117), one gets the final governing partial differential equations as follows:

$$\begin{aligned} c_{11}\frac{\partial^2 u}{\partial x^2}+c_{44}\frac{\partial^2 u}{\partial z^2}+(c_{13}+c_{44})\frac{\partial^2 w}{\partial x\partial z}+(e_{31}+e_{15})\frac{\partial^2 \phi}{\partial x\partial z}-\mu_{11}\frac{\partial^3 \phi}{\partial x^3} \\ -(\mu_{13}+\mu_{41}+\mu_{42})\frac{\partial^3 \phi}{\partial x\partial z^2}=0 \end{aligned} \tag{7.119}$$

$$\begin{aligned} (c_{13}+c_{44})\frac{\partial^2 u}{\partial x\partial z}+c_{44}\frac{\partial^2 w}{\partial x^2}+c_{33}\frac{\partial^2 w}{\partial z^2}+e_{15}\frac{\partial^2 \phi}{\partial x^2}+e_{33}\frac{\partial^2 \phi}{\partial z^2} \\ -(\mu_{31}+\mu_{41}+\mu_{42})\frac{\partial^3 \phi}{\partial x^2\partial z}-\mu_{33}\frac{\partial^3 \phi}{\partial z^3}=0 \end{aligned} \tag{7.120}$$

$$\begin{aligned} (e_{31}+e_{15})\frac{\partial^2 u}{\partial x\partial z}+\mu_{11}\frac{\partial^3 u}{\partial x^3}+(\mu_{13}+\mu_{41}+\mu_{42})\frac{\partial^3 u}{\partial x\partial z^2}+e_{15}\frac{\partial^2 w}{\partial x^2}+e_{33}\frac{\partial^2 w}{\partial z^2}+ \\ \mu_{33}\frac{\partial^3 w}{\partial z^3}+(\mu_{31}+\mu_{41}+\mu_{42})\frac{\partial^3 w}{\partial x^2\partial z}-\kappa_{11}\frac{\partial^2 \phi}{\partial x^2}+\kappa_{33}\frac{\partial^2 \phi}{\partial z^2}=0 \end{aligned} \tag{7.121}$$

where we have denoted

$$\kappa_{ii}=\kappa_0+\chi_{ii}, i=1,3. \tag{7.122}$$

As a check, by setting all flexoelectric coefficients to zero, the governing partial differential equations for a transversely isotropic piezoelectric material are recovered. It is particularly noted that due to the introduction of strain gradients and electric field gradients in the constitutive equations, the order of the resulting governing equations for piezoelectric materials with flexoelectricity is higher than that for materials without flexoelectricity. This is clear since the coefficients of the highest-order derivatives in Eqs. (7.119)-(7.121) only contain the flexoelectric coefficients, which vanish if the flexoelectricity is neglected, leading to the elimination of the highest-order derivatives. Therefore, in addition to the conventional boundary conditions, additional conditions must be supplemented to meet the solvability of the resulting governing equations, which will be given in the following section.

7.4.2 Elasticity solution

Now we consider a simply supported rectangular beam made of a piezoelectric material with the flexoelectric effect, as shown in Fig. 7.3. Its length is denoted as L, and thickness (or height) as h, respectively. The thickness direction is parallel to the polarization direction. A distributed mechanical load $q(x)$ per unit length is applied on its top surface, and the bottom surface is traction-free. In addition, in order to simulate a general case, an electric potential difference is exerted between the top and the bottom surfaces. Since an arbitrary function can be expanded in terms of trigonometric functions, without loss of generality, in the present study, we consider the following boundary conditions on the top and bottom surfaces.

$$\begin{cases} \sigma_{zz}\left(x, \dfrac{h}{2}\right) = \sum\limits_{n=1}^{\infty} q_n \sin \eta x \\ \sigma_{zz}\left(x, -\dfrac{h}{2}\right) = 0, \sigma_{xz}\left(x, \pm\dfrac{h}{2}\right) = 0 \end{cases} \tag{7.123}$$

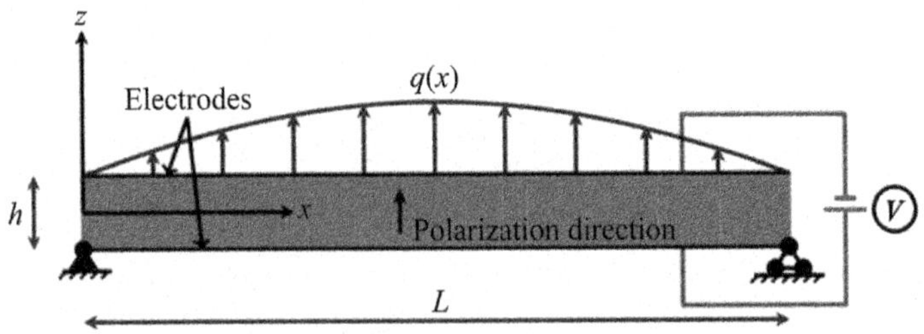

FIGURE 7.3 Schematic of a simply-supported beam with the piezoelectric and flexoelectric effects subjected to mechanical and electric voltage.

$$\begin{cases} \phi\left(x, \frac{h}{2}\right) = \sum_{n=1}^{\infty} V_n \sin \eta x, \\ \phi\left(x, -\frac{h}{2}\right) = 0 \end{cases} \tag{7.124}$$

where

$$\eta = \frac{n\pi}{L}, n = 1, 2, 3, \ldots \tag{7.125}$$

Obviously, the above relations provide six boundary conditions at the surfaces however, we need eight boundary conditions to solve the resulting governing equations. Therefore, two supplementary conditions are needed. In fact, we may require

$$\mu_{13}\varepsilon_{xx}\left(x, \pm \frac{h}{2}\right) + \mu_{33}\varepsilon_{zz}\left(x, \pm \frac{h}{2}\right) = 0 \tag{7.126}$$

the derivation of which is omitted here for brevity. This condition corresponds to the vanishing of electric quadrupoles at the top and bottom surfaces. Thus, Eqs. (7.123), (7.124) and (7.126) form eight conditions, which are employed to solve the problem under scrutiny.

For a simply supported beam, the supported conditions at the ends $x = 0, L$ require

$$\sigma_{xx} = 0, \ w = 0, \phi = 0 \tag{7.127}$$

In view of Eq. (7.127), an appropriate solution may be assumed to have the form

$$u(x, z) = \sum_{n=1}^{\infty} U_n e^{sz} \cos \eta x, \ w(x, z) = \sum_{n=1}^{\infty} W_n e^{sz} \sin \eta x, \phi(x, z) = \sum_{n=1}^{\infty} \Phi_n e^{sz} \sin \eta x \tag{7.128}$$

where U_n, W_n, and Φ_n are unknown constants to be determined and s is an unknown parameter. By substituting the relations in Eq. (7.128) into the governing Eqs. (7.119)-(7.121), we get

$$\begin{aligned} \left(c_{44}s^2 - \eta^2 c_{11}\right) U_n + (c_{13} + c_{44})\eta s W_n + \left[(e_{31} + e_{15})\eta s + \mu_{11}\eta^3 \right. \\ \left. - (\mu_{13} + \mu_{41} + \mu_{42})\eta s^2\right] \Phi_n = 0 \end{aligned} \tag{7.129}$$

$$\begin{aligned} -(c_{13} + c_{44})\eta s U_n + \left(c_{33}s^2 - c_{44}\eta^2\right) W_n + \left[e_{33}s^2 - e_{15}\eta^2 \right. \\ \left. + (\mu_{31} + \mu_{41} + \mu_{42})\eta^2 s - \mu_{33}s^3\right] \Phi_n = 0 \end{aligned} \tag{7.130}$$

$$-\left[(e_{31}+e_{15})\eta s+\mu_{11}\eta^3-(\mu_{13}+\mu_{41}+\mu_{42})\eta s^2\right]U_n+\left[e_{33}s^2-e_{15}\eta^2+\mu_{33}s^3\right.$$
$$\left.-(\mu_{31}+\mu_{41}+\mu_{42})\eta^2 s\right]W_n+\left(\kappa_{11}\eta^2-\kappa_{33}s^2\right)\Phi_n=0 \tag{7.131}$$

or in compact form

$$\begin{bmatrix} a_{11} & a_{12} & a_{13} \\ a_{21} & a_{22} & a_{23} \\ a_{31} & a_{32} & a_{33} \end{bmatrix}\begin{Bmatrix} U_n \\ W_n \\ \Phi_n \end{Bmatrix}=\begin{Bmatrix} 0 \\ 0 \\ 0 \end{Bmatrix} \tag{7.132}$$

where

$$\begin{cases} a_{11}=c_{44}s^2-\eta^2 c_{11} \\ a_{12}=(c_{13}+c_{44})\eta s \\ a_{13}=(e_{31}+e_{15})\eta s+\mu_{11}\eta^3-(\mu_{13}+\mu_{41}+\mu_{42})\eta s^2 \\ a_{21}=-(c_{13}+c_{44})\eta s \\ a_{22}=c_{33}s^2-c_{44}\eta^2 \\ a_{23}=e_{33}s^2-e_{15}\eta^2+(\mu_{31}+\mu_{41}+\mu_{42})\eta^2 s-\mu_{33}s^3 \\ a_{31}=-(e_{31}+e_{15})\eta s+\mu_{11}\eta^3-(\mu_{13}+\mu_{41}+\mu_{42})\eta s^2 \\ a_{32}=e_{33}s^2-e_{15}\eta^2+\mu_{33}s^3-(\mu_{13}+\mu_{41}+\mu_{42})\eta^2 s \\ a_{33}=\kappa_{11}\eta^2+\kappa_{33}s^2 \end{cases} \tag{7.133}$$

To obtain a nontrivial solution of U_n, W_n, and Φ_n, the determinant of the coefficient matrix in Eq. (7.132) must vanish. Expanding the determinant leads to an eight-degree polynomial equation

$$s^8+a_7s^7+a_6s^6+a_5s^5+a_4s^4+a_3s^3+a_2s^2+a_1s+a_0=0 \tag{7.134}$$

where the coefficients a_j ($j=0, 1, \ldots, 7$) are readily given and omitted here for brevity. Eq. (7.134) is a general polynomial equation with real coefficients, the roots of which can be sought with the help of commercial software, such as Mathematica. Solving the above polynomial equation gives the eigenvalues, which may be either real, γ, or appear as complex conjugate pairs, $\gamma=\alpha\pm i\beta$. These corresponds to eigenfunctions $e^{\gamma z}$ or $e^{\alpha z}(A_1\cos\beta z+A_2\sin\beta z)$, respectively. If a multiple real root γ occurs such as $\gamma_1=\gamma_2(-\gamma,\text{say})$, it is only to appeal to eigenfunctions $(A_1+A_2z)e^{\gamma z}$. By superposing all eigenfunctions, an appropriate solution can be constructed. A typical case is that the eigenvalues contain both a set of unequal real roots and a pair of complex roots. For example, for a given η value, if Eq. (7.134) has roots $s_1, s_2=\pm\gamma_1, s_3, s_4=\pm\gamma_2, s_5, s_6=\alpha\pm i\beta, s_7, s_8=-\alpha\pm i\beta$, the relations from Eqs. (7.130) and (7.131),

$$W_n = f_1(s)U_n, \Phi_n = f_2(s)U_n \tag{7.135}$$

hold, in which

$$f_1(s) = \frac{a_{31}a_{23} - a_{21}a_{33}}{a_{22}a_{33} - a_{32}a_{23}}, \ f_2(s) = \frac{a_{21}a_{32} - a_{31}a_{22}}{a_{22}a_{33} - a_{32}a_{23}} \tag{7.136}$$

With the above results at hand, the displacement components u, w, and the electric potential ϕ can be given by

$$u(x,z) = \sum_{n=1}^{\infty} \cos \eta x \left[\sum_{j=1}^{4} U_j e^{s_j z} + e^{\alpha z}(U_5 \cos \beta z + U_6 \sin \beta z) \right. \\ \left. + e^{-\alpha z}(U_7 \cos \beta z + U_8 \sin \beta z) \right] \tag{7.137}$$

$$w(x,z) = \sum_{n=1}^{\infty} \sin \eta x \left[\sum_{j=1}^{4} R_j U_j e^{s_j z} + U_5 e^{\alpha z}(R_5 \cos \beta z - R_6 \sin \beta z) \right. \\ \left. - U_6 e^{\alpha z}(R_6 \cos \beta z + R_5 \sin \beta z) \right] \\ + \sum_{n=1}^{\infty} \sin \eta x [U_7 e^{-\alpha z}(R_7 \cos \beta z - R_8 \sin \beta z) + U_8 e^{-\alpha z} \\ (R_8 \cos \beta z + R_7 \sin \beta z)] \tag{7.138}$$

$$\phi(x,z) = \sum_{n=1}^{\infty} \sin \eta x \left[\sum_{j=1}^{4} H_j U_j e^{s_j z} + U_5 e^{\alpha z}(H_5 \cos \beta z - H_6 \sin \beta z) \right. \\ \left. + U_6 e^{\alpha z}(H_6 \cos \beta z + H_5 \sin \beta z) \right] + \sum_{n=1}^{\infty} \sin \eta x [U_7 e^{-\alpha z} \\ (H_7 \cos \beta z - H_8 \sin \beta z) + U_8 e^{-\alpha z}(H_8 \cos \beta z + H_7 \sin \beta z)] \tag{7.139}$$

with

$$R_j = f_1(s_j), H_j = f_2(s_j), \ j = 1, 2, 3, 4 \tag{7.140}$$

$$R_5 = \mathrm{Re}[f_1(s_5)], R_6 = \mathrm{Im}[f_1(s_5)], H_5 = \mathrm{Re}[f_2(s_5)], H_6 = \mathrm{Im}[f_2(s_5)] \tag{7.141}$$

$$R_7 = \mathrm{Re}[f_1(s_7)], R_8 = \mathrm{Im}[f_1(s_7)], H_7 = \mathrm{Re}[f_2(s_7)], H_8 = \mathrm{Im}[f_2(s_7)] \tag{7.142}$$

where U_j ($j = 1, 2, \ldots, 8$) are unknown constants to be determined through prescribed boundary conditions on the top and the bottom surfaces.

Next, inserting the expresses (7.137)-(7.139) for the displacement components and electric potential into the constitutive relations and electric potential into the constitutive relations Eqs. (7.108)-(7.110)results in

$$\begin{aligned}\sigma_{xx} = \sum_{n=1}^{\infty} \sin \eta x \Bigg[\sum_{j=1}^{4} T_{1j} U_j e^{s_j z} + U_5 e^{\alpha z}(T_{15} \cos \beta z + T_{16} \sin \beta z) + \\ U_6 e^{\alpha z}(T_{16} \cos \beta z + T_{15} \sin \beta z) + \\ U_7 e^{-\alpha z}(T_{17} \cos \beta z + T_{18} \sin \beta z) - U_8 e^{-\alpha z}(T_{18} \cos \beta z - T_{17} \sin \beta z) \Bigg]\end{aligned} \tag{7.143}$$

$$\begin{aligned}\sigma_{zz} = \sum_{n=1}^{\infty} \sin \eta x \Bigg[\sum_{j=1}^{4} T_{2j} U_j e^{s_j z} + U_5 e^{\alpha z}(T_{25} \cos \beta z - T_{26} \sin \beta z) + \\ U_6 e^{\alpha z}(T_{26} \cos \beta z + T_{25} \sin \beta z) \\ U_7 e^{-\alpha z}(T_{27} \cos \beta z + T_{28} \sin \beta z) - U_8 e^{-\alpha z}(T_{28} \cos \beta z - T_{27} \sin \beta z) \Bigg]\end{aligned} \tag{7.144}$$

$$\begin{aligned}\sigma_{xz} = \sum_{n=1}^{\infty} \cos \eta x \Bigg[\sum_{j=1}^{4} T_{3j} U_j e^{s_j z} + U_5 e^{\alpha z}(T_{35} \cos \beta z - T_{36} \sin \beta z) + \\ U_6 e^{\alpha z}(T_{36} \cos \beta z + T_{35} \sin \beta z) - \\ U_7 e^{-\alpha z}(T_{37} \cos \beta z + T_{38} \sin \beta z) + U_8 e^{-\alpha z}(T_{38} \cos \beta z - T_{37} \sin \beta z) \Bigg]\end{aligned} \tag{7.145}$$

where the coefficients T_{ij} ($i = 1, 2, 3; j = 1, 2, \ldots, 8$) are given in appendix A of reference [91]. In addition, with the aid of Eq. (7.113) we also give the expressions for the strains ε_{xx} and ε_{zz} through Eqs. (7.137) and (7.138), namely

$$\begin{aligned}\varepsilon_{xx} = -\sum_{n=1}^{\infty} \eta \sin \eta x \Bigg[\sum_{j=1}^{4} U_j e^{s_j z} + e^{\alpha z}(U_5 \cos \beta z + U_6 \sin \beta z) \\ + e^{-\alpha z}(U_7 \cos \beta z + U_8 \sin \beta z) \Bigg]\end{aligned} \tag{7.146}$$

$$\begin{aligned}\varepsilon_{zz} = \sum_{n=1}^{\infty} \sin \eta x \Bigg\{ \sum_{j=1}^{4} s_j R_j U_j e^{s_j z} + U_5 e^{\alpha z}[(\alpha R_5 - \beta R_6)\cos \beta z - (\alpha R_6 + \beta R_5) \\ \sin \beta z] + U_6 e^{\alpha z}[(\alpha R_6 + \beta R_5)\cos \beta z + (\alpha R_5 - \beta R_6)\sin \beta z] + U_7 e^{-\alpha z} \\ [-(\alpha R_7 + \beta R_8)\cos \beta z + (\alpha R_8 - \beta R_7)\sin \beta z] + U_8 e^{-\alpha z}[(-\alpha R_8 + \beta R_7) \\ \cos \beta z - (\alpha R_7 + \beta R_8)\sin \beta z] \Bigg\}\end{aligned} \tag{7.147}$$

We substitute the above strains ε_{xx} and ε_{zz} into Eq. (7.126) to obtain two independent equations. Substitution of the above expressions (7.139), (7.144)-(7.147) into the boundary conditions Eqs. (7.123) and (7.124) along with Eq.

(7.126) forms eight equations. Thus, after combining them together and collection, one gets eight linear equations in matrix form

$$[\boldsymbol{K}_{ij}][\boldsymbol{U}_j] = \{\boldsymbol{F}_j\} \tag{7.148}$$

with

$$\{\boldsymbol{U}_j\} = \{U_1, U_2, U_3, U_4, U_5, U_6, U_7, U_8\}^T \tag{7.149}$$

$$\{\boldsymbol{F}_j\} = \{q_n, 0, 0, 0, V_n, 0, 0, 0\}^T \tag{7.150}$$

where the superscript T denotes the transpose of a matrix (vector), and the elements in the matrix $[\boldsymbol{K}_{ij}]$ are given in appendix B of reference [91]. The solution of the above algebraic equation is easily obtained. Once all the constants U_j ($j = 1, \ldots, 8$) are determined, all field variables are easily given. The stresses distribution in the beam can be determined by Eqs. (7.143)-(7.145), whereas the electric potential is given by Eq. (7.139). Electric polarization P_z can be evaluated by Eq. (7.112).

7.5 The coupled stress effects

Presented in this sections are the derivation procedure of higher-order beams on the basis of the couple stress theory. The couple stress theory [92] can explain size dependent effects, which cannot be described by classical elasticity. In the couple stress theories for linear elastic materials [93], the applied loads on the material particle include not only a force that causes translation but also a couple stress that induces rotation. The classical couple stress theory includes four elastic constants and provides another constitutive relation between the couple stress and curvature. For convenience of application, two simplified versions of the couple stress theory are presented, namely, modified couple stress theory [94] and consistent couple stress theory [95]. What they have in common is that they all contain only three elastic parameters, that is, two Levy constants and length-scale parameter characterizing size effects. They differ in that the couple stress and curvature tensors are symmetric in the former, but antisymmetric in the latter. However, using different couple stress theories will yield the same result for the study of beams. In this section, the constitutive equations of the higher-order shear beam based on couple stress theory are presented in literature [96].

7.5.1 The modified couple stress theory

According to the modified couple stress theory proposed byRef. [94], the strain energy density is a function of both strain tensor (conjugated with stress tensor) and curvature tensor (conjugated with couple stress tensor). Then, the strain energy of a deformed isotropic linear elastic body occupying a volume Ω can be written as

$$U_s = \frac{1}{2}\int_{\Omega} (\boldsymbol{\sigma} : \boldsymbol{\varepsilon} + \boldsymbol{m} : \boldsymbol{\chi})\mathrm{d}V \tag{7.151}$$

where $\boldsymbol{\sigma}$ is the stress tensor, $\boldsymbol{\varepsilon}$ is the strain tensor, $\boldsymbol{m}$ is the deviatoric part of the couple stress tensor, and $\boldsymbol{\chi}$ is the symmetric curvature tensor. These tensors are defined by

$$\boldsymbol{\sigma} = \lambda \mathrm{tr}(\boldsymbol{\varepsilon})\boldsymbol{I} + 2\mu\boldsymbol{\varepsilon} \tag{7.152}$$

$$m = 2l^2\mu\chi \tag{7.153}$$

$$\boldsymbol{\varepsilon} = \frac{1}{2}\left[\nabla \boldsymbol{u} + (\nabla \boldsymbol{u})^T\right] \tag{7.154}$$

$$\chi = \frac{1}{2}\left[\nabla \boldsymbol{\theta} + (\nabla \boldsymbol{\theta})^T\right] \tag{7.155}$$

where $\boldsymbol{u}$ is the displacement vector, λ and μ are Láme's constants, l is the material length scale parameter which reflects the effect of couple stress, and θ is the rotation vector that can be expressed as

$$\boldsymbol{\theta} = \frac{1}{2}\mathrm{curl}\ \boldsymbol{u} \tag{7.156}$$

Here a unified higher-order theory that contains various other theories is used. In this theory, the axial displacement, u_x, and the transverse displacement, u_z, are given as

$$\begin{aligned} u_z(x, z, t) &= u(x, t) - z\frac{\partial w(x, t)}{\partial x} + \Phi(z)\gamma(x, t) \\ u_y(x, z, t) &= 0 \\ u_z(x, z, t) &= w(x, t) \end{aligned} \tag{7.157}$$

where u and w are the axial and the transverse displacement of any point on the neutral axis, γ is the transverse shear strain of any point on the neutral axis

$$\gamma(x, t) = \frac{\partial w(x, t)}{\partial x} - \phi(x, t) \tag{7.158}$$

where ϕ is the total bending rotation of the cross sections at any point on the neutral axis, t denotes time. In Eq. (7.157), Φ is a function of z that characterizes the transverse shear and stress distribution along the thickness of the beam. The displacement fields of different beam theories can be obtained by appropriately choosing $\Phi(z)$. Using Eqs. (7.152) and (7.157), the components of the strain tensor are obtained as

$$\varepsilon_{xx} = \frac{\partial u}{\partial x} - z\frac{\partial w}{\partial x^2} + \Phi(z)\frac{\partial \gamma}{\partial x}, \varepsilon_{yy} = \varepsilon_{zz} = \varepsilon_{xy} = \varepsilon_{yz} = 0$$
$$\gamma_{xz} = 2\varepsilon_{xz} = \Phi_{,z}\gamma \tag{7.159}$$

where the subscript " z" indicates the derivative with respect to z. Also, Eqs. (7.154) and (7.157) give

$$\theta_y = \frac{1}{2}\Phi_{,z}\gamma - \frac{\partial w}{\partial x}, \theta_x = \theta_z = 0 \tag{7.160}$$

Substitution of Eq. (7.160) into Eq. (7.153) yields the following expression for the nonzero components of the symmetric curvature tensor

$$\chi_{xy} = \frac{1}{4}\Phi_{,z}\frac{\partial \gamma}{\partial x} - \frac{1}{2}\frac{\partial^2 w}{\partial x^2}, \chi_{yz} = \frac{1}{4}\Phi_{,zz}\gamma, \chi_{xx} = \chi_{yy} = \chi_{zz} = \chi_{xz} = 0 \tag{7.161}$$

7.5.2 The governing equations and boundary conditions

In this study, the equations of motion are derived by using Hamilton's principle [97]. This principle can be expressed as

$$\delta\int_0^T [U_k - (U_s - W)]\mathrm{d}t = 0 \tag{7.162}$$

where U_k is the kinetic energy, U_s is the strain energy and W is the work done by the external applied forces. The first variation of the strain energy on the time interval $[0, T]$ can be written as

$$\begin{aligned}
\delta\int_0^T U_s\mathrm{d}t &= \int_0^T \int_\Omega \left(\sigma_{ij}\delta\varepsilon_{ij} + m_{ij}\delta\chi_{ij}\right)\mathrm{d}V\mathrm{d}t \\
&= \int_0^T \int_\Omega \left(\sigma_{xx}\delta\varepsilon_{xx} + 2\sigma_{xz}\delta\varepsilon_{xz} + 2m_{xy}\delta\chi_{xy} + 2m_{xz}\delta\chi_{xz}\right)\mathrm{d}V\mathrm{d}t \\
&= \int_0^T \int_0^L \left[N\frac{\partial\delta u}{\partial x} - (M_z + Y_1)\frac{\partial^2\delta w}{\partial x^2} + \left(M_{nc} + \frac{1}{2}Y_2\right)\frac{\partial^2\delta\gamma}{\partial x^2}\right. \\
&\qquad \left. + \left(Q + \frac{1}{2}Y_3\right)\delta\gamma\right]\mathrm{d}x\mathrm{d}t
\end{aligned} \tag{7.163}$$

where L is the length of the microbeam and the following stress resultants are defined as

$$N = \int_A \sigma_{xx}\mathrm{d}A = A_{xx}\frac{\partial u}{\partial x} - B_{xx}\frac{\partial^2 w}{\partial x^2} + E_{xx}\frac{\partial \gamma}{\partial x} \tag{7.164}$$

$$M_z = \int_A z\sigma_{xx}\mathrm{d}A = B_{xx}\frac{\partial u}{\partial x} - D_{xx}\frac{\partial^2 w}{\partial x^2} + F_{xx}\frac{\partial \gamma}{\partial x} \tag{7.165}$$

$$M_{nc} = \int_A \Phi(z)\sigma_{xx}\mathrm{d}A = E_{xx}\frac{\partial u}{\partial x} - F_{xx}\frac{\partial^2 w}{\partial x^2} + H_{xx}\frac{\partial \gamma}{\partial x} \tag{7.166}$$

$$Q = \int_A \Phi_{,z}(z)\sigma_{xz}\mathrm{d}A = B_{xz}\gamma \tag{7.167}$$

$$Y_1 = \int_A m_{xy}\mathrm{d}A = \int_A 2\mu(z)l^2\chi_{xy}\mathrm{d}A = \frac{1}{2}D_{xy}l^2\frac{\partial \gamma}{\partial x} - A_{xz}l^2\frac{\partial^2 w}{\partial x^2} \tag{7.168}$$

$$Y_2 = \int_A \Phi_{,z}(z)m_{xy}\mathrm{d}A = \int_A 2\Phi_{,z}(z)\mu(z)l^2\chi_{xy}\mathrm{d}A = \frac{1}{2}B_{xy}l^2\frac{\partial \gamma}{\partial x} - D_{xz}l^2\frac{\partial^2 w}{\partial x^2} \tag{7.169}$$

$$Y_3 = \int_A \Phi_{,zz}(z)m_{yz}\mathrm{d}A = \int_A 2\Phi_{,zz}(z)\mu(z)l^2\chi_{yz}\mathrm{d}A = \frac{1}{2}E_{xz}l^2\gamma \tag{7.170}$$

where

$$(A_{xx}, B_{xx}, D_{xx}) = \int_A \frac{E(z)[1-\nu(z)]}{[1+\nu(z)][1-2\nu(z)]}\left(1, z, z^2\right)\mathrm{d}A \tag{7.171}$$

$$(E_{xx}, F_{xx}, H_{xx}) = \int_A \frac{E(z)[1-\nu(z)]}{[1+\nu(z)][1-2\nu(z)]}\Phi(z)[1, z, \Phi(z)]\mathrm{d}A \tag{7.172}$$

$$A_{xz} = \int_A \mu(z)\mathrm{d}A = \int_A \frac{E(z)}{2[1+\nu(z)]}\mathrm{d}A \tag{7.173}$$

$$B_{xz} = \int_A \mu(z)\left[\Phi_{,z}(z)\right]^2\mathrm{d}A \tag{7.174}$$

$$D_{xz} = \int_A \mu(z)\Phi_{,z}(z)\mathrm{d}A \tag{7.175}$$

$$E_{xz} = \int_A \mu(z)\left[\Phi_{,zz}(z)\right]^2\mathrm{d}A \tag{7.176}$$

The kinetic energy of the beam is given by

$$U_k = \frac{1}{2}\int_\Omega \rho(z)\left[\left(\frac{\partial u_x}{\partial t}\right)^2 + \left(\frac{\partial u_y}{\partial t}\right)^2 + \left(\frac{\partial u_z}{\partial t}\right)^2\right]\mathrm{d}V \tag{7.177}$$

where $\rho(z)$ is the mass density of FG microbeam. By using Eqs. (7.157) and (7.177), the first variation of the kinetic energy on the time interval $[0, T]$ is given by

$$\int_0^T \delta U_k \mathrm{d}t = \int_0^T \int_0^L \left[I_A \left(\frac{\partial u}{\partial t}\frac{\partial \delta u}{\partial t} + \frac{\partial w}{\partial t}\frac{\partial \delta w}{\partial t} \right) - I_B \left(\frac{\partial u}{\partial t}\frac{\partial^2 \delta w}{\partial x \partial t} + \frac{\partial \delta u}{\partial t}\frac{\partial^2 w}{\partial x \partial t} \right) \right.$$
$$+ I_D \frac{\partial^2 w}{\partial x \partial t}\frac{\partial^2 \delta w}{\partial x \partial t} + I_E \left(\frac{\partial u}{\partial t}\frac{\partial \delta \gamma}{\partial t} + \frac{\partial \delta u}{\partial t}\frac{\partial \gamma}{\partial t} \right)$$
$$\left. - I_F \left(\frac{\partial^2 w}{\partial x \partial t}\frac{\partial \delta \gamma}{\partial t} + \frac{\partial^2 \delta w}{\partial x \partial t}\frac{\partial \gamma}{\partial t} \right) + I_H \frac{\partial^2 \gamma}{\partial t}\frac{\partial \delta \gamma}{\partial t} \right] \mathrm{d}x\mathrm{d}t \tag{7.178}$$

where

$$(I_A, I_B, I_D) = \int_A \rho(z)\left(1, z, z^2\right)\mathrm{d}A \tag{7.179}$$

$$I_E, I_F, I_H = \int_A \rho(z)\Phi(z)[1, z, \Phi(z)]\mathrm{d}A \tag{7.180}$$

General form of the first variation of work done by the external applied forces on the time interval $[0, T]$ is given by [98]

$$\delta \int_0^T W \mathrm{d}t = \int_0^T \int_\Omega (\widehat{f}\delta u + \widehat{c}\delta\theta)\mathrm{d}V\mathrm{d}t + \int_0^T \int_S (\widehat{t}\delta u + \widehat{s}\delta\theta)\mathrm{d}S\mathrm{d}t \tag{7.181}$$

where $\widehat{f}, \widehat{c}, \widehat{t}, \widehat{s}$ are, respectively, body force per unit volume, body couple per unit volume, traction per unit area and surface couple per unit area, and S is the surface of Ω. By neglecting the traction and the surface couple, the first variation of work done by the external applied forces on the time interval $[0, T]$ is

$$\int_0^T \delta W \mathrm{d}t = \int_0^T \int_0^L \left[f\delta u + \left(q + \frac{\partial c}{\partial x} \right)\delta w + \frac{1}{2} a c \delta\gamma \right] \mathrm{d}x\mathrm{d}t - \int_0^T (c\delta w)\Big|_0^L \mathrm{d}t \tag{7.182}$$

where f and q are the x and z components of the body forces per unit length, c is the y component of the body couple per unit length, and $a = \int_0^L \Phi_{,z}(z)\mathrm{d}A$. Substituting Eqs. (7.163), (7.178) and (7.182) into Eq. (7.162), integrating by parts and setting the coefficients $\delta u, \delta w$ and $\delta\gamma$ to zero lead to the following equations of motion:

$$\frac{\partial N}{\partial x} + f = I_A \frac{\partial^2 u}{\partial t^2} - I_B \frac{\partial^3 w}{\partial x \partial t^2} + I_E \frac{\partial^2 \gamma}{\partial t^2} \tag{7.183}$$

$$\frac{\partial^2 M_z}{\partial x^2} + \frac{\partial^2 Y_1}{\partial x^2} + \frac{\partial c}{\partial x} + q = I_A \frac{\partial^2 w}{\partial t^2} + I_B \frac{\partial^3 u}{\partial x \partial t^2} - I_D \frac{\partial^4 w}{\partial x^2 \partial t^2} + I_F \frac{\partial^3 \gamma}{\partial x \partial t^2} \tag{7.184}$$

$$\frac{\partial M_{nc}}{\partial x}+\frac{1}{2}\frac{\partial Y_2}{\partial x}-\frac{1}{2}Y_3-Q+\frac{1}{2}ac=I_E\frac{\partial^2 u}{\partial t^2}-I_F\frac{\partial^3 w}{\partial x\partial t^2}+I_H\frac{\partial^2 \gamma}{\partial t^2} \tag{7.185}$$

The corresponding boundary conditions at the beam ends (at $x=0$ and $x=L$) are obtained

$$\text{either } N=0 \text{ or } u=0 \text{ at } x=0 \text{ and } x=L \tag{7.186}$$

$$\begin{aligned}\text{either } &\frac{\partial M_c}{\partial x}+\frac{\partial Y_1}{\partial x}+c-I_B\frac{\partial^2 u}{\partial t^2}+I_D\frac{\partial^3 w}{\partial x\partial t^2}-I_F\frac{\partial^2 \gamma}{\partial t^2}\\ &=0 \text{ or } w=0 \text{ at } x=0 \text{ and } x=L\end{aligned} \tag{7.187}$$

$$\text{either } M_{nc}+\frac{1}{2}Y_2=0 \text{ or } \gamma=0 \text{ at } x=0 \text{ and } x=L \tag{7.188}$$

$$\text{either } M_c+Y_1=0 \text{ or } \frac{\partial w}{\partial x}=0 \text{ at } x=0 \text{ and } x=L \tag{7.189}$$

By using Eqs. (7.164)-(7.170) and (7.183)-(7.185), the equations of motion of FG microbeam in terms of the displacements are obtained as

$$A_{xx}\frac{\partial^2 u}{\partial x^2}-B_{xx}\frac{\partial^3 w}{\partial x^3}+E_{xx}\frac{\partial^2 \gamma}{\partial x^2}-I_0\frac{\partial^2 u}{\partial t^2}+I_8\frac{\partial^3 w}{\partial x\partial t^2}-I_9\frac{\partial^2 \gamma}{\partial t^2}+f=0 \tag{7.190}$$

$$\begin{aligned}&-B_{xx}\frac{\partial^2 u}{\partial x^2}+\left(D_{xx}+A_{xz}l^2\right)\frac{\partial^4 w}{\partial x^4}-\left(F_{xx}+\frac{1}{2}D_{xz}l^2\right)\frac{\partial^3 \gamma}{\partial x^3}+I_8\frac{\partial^2 u}{\partial t^2}\\ &+I_9\frac{\partial^4 w}{\partial x^2\partial t^2}-I_9\frac{\partial^3 \gamma}{\partial x\partial t^2}-\frac{\partial^2 u}{\partial t^2}-I_7\frac{\partial^2 \gamma}{\partial t^2}-\frac{\partial c}{\partial x}-q=0\end{aligned} \tag{7.191}$$

$$\begin{aligned}&E_{xx}\frac{\partial^2 u}{\partial x^2}-\left(F_{xx}+\frac{1}{2}D_{xz}l^2\right)\frac{\partial^3 w}{\partial x^3}+\left(H_{xx}+\frac{1}{4}B_{xz}l^2\right)\frac{\partial^2 \gamma}{\partial x^2}-\left(B_{xz}+\frac{1}{4}E_{xz}l^2\right)\gamma\\ &-I_7\frac{\partial^2 u}{\partial t^2}+I_5\frac{\partial^3 w}{\partial x\partial t^2}-I_{11}\frac{\partial^2 \gamma}{\partial t^2}+\frac{1}{2}ac=0\end{aligned} \tag{7.192}$$

It should be noted that when the scale parameter l and the Poisson's ratio are set to zero in Eqs. (7.190)-(7.192), the governing equations reduce to those of the classical higher-order beam model.

References

[1] Zener C. Internal friction in solids. I. Theory of internal friction in reeds. Phys Rev 1937;52 (3):230.

[2] Zener C. Internal friction in solids II. General theory of thermoelastic internal friction. Phys Rev 1938;53(1):90–9.

[3] Shi S, Jin F, He T, et al. Thermoelastic damping analysis model of transversely isotropic micro/nano- resonators based on dual-phase-lag heat conduction model and surface effect. Compos Struct 2022;292:115664.

[4] Shi S, He T, Jin F. Thermoelastic damping analysis of size-dependent nano-resonators considering dual-phase-lag heat conduction model and surface effect. Int J Heat Mass Tran 2021;170:120977.

[5] Kumar H, Mukhopadhyay S. Thermoelastic damping in micro and nano-mechanical resonators utilizing entropy generation approach and heat conduction model with a single delay term. Int J Mech Sci 2020;165:105211.

[6] Wang YW, Li XF. Synergistic effect of memory-size-microstructure on thermoelastic damping of a micro-plate. Int J Heat Mass Tran 2021;181:122031.

[7] Kim JH, Kim JH. Thermoelastic damping effect of the micro-ring resonator with irregular mass and stiffness. J Sound Vib 2016;369:168–77.

[8] Zhou H, Li P, Fang Y. Single-phase-lag thermoelastic damping models for rectangular cross-sectional micro-and nano-ring resonators. Int J Mech Sci 2019;163:105132.

[9] Li M, Cai Y, Fan R, et al. Generalized thermoelasticity model for thermoelastic damping in asymmetric vibrations of nonlocal tubular shells. Thin-Walled Struct 2022;174:109142.

[10] Lu P, Lee HP, Lu C, et al. Thermoelastic damping in cylindrical shells with application to tubular oscillator structures. Int J Mech Sci 2008;50(3):501–12.

[11] Li SR, Zhang F, Batra RC. Thermoelastic damping in high frequency resonators using higher-order shear deformation theories. Thin-Walled Struct 2023;188:110778.

[12] Kim JH, Kim JH. Phase-lagging of the thermoelastic dissipation for a tubular shell model. Int J Mech Sci 2019;163:105094.

[13] Gu B, He T, Ma Y. Thermoelastic damping analysis in micro-beam resonators considering nonlocal strain gradient based on dual-phase-lag model. Int J Heat Mass Tran 2021;180:121771.

[14] Auriault JL. Cattaneo–Vernotte equation versus Fourier thermoelastic hyperbolic heat equation. Int J Eng Sci 2016;101:45–9.

[15] Maillet D. A review of the models using the Cattaneo and Vernotte hyperbolic heat equation and their experimental validation. Int J Therm Sci 2019;139:424–32.

[16] Alihemmati J, Beni YT. Generalized thermoelasticity of microstructures: Lord-Shulman theory with modified strain gradient theory. Mech Mater 2022;172:104412.

[17] Guyer RA, Krumhansl JA. Solution of the linearized phonon Boltzmann equation. Phys Rev 1966;148(2):766.

[18] Guyer RA, Krumhansl JA. Thermal conductivity, second sound, and phonon hydrodynamic phenomena in nonmetallic crystals. Phys Rev 1966;148(2):778.

[19] Wang YW, Zhang XY, Li XF. Thermoelastic damping in moderately thick microplate resonators based on the fractional heat conduction model. Int J Heat Mass Tran 2025;237:126394.

[20] Deng W, Li L, Hu Y, et al. Thermoelastic damping of graphene nanobeams by considering the size effects of nanostructure and heat conduction. J Therm Stress 2018;41(9):1182–200.

[21] Sellitto A, Jou D, Bafaluy J. Non-local effects in radial heat transport in silicon thin layers and graphene sheets. Proc R Soc A 2012;468(2141):1217–29.

[22] Ma Y. Size-dependent thermal conductivity in nanosystems based on non-Fourier heat transfer. Appl Phys Lett 2012;101(21).

[23] Ma WL, Jiang ZC, Li XF. Effect of warping shape on buckling of circular and rectangular columns under axial compression. Appl Math Model 2021;89:1475–90.

[24] Kasuya A, Sasaki Y, Saito Y, et al. Evidence for size-dependent discrete dispersions in single-wall nanotubes. Phys Rev Lett 1997;78(23):4434.

[25] Poncharal P, Wang ZL, Ugarte D, et al. Electrostatic deflections and electromechanical resonances of carbon nanotubes. Science 1999;283(5407):1513–6.

[26] Chen CQ, Shi Y, Zhang YS, et al. Size dependence of Young's modulus in ZnO nanowires. Phys Rev Lett 2006;96(7):075505.

[27] Peigney A, Laurent C, Flahaut E, et al. Specific surface area of carbon nanotubes and bundles of carbon nanotubes. Carbon 2001;39(4):507–14.

[28] Li X, Gao H, Murphy CJ, et al. Nanoindentation of silver nanowires. Nano Lett 2003;3 (11):1495–8.

[29] Stan G, Ciobanu CV, Parthangal PM, et al. Diameter-dependent radial and tangential elastic moduli of ZnO nanowires. Nano Lett 2007;7(12):3691–7.

[30] Zijlstra P, Tchebotareva AL, Chon JWM, et al. Acoustic oscillations and elastic moduli of single gold nanorods. Nano Lett 2008;8(10):3493–7.

[31] Bernal RA, Agrawal R, Peng B, et al. Effect of growth orientation and diameter on the elasticity of GaN nanowires. A combined in situ TEM and atomistic modeling investigation. Nano Lett 2011;11(2):548–55.

[32] Tang DM, Ren CL, Wang MS, et al. Mechanical properties of Si nanowires as revealed by in situ transmission electron microscopy and molecular dynamics simulations. Nano Lett 2012;12(4):1898–904.

[33] Liu F, Tang DM, Gan H, et al. Individual boron nanowire has ultra-high specific young's modulus and fracture strength as revealed by in situ transmission electron microscopy. ACS Nano 2013;7(11):10112–20.

[34] Yu H, Sun C, Zhang WW, et al. Study on size-dependent young's modulus of a silicon nanobeam by molecular dynamics simulation. J Nanomater 2013;2013(1):319302.

[35] Ding W, Calabri L, Chen X, et al. Mechanics of crystalline boron nanowires. Compos Sci Technol 2006;66(9):1112–24.

[36] Babaei Gavan K, Westra HJR, van der Drift EWJM, et al. Size-dependent effective Young's modulus of silicon nitride cantilevers. Appl Phys Lett 2009;94(23).

[37] Zeng D, Wei X, Liu JZ, et al. Tunable resonant frequencies for determining Young's moduli of nanowires. J Appl Phys 2009;105(11).

[38] Wu B, Heidelberg A, Boland JJ. Mechanical properties of ultrahigh-strength gold nanowires. Nat Mater 2005;4(7):525–9.

[39] Ngo LT, Almécija D, Sader JE, et al. Ultimate-strength germanium nanowires. Nano Lett 2006;6(12):2964–8.

[40] Agrawal R, Peng B, Espinosa HD. Experimental-computational investigation of ZnO nanowires strength and fracture. Nano Lett 2009;9(12):4177–83.

[41] Huang JY, Zheng H, Mao SX, et al. In situ nanomechanics of GaN nanowires. Nano Lett 2011;11(4):1618–22.

[42] Wong EW, Sheehan PE, Lieber CM. Nanobeam mechanics: elasticity, strength, and toughness of nanorods and nanotubes. Science 1997;277(5334):1971–5.

[43] Kramer D, Viswanath RN, Weissmüller J. Surface-stress induced macroscopic bending of nanoporous gold cantilevers. Nano Lett 2004;4(5):793–6.

[44] Jing GY, Duan HL, Sun XM, et al. Surface effects on elastic properties of silver nanowires: contact atomic-force microscopy. Phys Rev B Condens Matter 2006;73(23):235409.

[45] Guo JG, Zhao YP. The size-dependent bending elastic properties of nanobeams with surface effects. Nanotechnology 2007;18(29):295701.

[46] He J, Lilley CM. Surface effect on the elastic behavior of static bending nanowires. Nano Lett 2008;8(7):1798–802.

[47] Yun G, Park HS. Surface stress effects on the bending properties of fcc metal nanowires. Phys Rev B Condens Matter 2009;79(19):195421.

[48] Wang GF, Feng XQ. Surface effects on buckling of nanowires under uniaxial compression. Appl Phys Lett 2009;94(14).

[49] Li XF, Zhang H, Lee KY. Dependence of Young's modulus of nanowires on surface effect. Int J Mech Sci 2014;81:120–5.

[50] Wang GF, Feng XQ. Effects of surface elasticity and residual surface tension on the natural frequency of microbeams. Appl Phys Lett 2007;90(23).

[51] Wang GF, Feng XQ. Timoshenko beam model for buckling and vibration of nanowires with surface effects. J Phys Appl Phys 2009;42(15):155411.

[52] Farshi B, Assadi A, Alinia-Ziazi A. Frequency analysis of nanotubes with consideration of surface effects. Appl Phys Lett 2010;96(9).

[53] Lee HL, Chang WJ. Surface effects on frequency analysis of nanotubes using nonlocal Timoshenko beam theory. J Appl Phys 2010;108(9).

[54] He Q, Lilley CM. Resonant frequency analysis of Timoshenko nanowires with surface stress for different boundary conditions. J Appl Phys 2012;112(7).

[55] Lei X, Natsuki T, Shi J, et al. Surface effects on the vibrational frequency of double-walled carbon nanotubes using the nonlocal Timoshenko beam model. Compos B Eng 2012;43 (1):64–9.

[56] Askari H, Esmailzadeh E, Zhang D. Nonlinear vibration analysis of nonlocal nanowires. Compos B Eng 2014;67:607–13.

[57] Cammarata RC. Surface and interface stress effects on interfacial and nanostructured materials. Mater Sci Eng A 1997;237(2):180–4.

[58] Gurtin ME, Weissmüller J, Larche F. A general theory of curved deformable interfaces in solids at equilibrium. Philos Mag A 1998;78(5):1093–109.

[59] Karnovsky IA, Lebed OI. Formulas for structural dynamics: tables, graphs and solutions. McGraw-Hill; 2001.

[60] Harden J, Mbanga B, Éber N, et al. Giant flexoelectricity of bent-core nematic liquid crystals. Phys Rev Lett 2006;97(15):157802.

[61] Cross LE. Flexoelectric effects: Charge separation in insulating solids subjected to elastic strain gradients. J Mater Sci 2006;41(1):53–63.

[62] Petrov AG. Electricity and mechanics of biomembrane systems: flexoelectricity in living membranes. Anal Chim Acta 2006;568(1–2):70–83.

[63] Harris P. Mechanism for the shock polarization of dielectrics. J Appl Phys 1965;36 (3):739–41.

[64] Mindlin RD. Continuum and lattice theories of influence of electromechanical coupling on capacitance of thin dielectric films. Int J Solid Struct 1969;5(11):1197–208.

[65] Tagantsev AK. Piezoelectricity and flexoelectricity in crystalline dielectrics. Phys Rev B 1986;34(8):5883.

[66] Ma W, Cross LE. Large flexoelectric polarization in ceramic lead magnesium niobate. Appl Phys Lett 2001;79(26):4420–2.

[67] Ma W, Cross LE. Flexoelectric effect in ceramic lead zirconate titanate. Appl Phys Lett 2005;86(7).

[68] Lee D, Yoon A, Jang SY, et al. Giant flexoelectric effect in ferroelectric epitaxial thin films. Phys Rev Lett 2011;107(5):057602.

[69] Maranganti R, Sharma ND, Sharma P. Electromechanical coupling in nonpiezoelectric materials due to nanoscale nonlocal size effects: Green's function solutions and embedded inclusions. Phys Rev B Condens Matter 2006;74(1):014110.

[70] Majdoub MS, Sharma P, Cagin T. Enhanced size-dependent piezoelectricity and elasticity in nanostructures due to the flexoelectric effect. Phys Rev B Condens Matter 2008;77 (12):125424.

[71] Shen S, Hu S. A theory of flexoelectricity with surface effect for elastic dielectrics. J Mech Phys Solid 2010;58(5):665–77.

[72] Yan Z, Jiang L. Size-dependent bending and vibration behaviour of piezoelectric nanobeams due to flexoelectricity. J Phys Appl Phys 2013;46(35):355502.

[73] Yan Z, Jiang LY. Flexoelectric effect on the electroelastic responses of bending piezoelectric nanobeams. J Appl Phys 2013;113(19).

[74] Liang X, Hu S, Shen S. Effects of surface and flexoelectricity on a piezoelectric nanobeam. Smart Mater Struct 2014;23(3):035020.

[75] Liang X, Hu S, Shen S. Size-dependent buckling and vibration behaviors of piezoelectric nanostructures due to flexoelectricity. Smart Mater Struct 2015;24(10):105012.

[76] Zhou ZD, Yang CP, Su YX, et al. Electromechanical coupling in piezoelectric nanobeams due to the flexoelectric effect. Smart Mater Struct 2017;26(9):095025.

[77] Yue YM, Xu KY, Chen T. A micro scale Timoshenko beam model for piezoelectricity with flexoelectricity and surface effects. Compos Struct 2016;136:278–86.

[78] Zhang R, Liang X, Shen S. A Timoshenko dielectric beam model with flexoelectric effect. Meccanica 2016;51(5):1181–8.

[79] Ray MC. Exact solutions for flexoelectric response in nanostructures. J Appl Mech 2014;81 (9):091002.

[80] Wang ZL, Song J. Piezoelectric nanogenerators based on zinc oxide nanowire arrays. Science 2006;312(5771):242–6.

[81] Jiang X, Huang W, Zhang S. Flexoelectric nano-generator: Materials, structures and devices. Nano Energy 2013;2(6):1079–92.

[82] Wang KF, Wang BL. An analytical model for nanoscale unimorph piezoelectric energy harvesters with flexoelectric effect. Compos Struct 2016;153:253–61.

[83] Liang X, Hu S, Shen S. Nanoscale mechanical energy harvesting using piezoelectricity and flexoelectricity. Smart Mater Struct 2017;26(3):035050.

[84] Managheb SAM, Ziaei-Rad S, Tikani R. Energy harvesting from vibration of Timoshenko nanobeam under base excitation considering flexoelectric and elastic strain gradient effects. J Sound Vib 2018;421:166–89.

[85] Yudin PV, Tagantsev AK. Fundamentals of flexoelectricity in solids. Nanotechnology 2013;24(43):432001.

[86] Le Quang H, He QC. The number and types of all possible rotational symmetries for flexoelectric tensors. Proc R Soc A 2011;467(2132):2369–86.

[87] Shu L, Wei X, Pang T, et al. Symmetry of flexoelectric coefficients in crystalline medium. J Appl Phys 2011;110(10).

[88] Ma W. A study of flexoelectric coupling associated internal electric field and stress in thin film ferroelectrics. Phys status Solidi (B) 2008;245(4):761–8.

[89] Mao S, Purohit PK. Insights into flexoelectric solids from strain-gradient elasticity. J Appl Mech 2014;81(8):081004.

[90] Abdollahi A, Millán D, Peco C, et al. Revisiting pyramid compression to quantify flexoelectricity: A three-dimensional simulation study. Phys Rev B 2015;91(10):104103.

[91] Xiang S, Li XF. Elasticity solution of the bending of beams with the flexoelectric and piezoelectric effects. Smart Mater Struct 2018;27(10):105023.

[92] Toupin RA. Elastic materials with couple-stresses. Arch Ration Mech Anal 1962;11(1):385–414.

[93] Mindlin RD, Tiersten HF. Effects of couple-stresses in linear elasticity. Arch Ration Mech Anal 1962;11(1):415–48.

[94] Yang F, Chong ACM, Lam DCC, et al. Couple stress based strain gradient theory for elasticity. Int J Solid Struct 2002;39(10):2731–43.

[95] Hadjesfandiari AR, Dargush GF. Couple stress theory for solids. Int J Solid Struct 2011;48(18):2496–510.

[96] Simsek M, Reddy JN. Bending and vibration of functionally graded microbeams using a new higher order beam theory and the modified couple stress theory. Int J Eng Sci 2013;64:37–53.

[97] Reddy JN. Energy principles and variational methods in applied mechanics. John Wiley and Sons; 2017.

[98] Ma HM, Gao XL, Reddy JN. A microstructure-dependent Timoshenko beam model based on a modified couple stress theory. J Mech Phys Solid 2008;56(12):3379–91.

Index

Note: Page numbers followed by *f* indicate figures and *t* indicate tables.

Printed and bound by CPI Group (UK) Ltd, Croydon, CR0 4YY
11/09/2026
02215549-0001